Applied Probability and Statistics (Continued)

BENNETT and FRANKLIN · Statistical Analysis in Chemistry and the Chemical Industry

BHAT · Elements of Applied Stochastic Processes

BOX and DRAPER · Evolutionary Operation: A Statistical Method for Process Improvement

BROWNLEE · Statistical Theory and Methodology in Science and Engineering, *Second Edition*

CHERNOFF and MOSES · Elementary Decision Theory

CHIANG · Introduction to Stochastic Processes in Biostatistics

CLELLAND, deCANI, BROWN, BURSK, and MURRAY · Basic Statistics with Business Applications, *Second Edition*

COCHRAN · Sampling Techniques, *Second Edition*

COCHRAN and COX · Experimental Designs, *Second Edition*

COX · Planning of Experiments

COX and MILLER · The Theory of Stochastic Processes, *Second Edition*

DANIEL and WOOD · Fitting Equations to Data

DAVID · Order Statistics

DEMING · Sample Design in Business Research

DODGE and ROMIG · Sampling Inspection Tables, *Second Edition*

DRAPER and SMITH · Applied Regression Analysis

DUNN and CLARK · Applied Statistics: Analysis of Variance and Regression

ELANDT-JOHNSON · Probability Models and Statistical Methods in Genetics

FLEISS · Statistical Methods for Rates and Proportions

GOLDBERGER · Econometric Theory

GROSS and HARRIS · Fundamentals of Queueing Theory

GUTTMAN, WILKS and HUNTER · Introductory Engineering Statistics, *Second Edition*

HAHN and SHAPIRO · Statistical Models in Engineering

HALD · Statistical Tables and Formulas

HALD · Statistical Theory with Engineering Applications

HOEL · Elementary Statistics, *Third Edition*

HOLLANDER and WOLFE · Nonparametric Statistical Methods

HUANG · Regression and Econometric Methods

JOHNSON and KOTZ · Distributions in Statistics

Discrete Distributions

Continuous Univariate Distributions-1

Continuous Univariate Distributions-2

Continuous Multivariate Distributions

JOHNSON and LEONE · Statistics and Experimental Design: In Engineering and the Physical Sciences, Volumes I and II

LANCASTER · The Chi Squared Distribution

LANCASTER · An Introduction to Medical Statistics

LEWIS · Stochastic Point Processes

MANN, SCHAFER and SINGPURWALLA · Methods for Statistical Analysis of Reliability and Life Data

MILTON · Rank Order Probabilities: Two-Sample Normal Shift Alternatives

continued on back

Introduction to
Probability Theory
and Statistical Inference

Introduction to Probability Theory and Statistical Inference

SECOND EDITION

Harold J. Larson
Naval Postgraduate School
Monterey, California

John Wiley & Sons
New York   London   Sydney   Toronto

Library of Congress Cataloging in Publication Data:

Larson, Harold J. 1934–
Introduction to probability theory and statistical inference.

Bibliography: p.
1. Probabilities. 2. Mathematical statistics.
I. Title. II. Title: Probability theory and statistical inference.

QA273.L352 1974   519.2   73–19852
ISBN 0-471-51781-X

Printed in the United States of America

10 9 8 7 6 5 4 3 2

for Marie

Preface

I have been quite gratified by the reception afforded the first edition of this book. This second edition is, I believe, a considerable improvement over the first. The spirit of the first edition has been maintained, to give the student a more rigorous, but no more difficult, introduction to the commonly applied techniques of probability theory and statistical inference. A one-year course in calculus is a necessary prerequisite for the material; the many examples discussed are purposely selected from many different fields and are not slanted toward any particular applied area. I feel that this type of presentation is appropriate for students in many different disciplines.

As with the first edition, the first five chapters give an introduction to set theory and to probability theory. Both discrete and continuous probability distributions are covered; specialized discussions are given for the more common distributions which are used frequently in applications. Capital letters are used for random variables and lower case letters for their observed values; this notation is carried through the chapters on statistical inference to draw clear-cut distinctions between estimators, estimates, and parameters.

The material in this edition is presented in essentially the same order as in the first edition. The few peripheral topics which were presented in separate appendices in the first edition have been absorbed into the body of the text. The discussion of functions in Chapter 1 has been rewritten, as has the material on functions of a random variable in Chapter 3. Numerous changes have been made in Chapter 5 on jointly distributed random variables. Chapter 6 has been completely rewritten, to more smoothly make the transition from probability theory to problems of statistical inference; this chapter includes a discussion of populations and the concept of sampling from populations. Discussions of sufficient statistics and their uses are included in Chapters 7 and 8. The discussion of Bayesian methods has been expanded, as has the discussion of nonlinear models in Chapter 10. Numerous other minor changes have also been effected throughout the text. This second edition includes 202 numbered examples; the number of exercises has been increased to 684. As with the first edition, answers to all the exercises are given at the back of the book. Tables of the distribution functions for the binomial, Poisson, standard normal, χ^2, and t random variables are also given at the back of the book. I would like to thank D. Van Nostrand Co., Inc., the editor of Biometrika, and Oliver & Boyd, Ltd., for permission to reproduce parts of previously published tables.

None of the methods described in this text were originated by the author; my indebtedness to those that originally did devise them is apparent. I would like to thank the many people that have given me comments about the first edition of this book. I look forward to receiving comments about this second edition. I am grateful to Sylvia Vaden for most of the typing for the original version of the book and to Mary Wright for typing the revision.

Pacific Grove, California   *Harold J. Larson*

Contents

1 Set Theory **1**

1.1 Set Notation, Equality and Subsets 1
1.2 Algebra of Sets 4
1.3 Functions 11

2 Probability **17**

2.1 Sample Space; Events 19
2.2 Probability Axioms 26
2.3 Single-Element Events and the Equally Likely Case 31
2.4 Counting Techniques 36
2.5 Some Particular Probability Problems 47
2.6 Conditional Probability 53
2.7 Independent Events 59
2.8 Discrete and Continuous Sample Spaces 65

3 Random Variables and Distribution Functions **75**

3.1 Random Variables 75
3.2 Distribution Functions and Density Functions 81
3.3 Expected Values and Summary Measures 93
3.4 Moments and Generating Functions 104
3.5 Functions of a Random Variable 114

4 Some Standard Distributions **125**

4.1 Binomial and Bernoulli Random Variables 125
4.2 Geometric and Hypergeometric Random Variables 132
4.3 The Poisson Random Variable 143

4.4   Uniform and Exponential Random Variables   151
4.5   Normal Random Variable   160

5   Jointly Distributed Random Variables   **169**

5.1   Two-dimensional Random Variables   169
5.2   Expected Values and Moments   176
5.3   Conditional Distributions and Independent Random Variables   187
5.4   n-Dimensional Random Variables and the Multinomial Random
      Variable   197
5.5   The Chebychev Inequality and the Law of Large Numbers   209
5.6   The Central Limit Theorem; Approximations   215

6   Sampling and Statistics   **225**

6.1   Populations and Samples   226
6.2   Statistics and Distributions   232

7   Statistical Inference: Estimation   **249**

7.1   The Method of Moments and Maximum Likelihood   250
7.2   Properties of Estimators   260
7.3   Confidence Intervals   272

8   Statistical Inference: Tests of Hypotheses   **285**

8.1   Hypotheses and Tests   285
8.2   Simple Hypotheses   292
8.3   Composite Hypotheses   303
8.4   Goodness-of-fit Tests   313
8.5   Contingency Tables   323

9   Bayesian Methods   **333**

9.1   Prior and Posterior Distributions   334
9.2   Bayes Estimators   341
9.3   Interval Estimation   352

10   Least Squares and Regression Theory   **359**

10.1   Least Squares Estimation   359
10.2   Interval Estimation and Tests of Hypotheses   374
10.3   Other Models   383

References   387

Appendix: Tables   389

Answers to Exercises   403

Index   425

1

Set Theory

In the study of probability theory and statistics an exact medium of communication is extremely important; if the meaning of the question that is asked is confused by semantics, the solution is all the more difficult, if not impossible, to find. The usual exact language employed to state and solve probability problems is that of set theory. The amount of set theory that is required for relative ease and comfort in probability manipulations is easily acquired. We shall take a brief look at some of the simpler definitions, operations, and concepts of set theory, not because these ideas are necessarily a part of probability theory but rather because the time needed to master them is more than compensated for by later simplifications in the study of probability.

1.1. Set Notation, Equality, and Subsets

A set is a collection of objects. The objects themselves can be anything from numbers to battleships. An object that belongs to a particular set is called an element of that set. We shall commonly use capital letters from the beginning of the alphabet to denote sets (A, B, C, etc.) and lower case letters from the end of the alphabet to denote elements of sets (x, y, z, etc.).

To specify that certain objects belong to a given set, we shall use braces { } (commonly called set builders) and either the *roster* (complete listing of all elements) or the *rule* method. For example, if we want to write that the set A consists of the letters a, b, c and that the set B consists of the first 10

1

integers, we may write

$A = \{a, b, c\}$ (*roster* method of specification)
$B = \{x : x = 1, 2, 3, \ldots, 10\}$ (*rule* method of specification).

The above two sets can easily be read as "A is the set of elements a, b, c" and "B is the set of elements x such that $x = 1$ or $x = 2$ or $x = 3$ and so on up to or $x = 10$." We shall use the symbol $\in$ as shorthand for "belongs to" and thus can write for the two sets defined above that $a \in A$, $7 \in B$. Just as a line drawn through an equals sign is taken as negation of the equality, we shall use $\notin$ to mean "does not belong to"; thus $a \notin B$, $9 \notin A$, $f \notin A$, $102 \notin B$, etc., where A and B are the sets defined above.

DEFINITION 1.1.1. Two sets A and B are *equal* if and only if every element that belongs to A also belongs to B and every element that belongs to B also belongs to A.

Then two sets are equal only if both contain exactly the same elements. Notice in particular that the order of listing of elements of a set is of no importance and that the sets

$$A = \{1, 2, a, 3\}, \qquad B = \{a, 1, 2, 3\}$$

are equal. Also, the number of times that an element is listed in the roster specifying the set is of no concern; the two sets

$$C = \{1, 2, 3\}, \qquad D = \{1, 2, 2, 1, 3, 1\}$$

are equal. We shall have no use for the redundancy exhibited in the roster of D and thus shall generally assume that if the roster of a particular set contains n elements, the elements are distinguishable and that the same element does not occur more than once.

DEFINITION 1.1.2 A is a *subset* of B (written $A \subset B$) if and only if every element that belongs to A also belongs to B.

(Actually, two different types of subsets can be distinguished, proper and improper subsets. A is a *proper* subset of B if it is a subset of B and there exists at least one element of B which does not belong to A; otherwise A is an *improper* subset of B. We shall have no need to this distinction.) In a sense, A is a subset of B if A is contained in B. For example, $\{a\}$ is a subset of $\{a, z\}$ and both of these are subsets of $\{a, b, z\}$. The set of all people residing in California is a subset of the set of all people residing in the United States; the set of all pine trees is a subset of the set of all trees.

A word regarding the difference between something belonging to a set A and being a subset of A may be in order. For example, define

$$A = \{1, 2, 3\}, \qquad B = \{1, 3\}, \qquad C = \{1\}.$$

Then it is correct to say $B \subset A$ and $C \subset A$ since every element that belongs to B also belongs to A, as does every element that belongs to C. However, it is not correct to say that $B \in A$ or that $C \in A$ since B and C are not specified in the roster of elements that belong to A. Similarly, it is correct to say that $1 \in A$ and $1 \in C$, not $1 \subset A$ and $1 \subset C$ since 1 is not a set. Thus we can say that the set of all married people is a subset of the set of all people, but we do not say that the set of all married people belongs to the set of all people. A particular married person belongs to the set of all married people (as well as to the set of all people), but this person is not a subset of either (since a person is not a set).

An alternative definition of set equality can be given by saying that $A = B$ if and only if $A \subset B$ and $B \subset A$. This, in fact, provides a very useful way of demonstrating that two sets are equal, as we shall see. Most simple set equalities can be rigorously and easily proved in this way.

From the definitions, it is easily seen that every set is equal to itself and that every set is a subset of itself. That is, we can say

$$A = A, \quad \text{for all } A,$$
$$A \subset A, \quad \text{for all } A.$$

In textbooks on algebra, these two statements are summarized by saying that both the equality and the subset relations are *reflexive*.

EXERCISE 1.1

1. Show that the subset relationship is *transitive*; that is, $A \subset B$ and $B \subset C$ imply $A \subset C$.

2. Show that the subset relationship is not *symmetric*; that is, $A \subset B$ and $B \subset A$ are logically different statements and one is not implied by the other. Give an example illustrating this nonsymmetry.

3. Give an example showing that if $E \subset F$ and $D \subset F$, we do not necessarily have $D \subset E$ or $E \subset D$.

4. Show that set equality is transitive. (Transitivity is defined in problem 1 above.)

5. If $A = \{x : x = 1, 0, 1\}$, $B = \{-1, 0, 1\}$, $C = \{0, 1, -1\}$, $D = \{-1, 1, -1\}$ mark each of the following either true or false.

   (a) $A = B$
   (b) $A = C$
   (c) $B = C$
   (d) $A \subset B$
   (e) $A \subset C$
   (f) $B \subset C$
   (g) $C \subset A$

   (h) $C \subset B$
   (i) $0 \in B$
   (j) $0 \subset C$
   (k) $A = D$
   (l) $D \subset A$
   (m) $B \subset D$
   (n) $D \in C$.

6. Define $E = \{x: 0 \leq x \leq 1\}$, $F = \{y: 0 \leq y \leq 1\}$, $G = \{x: 0 < x < 1\}$, $H = \{1\}$. Mark each of the following either true or false.
   (a) $E = F$
   (b) $F = G$
   (c) $F \subset G$
   (d) $G \subset F$
   (e) $H \subset G$
   (f) $H \subset E$
   (g) $H \in F$.

7. Is the set of all students a subset of the set of all people? Is it a subset of the set of all people under 36 years of age? Under 50 years of age?

8. Consider the set of all people in your family.
   (a) Is this set well defined? What sort of questions must be considered to make it well defined?
   (b) Are you an element of this set or a subset of this set?

9. Is the set of all automobiles produced in Detroit in a calendar year well defined? What sort of questions must be considered to make it well defined?

10. Is the set of all living United States citizens, at a given instant of time, equal to the set of all people residing in the United States at the same instant of time?

1.2.   Algebra of Sets

We shall now study some of the algebra of sets. Just as the algebra of real numbers is concerned with operations performed with numbers and their consequences, the algebra of sets is concerned with operations performed with sets and their consequences. The first set operation we define is called the union of two sets.

DEFINITION 1.2.1.   The *union* of A and B (written $A \cup B$) is the set which consists of all the elements that belong to A *or* to B or to both; i.e., $A \cup B = \{x: x \in A \text{ or } x \in B\}$.

For example, if $A = \{1, 2\}$, $B = \{1, 3\}$, $C = \{0\}$, then $A \cup B = \{1, 2, 3\}$, $A \cup C = \{0, 1, 2\}$, and $B \cup C = \{1, 3, 0\}$. Notice from the definition of the union that $A \cup B$ and $B \cup A$ are identical sets (the operation is *commutative*), since the collection of objects that belong to A or to B is the same; it does not matter whether we first list those that belong to A or those that belong to B. We also note that the operation of forming unions is *associative*; i.e.,

$$(A \cup B) \cup C = A \cup (B \cup C).$$

For this reason it is not necessary to place parentheses about pairs of sets when we take the union of several sets; it doesn't matter which union is formed first.

We notice a number of interesting relationships linking unions of sets and the concept of subsets. For example, for any A and B we would know that $A \subset A \cup B$ and $B \subset A \cup B$ (verification of these facts is asked for in problem 1 below). Not quite so obviously, $A \cup B = A$ if $B \subset A$, since the set which consists of all those elements that belong to A or to B or to both would merely consist of those elements that belong to A originally if every element belonging to B also belongs to A. Thus the union of the set of people living in California with the set of people living in the United States is the set of people living in the United States; the union of the set of all cows with the set of all animals is the set of all animals.

A second set operation which we shall find useful is called the intersection of two sets.

> DEFINITION 1.2.2. The *intersection* of A and B (written $A \cap B$) is that set which consists of all elements that belong *both* to A and to B; i.e.,
> $A \cap B = \{x : x \in A \text{ and } x \in B\}$.

For example, if $A = \{0, 1\}$, $B = \{1, 3\}$, $C = \{0, 1, 3\}$, then $A \cap B = \{1\}$, $A \cap C = \{0, 1\}$, $B \cap C = \{1, 3\}$. If A is the set of married people and B is the set of people living in California, then $A \cap B$ is the set of married people living in California.

It is easy to see that the intersection operation is also commutative and associative; that is,

$$A \cap B = B \cap A \quad \text{and} \quad A \cap (B \cap C) = (A \cap B) \cap C.$$

We note as well that

$$A \cap B \subset B, \quad A \cap B \subset A,$$

no matter what the sets A and B may be. Similarly, if $B \subset A$, then

$$A \cap B = B.$$

Thus the intersection of the set of all buildings in California with the set of all buildings in the United States is the set of all buildings in California.

A handy way of picturing sets, relationships between sets, and operations with sets is to employ a Venn diagram. Various geometric shapes can be drawn on a plane and, if we assume that the points interior to and on the boundary of each figure constitute a set, many of the concepts already discussed can be visually displayed. In Figure 1.2.1 are given sets A, B, C, D, and $A \cup B$ and $C \cap D$.

There are two *distributive laws* linking the operations of unions and intersections. These are:

$$A \cap (B \cup C) = (A \cap B) \cup (A \cap C) \tag{1.1}$$

$$A \cup (B \cap C) = (A \cup B) \cap (A \cup C). \tag{1.2}$$

Figure 1.2.2 gives a Venn diagram which the reader can use to verify that both of these statements are true.

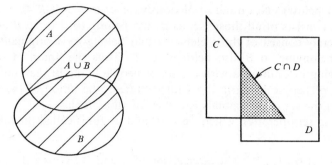

FIGURE 1.2.1   Union and intersection.

Note that there exists a rather natural analog between unions of sets and addition of numbers since, in some senses, the set $A \cup B$ is the result of "adding" the set B to the set A. If we also pretend that there is an analog between set intersection and multiplication of numbers, then distributive law 1.1 above is analogous to $a(b + c) = ab + ac$, which is true for all real numbers a, b, and c. However, distributive law 1.2 would be analogous to $a + b \cdot c = (a + b) \cdot (a + c)$, which is not true for all a, b, and c.

Generally, all of the sets entering into particular discussions will have certain things in common. For example, a group of botanists might discuss sets of trees, in which event all of the sets entering into their discussion would have trees as elements. A group of mathematicians might discuss sets of real numbers, in which event all the sets entering into their discussion would have real numbers as elements. The *universal set* for a particular discussion is that set which has as elements all elements of every set entering into the discussion. Thus the universal set for the botanists' discussion could be the set of all trees; the universal set for the mathematicians' discussion might be

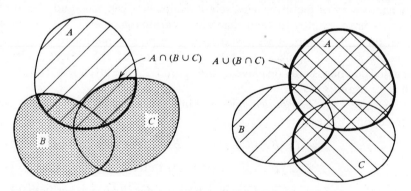

FIGURE 1.2.2   Distributive laws.

the set of all real numbers. Note that different discussions may have different universal sets.

If we have in mind a particular universal set, then definition of a set A automatically also specifies a second set $\bar{A}$, called the complement of A, which is defined as follows.

DEFINITION 1.2.3. The *complement* of A (with respect to a given universal set) is the set of all elements *not* belonging to A; i.e.,

$$\bar{A} = \{x : x \notin A\}.$$

If, for example, our universal set is

$$U = \{x : 1 \leq x \leq 10\}$$

and we define

$$A = \{x : 1 \leq x \leq 2\}$$

$$B = \{1, 10\}$$

then

$$\bar{A} = \{x : 2 < x \leq 10\}$$

$$\bar{B} = \{x : 1 < x < 10\}.$$

Or, if our universal set is the set of all people living in the United States and we define A as the set of people living in California and B as the set of people living in San Francisco, then $\bar{A}$ is the set of people living in the United States outside of California and $\bar{B}$ is the set of people living in the United States outside of San Francisco.

Figure 1.2.3 gives a Venn diagram illustrating a universal set U, a set A, a

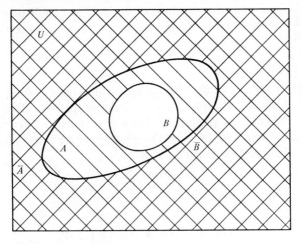

FIGURE 1.2.3   Set complements.

set $B \subset A$, $\bar{A}$, and $\bar{B}$. Note that $A \cup \bar{A} = U$ and, as is always the case, that if $B \subset A$, then $\bar{A} \subset \bar{B}$.

The null (or empty) set is frequently of use and is defined as follows:

DEFINITION 1.2.4. The *null set* $\varnothing$, is the set with no elements; i.e.,

$$\varnothing = \{\ \ \}.$$

$\varnothing$ occurs as the intersection of two sets that have no elements in common. For example, if A is the set of all frogs and B is the set of all people, then $A \cap B = \varnothing$. More generally, from the definition of $\bar{A}$, $A \cap \bar{A} = \varnothing$ for any A.

The empty set plays a role analogous to the real number 0 in some (not all) manipulations. If a is any real number, we know that

$$a + 0 = a, \qquad a \cdot 0 = 0.$$

If A is any set, we can easily see that

$$A \cup \varnothing = A, \qquad A \cap \varnothing = \varnothing.$$

Note, however, that we can have $A \cap B = \varnothing$ where neither A nor B is empty, whereas $a \cdot b = 0$ implies that a or b (or both) is 0. From our definition of subsets, we must admit that $\varnothing \subset A$ for all A, since every element belonging to $\varnothing$ (none) also belongs to A. Alternatively, every x belonging to $\varnothing$ also belongs to A is a true statement, vacuously, since there is no x belonging to $\varnothing$ to make the statement false. Unfortunately, $\varnothing$ cannot be pictured on a Venn diagram like other sets, because if we draw any shape at all we don't have a picture of $\varnothing$ since our shape is not empty.

As was discussed earlier, a set of elements is unchanged if we merely rearrange the ordering of its elements. To have at our disposal collections of n elements where order is of importance, we define an n-tuple as follows.

DEFINITION 1.2.5. An n-tuple is an *ordered* array of n components written $(x_1, x_2, \ldots, x_n)$.

For example, $(1, 2)$, $(2, 1)$, $(0, 100)$, (a, b) are all 2-tuples; (a, b, c), (b, a, c), $(1, 1, 1)$, $(2, 1, 2)$ are all 3-tuples; etc. Two n-tuples are different, even if they contain the same components, if they are written in a different order and, of course, an n-tuple and an $(n + 1)$-tuple cannot be equal since they have different numbers of components. Thus

$$(1, 2) \neq (2, 1)$$
$$(1, 3, 1) \neq (1, 1, 3)$$
$$(1, 1) \neq (1, 1, 1)$$
$$\{(1, 2), (1, 3)\} \neq \{(2, 1), (1, 3)\}$$
$$\{(4, 2), (2, 5)\} = \{(2, 5), (4, 2)\}.$$

We shall have use for *sets* of *n*-tuples as we proceed. For example, if we want to discuss the set of married couples living in Texas, then we want each couple to be an element of the set; for clarity we list the husband's name first when specifying the couple. Each element of the set then is a couple or a 2-tuple. Note that a married individual living in Texas does not belong to the set; he is a component of one of the 2-tuples that belongs to the set. As a second example, the set of points lying in the first quadrant of the usual Cartesian plane is a set of 2-tuples. Each point is represented by a 2-tuple (x, y) where x is the horizontal coordinate and y the vertical coordinate.

One other set operation, the Cartesian product of two sets, will be of interest to us. It is defined as follows.

DEFINITION 1.2.6. The *Cartesian product* $A \times B$ of A and B is the set of all possible 2-tuples (x_1, x_2) where $x_1 \in A$, $x_2 \in B$. That is,

$$A \times B = \{(x_1, x_2): x_1 \in A, x_2 \in B\}.$$

Let us consider a number of examples of Cartesian products. Suppose that $A = \{0, 1, 2\}$, $B = \{3, 5\}$, $C = \{0\}$. Then

$A \times B = \{(0, 3), (0, 5), (1, 3), (1, 5), (2, 3), (2, 5)\}$
$A \times C = \{(0, 0), (1, 0), (2, 0)\}$
$B \times C = \{(3, 0), (5, 0)\}$
$B \times A = \{(3, 0), (3, 1), (3, 2), (5, 0), (5, 1), (5, 2)\}$
$C \times A = \{(0, 0), (0, 1), (0, 2)\}$
$C \times B = \{(0, 3), (0, 5)\}$
$A \times A = \{(0, 0), (0, 1), (0, 2), (1, 0), (1, 1), (1, 2), (2, 0), (2, 1), (2, 2)\}$
$B \times B = \{(3, 3), (3, 5), (5, 3), (5, 5)\}$
$C \times C = \{(0, 0)\}.$

If D is the set of all positive real numbers, then $D \times D$ is the set of points in the first quadrant of the plane.

Notice that $A \times B$ and $B \times A$ are not equal in general. The operation is *not commutative*. Problem 19 below asks you to specify when these two would be equal. We can form the Cartesian product of any number of sets; $A \times B \times C$ is a set of 3-tuples, $A_1 \times A_2 \times \cdots \times A_n$ is a set of *n*-tuples. Taking A, B, and C as defined above,

$A \times B \times C = \{(0, 3, 0), (0, 5, 0), (1, 3, 0), (1, 5, 0), (2, 3, 0), (2, 5, 0)\}$
$A \times C \times B = \{(0, 0, 3), (0, 0, 5), (1, 0, 3), (1, 0, 5), (2, 0, 3), (2, 0, 5)\}.$

Strictly speaking, we should denote the first element listed in $A \times B \times C$ above as $((0, 3), 0)$ if we are thinking of $(A \times B) \times C$, or as $(0, (3, 0))$ if we are thinking of $A \times (B \times C)$. We have no reason to distinguish between $((0, 3), 0)$, $(0, (3, 0))$, and $(0, 3, 0)$, so we simply define $A \times B \times C$, as noted, to be the collection of 3-tuples which results when the innermost pair of parentheses is deleted. Thus Cartesian products are *associative*, with this convention.

EXERCISE 1.2

1. Show that $A \cap B \subset A \subset A \cup B$ and that $A \cap B \subset B \subset A \cup B$.
2. If $A = \{1, 0\}, B = \{x: 0 < x < 1\}, C = \{\frac{1}{2}\}$, compute $A \cup B, A \cup C, B \cup C$, $A \cap B, A \cap C, B \cap C$.
3. Show that $B \cup B = B \cap B = B$ for any B.
4. What is the implication of the equation $E \cap F = F$?
5. What is the implication of the equation $E \cup F = E$?
6. Define:
$$A = \{x: x = 1, 2, 3, \ldots, 10\}$$
$$B = \{x: 1 \leq x \leq 10\}$$
$$C = \{x: x = 0, 1, 2, 3, 4, 5, 6\}$$
$$D = \{0, 10, 20, 30\}$$
and compute

(a) $A \cup B$	(g) $B \cup C$	(m) $A \cup B \cup C$
(b) $A \cap B$	(h) $B \cap C$	(n) $A \cap (B \cup C)$
(c) $A \cup C$	(i) $B \cup D$	(o) $A \cup (B \cap C)$
(d) $A \cap C$	(j) $B \cap D$	(p) $A \cap B \cap C$
(e) $A \cup D$	(k) $C \cup D$	(q) $C \cup (A \cap D)$
(f) $A \cap D$	(l) $C \cap D$	(r) $(A \cup B) \cap (C \cup B)$.

7. Show that, algebraically, both distributive laws 1.1 and 1.2 must hold. (Hint: Show that if $x \in A \cap (B \cup C)$ then $x \in (A \cap B) \cup (A \cap C)$ and vice versa.)
8. Define A to be the set of safecrackers in the United States and define B to be the set of prison inmates in the United States. What, in words, is $A \cap B$?
9. Let A be the set of female sociologists in the United States and let B be the set of male sociologists in the United States. What, in words, is $A \cup B$?
10. Can you find any real numbers, a, b, c such that $(a + b)c = (a + b) \cdot (a + c)$? (This is the analog of distributive law 1.2.)
11. Name two distinct universal sets for a discussion involving sets whose elements are all people residing in a particular town.
12. Given the universal set
$$U = \{1, 2, 3, \ldots, n\}$$
and the sets
$$A = \{1, 2, 3\}, \qquad B = \{2, 3, n\},$$
compute $\bar{A}, \bar{B}$, and $\overline{\bar{A} \cup \bar{B}}$. Compare this latter set with $A \cap B$.
13. Can we define a set A such that $A = \bar{A}$? Such that $A \subset \bar{A}$?
14. Show that $\bar{\bar{A}} = A$.

15. Show that $\overline{A \cup B} = \bar{A} \cap \bar{B}$. (This is one of De Morgan's laws.)

16. Show that $\overline{A \cap B} = \bar{A} \cup \bar{B}$. (This is also one of De Morgan's laws.)
    Hint: In problem 15, replace A and B by their complements and then take the complement of both sides.

17. Draw Venn diagrams illustrating the sets $A \cap \bar{B}$, $\bar{A} \cap \bar{B}$, $A \cup B \cup C$, $A \cap B \cap C$, $(A \cup B) \cap C$, $(A \cap C) \cup (B \cap C)$.

18. Define $A = \{1, 2\}$, $B = \{2, 1\}$, $C = \{10, 12\}$, and form the Cartesian products $A \times B$, $A \times C$, $B \times C$, $B \times A$, $C \times A$, $C \times B$, $A \times A$, $B \times B$, $C \times C$, $A \times B \times C$, $C \times B \times A$, $C \times A \times B$.

19. What are the conditions under which $A \times B = B \times A$?

20. If A is the set of married men living in Texas and B is the set of married women living in Texas, is $A \times B$ the set of married couples living in Texas?

21. If A is the set of people with United States citizenship and B is the set of people with Canadian citizenship, does $A \cap B = \varnothing$?

22. How would you define $\varnothing \times \varnothing$?

23. Show that $(A \cup B) \times C = (A \times C) \cup (B \times C)$.

24. Show that $(A \times B) \cap (A \times C) = A \times (B \cap C)$.

25. For any universal set U, show that $\bar{U} = \varnothing$, $\overline{\varnothing} = U$.

26. Assume sets A and B are used in the same discussion (with universal set U). What would be a good universal set if Cartesian products are used? (What set are we sure $A \times B$ must be a subset of?)

27. Assume A and B are subsets of a universal set U; $U \times U$ then would serve as the universal set for any Cartesian products. Show that $\bar{A} \times \bar{B} \neq \overline{A \times B}$.
    Hint: Take $U = \{1, 2, 3, 4, 5,\}$, $A = \{1, 2\}$, $B = \{3, 4\}$ and examine $\bar{A} \times \bar{B}$ and $\overline{A \times B}$.

28. Express $\overline{A \times B}$ in terms of Cartesian products involving $\bar{A}$, A, $\bar{B}$ and B. (Consider taking unions of products.)

1.3. Functions

A function can be defined in many different ways. The reader is undoubtedly acquainted with the concept of a function from his study of analytic geometry and calculus. The definition of a function which we shall find most useful for our purposes is presented below.

DEFINITION 1.3.1. A *function* is a set of pairs (2-tuples), such that no two pairs have equal first components. The set which has as elements the *first* components of all the 2-tuples is called the *domain of definition* of the function. The set which has as elements the *second* components of all the 2-tuples is called the *range* of the function.

The reader is probably most familiar with real functions of a real variable. A real function of a real variable associates together pairs of real numbers. If we let x denote values on a horizontal axis, and let y denote values on a vertical axis, a real function f will associate values of y with values of x. Commonly, the y-value associated with a given value of x will be denoted by $f(x)$. Definition 1.3.1 displays this correspondence or association by listing pairs. The first element of each pair is the x-value and the second element is the y-value associated with it. Each different x value can have only one y associated with it. That is the reason for the requirement that each first element occur only once: the (unique) y-value associated with a given value of x is the second entry in the pair. The same y-value could occur more than one time. The following example discusses a few functions of a real variable, using Definition 1.3.1.

EXAMPLE 1.3.1. For simplicity of exposition, let us consider a real variable x which is constrained to lie between -5 and 5; thus we shall consider several functions of a real variable x where $-5 < x < 5$. First, consider the *constant* function:

$$y = f_1(x) = 2, \quad -5 < x < 5.$$

From Definition 1.3.1 we would write this function as

$$f_1 = \{(x, 2): -5 < x < 5\}.$$

to indicate that the same constant value, 2, is associated with each x. As a second example, consider the *identity* function

$$y = f_2(x) = x, \quad -5 < x < 5,$$

which associates each value of x with itself. This would be written

$$f_2 = \{(x, x): -5 < x < 5\},$$

to indicate that f_2 associates each value of x with itself (the second, associated value for each pair is equal to the first value). The *square* function

$$y = f_3(x) = x^2, \quad -5 < x < 5,$$

would be written

$$f_3 = \{(x, x^2): -5 < x < 5\}.$$

The *absolute value* function

$$y = f_4(x) = |x|, \quad -5 < x < 5$$

would be written

$$f_4 = \{(x, |x|): -5 < x < 5\}.$$

Figure 1.3.1 presents the graphs of f_1, f_2, f_3, and f_4. Since in every case we restricted x to lie between -5 and 5, the domain of definition for each of the 4 functions is

$$D_{f_1} = D_{f_2} = D_{f_3} = D_{f_4} = \{x: -5 < x < 5\}.$$

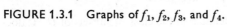

FIGURE 1.3.1   Graphs of f_1, f_2, f_3, and f_4.

The ranges, however, are not the same. It is easily verified that the ranges are

$$R_{f_1} = \{2\}$$
$$R_{f_2} = \{y: -5 < y < 5\}$$
$$R_{f_3} = \{y: 0 \le y < 25\}$$
$$R_{f_4} = \{y: 0 \le y < 5\}.$$

Note that we are using the same symbol (f with a subscript) to denote *both* the rule which associates a y-value with an x-value and the set of pairs. This redundancy in use should cause no difficulty; the primary usage will be to denote the rule.

Using Definition 1.3.1 it is easy to see that the definition of a function is appropriate for more than just real functions of a real variable. There is no requirement that the first entry in each pair should necessarily be a real number (nor that the second should be either, for that matter). In succeeding chapters we shall study *probability functions* and *random variables*; for each of these the first component (the quantity which will have a real number associated with it) in each pair is not a real number, although the second component will be. A probability function is a rule or function which associates a real number with every *subset* of a given set (and satisfies certain rules as well; see Section 2.2). Thus, for a probability function, the first component in each pair will be a subset of a given set; the second component will be a real number (called a probability). A random variable is a rule which associates a real number with every *element* of a given set. Thus, the first component of each pair will be an element of the given set and the second component will be a real number (the value of the random variable).

The following examples illustrate these two types of functions:

EXAMPLE 1.3.2   Define

$$U = \{1, 2, 3, 4, 5\}$$

and let $n(A) =$ number of elements in A, for $A \subset U$. Then n is a function which associates a number with each subset $A \subset U$; the number associated with each A is the number of elements belonging to A. Thus

$$n(\{1\}) = n(\{2\}) = n(\{4\}) = 1,$$
$$n(\{1, 2\}) = n(\{1, 5\}) = n(\{2, 3\}) = 2,$$
$$n(\{1, 2, 3\}) = n(\{1, 3, 5\}) = n(\{2, 3, 4\}) = 3$$
$$n(\{1, 2, 3, 4, 5\}) = 5, \qquad n(\varnothing) = 0$$

for example. n is called a set function defined on the class of subsets of U. It does not satisfy the axioms for a probability function, which is simply a special type of set function. Note that the domain of definition of n is the class of subsets of U and its range is $R_n = \{0, 1, 2, 3, 4, 5\}$.

As with functions of a real variable, a set function could use any rule for associating a real number with each subset of U. Instead of associating the number of elements in the subset, with each subset, it could associate the square root of that number, or the cube of that number, or use any arbitrary rule for the association.

EXAMPLE 1.3.3.   Let $A = \{1, 2, 3, 4\}$ and define $U = A \times A$. Then each element of U is a 2-tuple (u_1, u_2). Now define the function X by

$$X((u_1, u_2)) = u_1 + u_2$$

Then X associates a real number with each *element* belonging to U. In fact,

$$X((1, 1)) = 2$$
$$X((2, 2)) = 4$$
$$X((1, 4)) = 5$$
$$X((3, 4)) = 7, \text{ etc.}$$

The domain of definition for X is the set U and the range of X can be verified to be

$$R_X = \{2, 3, 4, 5, 6, 7, 8\}.$$

The function defined in Example 1.3.3 is what will be called a random variable, a function that associates a real number with every element of a certain type of set.

EXERCISE 1.3

1.   Let A be the set of individuals in your family and define the function $f(\omega) =$ age of ω for $\omega \in A$. Specify the range of f.

2.   Let

$$B = \{1, 2, 3, 4, 5, 6\}$$

$$C = B \times B$$

and define the function

$$g[(x_1, x_2)] = x_1 + x_2 \quad \text{for} \quad (x_1, x_2) \in C.$$

What is the range of g?

3.   Let $A = \{0, 1\}$ and define $B = A \times A \times A$. For each element $(x_1, x_2, x_3) \in B$, let

$$h((x_1, x_2, x_3)) = x_1 + x_2 + x_3.$$

What is the range of h? How many elements have the value 0 assigned to them? How many have the value 1 assigned to them?

4.   For the set U defined in Example 1.3.3, let $Y(u_1, u_2) = u_1/u_2$. What is the range of Y?

5. For the set C defined in Exercise 2 above, let $d(\,(x_1, x_2)\,) = x_1 - x_2$. What is the range of d?

6. In Exercise 5 let
$$a(\,(x_1, x_2)\,) = |d(\,(x_1, x_2)\,)|.$$
What is the range of a?

7. Let
$$U = \{1, 2, 3\}$$
and define the set function
$$m(A) = 1 \quad \text{if } 2 \in A$$
$$m(A) = 0 \quad \text{if } 2 \notin A$$

for every subset A of U. What is the range of m? How many subsets $A \subset U$ have $m(A) = 1$? (m is in fact a probability function.)

8. Refer to the set U and the set function $n(A)$ defined in Example 1.3.2. Let $p(A) = n(A)/5$ for all $A \subset U$. What is the range of p? (This is also a probability function.)

9. Let $U = \{1, 2, 3, 4\}$ and define
$$q(\varnothing) = 2$$
$$q(A) = \text{reciprocal of the number of elements in } A$$

for every nonempty subset $A \subset U$. What is the range of q?

10. Let U and n be as defined in Example 1.3.2. Show that

$$n(A) \leq n(B) \quad \text{if } A \subset B$$
$$n(A \cup B) = n(A) + n(B) \quad \text{if } A \cap B = \varnothing.$$

(The first of these says that n is *monotonic*, the second that n is *finitely additive*.)

11. Show that the set function m defined in Exercise 7 is monotonic and finitely additive. (See Exercise 10.)

12. Show that the set function q defined in Exercise 9 is not monotonic and is not finitely additive. (These terms are defined in Exercise 10.)

2

Probability

We are now ready to begin our study of probability theory itself. The earliest known applications of probability theory occurred in the seventeenth century. A French nobleman of that time was interested in several games then played at Monte Carlo; he tried unsuccessfully to describe mathematically the relative proportion of the time that certain bets would be won. He was acquainted with two of the best mathematicians of the day, Pascal and Fermat, and mentioned his difficulties to them. This began a famous exchange of letters between the two mathematicians concerning the correct application of mathematics to the measurement of relative frequencies of occurrences in simple gambling games. Historians generally agree that this exchange of letters was the beginning of probability theory as we now know it.

For many years a simple *relative frequency* definition of probability was all that was known and was all that many felt was necessary. This definition proceeds roughly as follows. Suppose that a chance experiment is to be performed (some operation whose outcome cannot be predicted in advance); thus there are several possible outcomes which can occur when the experiment is performed. If an event A occurs with m of these outcomes, then the probability of A occurring is the ratio m/n where n is the total number of outcomes possible. Thus, if the experiment consists of one roll of a fair die and A is the occurrence of an even number, the probability of A is $\frac{3}{6}$.

There are many problems for which this definition is appropriate, but such a heuristic approach is not conducive to a mathematical treatment of the theory of probability. The mathematical advances in probability theory were relatively limited and difficult to establish on a firm basis until the Russian

17

mathematician A. N. Kolmogorov gave a simple set of three *axioms* or rules which probabilities are assumed to obey. Since the establishment of this firm axiomatic basis, there have been great strides made in the theory of probability and in the number of practical problems to which it is applied.

In this chapter we shall see what these three axioms are and why we might reasonably adopt these rules for probabilities to follow. The axioms do not give any unique value which the probability of an event must equal; rather they express internal rules which ensure consistency in our arbitrary assignment of probabilities. The relative frequency definition of probability which was already mentioned is only one way in which probabilities can be computed; it is discussed more fully in Section 2.3 and, as is shown there, does satisfy the axioms. This relative frequency definition has built within it certain assumptions about the outcomes of the experiment which are not always appropriate. Thus, this arbitrary way of assigning probabilities is not always applicable.

It might be appropriate to discuss briefly at this time the notion of a probability model. An experiment is a physical operation which, in the real world, can result in one of many possible outcomes. For example, if we roll a pair of dice one time the two numbers we might observe can range anywhere from a pair of ones to a pair of sixes. Or, if a particular individual is going to run one hundred yards as fast as he can, the elapsed time from when he starts until when he finishes might be anywhere from nine seconds to thirty seconds. Or, if a particular person lives in a given way in terms of rest and habits of various kinds, his total life span might lie anywhere from zero years to one hundred years. In each of these cases the particular outcome we might observe cannot be predicted in advance, but the total collection of outcomes can be assumed to be known (one of which will be observed when the experiment is completed). When building a probability model for an experiment, we are concerned with specifying: (1) what the total collection of outcomes could be and (2) the relative frequency of occurrence of these outcomes, based on an analysis of the experiment. We are in many senses idealizing the physical situation by restricting the set of possible outcomes; but, to the extent that this idealization does not affect the relative frequency of events of interest, we can profitably use the results of our computations as descriptions of the actual physical experiment. The probability model then consists of the assumed collection of possible outcomes and the assigned relative frequencies or probabilities of these outcomes. The axioms are used to assure consistency in this assignment of probabilities.

In the latter half of the book we shall study problems and methods of statistical inference. Probability theory is essentially deductive in nature; from the facts known about an experiment one wants to deduce the probabilities of occurrence of events of interest. As we shall see, statistics is more inductive in nature. By observing what is true in a few cases, the statistician is interested in making generalizations or inferences about what may also be true in cases not observed. The statistician may, for example, observe the outcome of an experiment one or more times and then ask whether these observations are consistent with one or another type of probability distribution. Or he may observe what has happened through time to a particular quantity, such as the price of a stock or the weight of a baby, and then ask what will happen to this quantity in the future. Or, in a science such as genetics or physics, a "scientific law" may predict what should be observed when a certain experiment is performed; by observing the actual experiment the statistician then is interested in deciding whether or not the "law" appears to be correct. In each of these cases an inference is to be made; probability theory proves very useful to the statistician in comparing different ways or methods of making the inference.

2.1. Sample Space; Events

As mentioned in Chapter 1, set theory provides a language ideally suited for the efficient study of probability theory. After the following two basic definitions, we shall see in what manner set theory can be utilized.

DEFINITION 2.1.1. An *experiment* is any operation whose outcome cannot be predicted with certainty.

DEFINITION 2.1.2. The *sample space S* of an experiment is the set of all possible outcomes for the experiment.

Let us examine some particular examples which utilize these two definitions.

EXAMPLE 2.1.1. We roll a single die one time (dice is the plural of die). Then the experiment is the roll of the die. A sample space for this experiment could be

$$S = \{1, 2, 3, 4, 5, 6\}$$

where each of the integers 1 through 6 is meant to represent the face having that many spots being uppermost when the die stops rolling.

EXAMPLE 2.1.2. We select 1 card at random from a standard deck of 52. The experiment is the selection of a card. We might assume that we have numbered the cards (in some specified order) from 1 to 52; then a sample space for this experiment would be

$$S = \{1, 2, 3, \ldots, 52\}$$

since the particular card that we select in performing the experiment must correspond to exactly one of these integers.

EXAMPLE 2.1.3.   We select a person at random from the student body of UCLA. The experiment is the selection of a student from the total student body. A sample space for the experiment would be the set of all students that could be selected; i.e., S would simply be the roster of students at UCLA at the time the selection is made.

Frequently the performance of an experiment naturally gives rise to more than one piece of information which we may want to record. If we observe two pieces of information every time the experiment is performed, we would reasonably want a sample space that is a collection of 2-tuples, the two positions corresponding to the two pieces of information. Or, if we observe three pieces of information, we would want a sample space of 3-tuples. Or, more generally, if we observe r pieces of information, we would want a sample space of r-tuples. In each of the three examples given above, a single piece of information was generated when performing the experiment; thus each of these sample spaces had 1-tuples as elements. The next three examples discuss experiments in which more than one piece of information is derived.

EXAMPLE 2.1.4.   Suppose our experiment consists of one roll of two dice, one red and the other green. A reasonable sample space for the experiment would be the collection of all possible 2-tuples (x_1, x_2) that could occur where the number in the first position of any 2-tuple corresponds to the number on the red die and the number in the second position corresponds to the number on the green die. Thus, we might use as our sample space

$$S = D \times D$$

where $D = \{1, 2, 3, 4, 5, 6\}$; i.e.,

$$S = \{(x_1, x_2): x_1 = 1, 2, 3, \ldots, 6; \ x_2 = 1, 2, 3, \ldots, 6\}.$$

EXAMPLE 2.1.5.   Doug, Joe, and Hugh match coins. The experiment they perform is one flip of three coins. A reasonable sample space for this experiment is the set of 3-tuples, each of which has H (head) or T (tail) in every position. The first position in the 3-tuple corresponds to the face on Doug's coin, the second position to the face on Joe's coin, and the third position to the face on Hugh's coin. This sample space S can be written

$$S = C \times C \times C$$

where $C = \{H, T\}$;

$$S = \{(x_1, x_2, x_3): x_1 = H \text{ or } T, x_2 = H \text{ or } T, x_3 = H \text{ or } T\}.$$

EXAMPLE 2.1.6.   We select 10 students, at random, from the student body of the University of Chicago. We would normally restrict the sampling method so that the same student does not occur more than once in the sample (this is called sampling without replacement; see Chapter 6 for a more complete discussion of sampling and its relation to probability). The experiment then is the selection of 10 students. A sample space for the experiment would be the total collection of 10-tuples that could occur, the first element in any 10-tuple corresponding to the first student selected, the second element corresponding to the second student selected, the third element to the third student, etc. For convenience, let us assume that the student body at the University of Chicago consists of 30,000 students and that they have been numbered from 1 to 30,000. Then this sample space may be written

$$S = \{(x_1, x_2, \ldots, x_{10}): x_i = 1, 2, 3, \ldots, 30{,}000;$$
$$i = 1, 2, \ldots, 10; \ x_i \neq x_j \text{ for all } i \neq j\}.$$

Although S has 10-tuples as elements, note that it is not the Cartesian product of the same set with itself 10 times.

One thing to notice about the sample space of an experiment is that it is not unique; that is, there generally is more than one reasonable way of specifying all possible outcomes of the experiment. For example, if we roll a pair of dice we could use the set of 2-tuples mentioned in Example 2.1.4 as the sample space, or we could reason that every time we roll two dice the sum of the two numbers that occur could be any of the integers 2 through 12 and adopt this set as the sample space. In Example 2.1.5, Doug, Joe, and Hugh will observe 0 or 1 or 2 or 3 heads when they flip their coins; therefore we could adopt the set having these integers as elements as our sample space rather than the set of 3-tuples mentioned above. It would be equally easy to mention an alternative sample space for Example 2.1.6. As we shall see in the sequel, one sample space is generally easier to use than another for a given experiment. Of course, we shall try to use the one that is easiest for our purposes. Which sample space is used has no effect on the values of probabilities of interest but does affect the ease of computation of such probabilities.

The following two definitions are very basic to the sequel. Misunderstanding of the definition of an event and of when an event has or has not occurred can lead to a great deal of difficulty in many problems.

DEFINITION 2.1.3. An *event* is a subset of the sample space. Every subset is an event.

DEFINITION 2.1.4. An event *occurs* if any one of its elements is the outcome of the experiment.

The sample space used for Example 2.1.1 was

$$S = \{1, 2, 3, 4, 5, 6\}.$$

Then each of the sets

$$A = \{1\}, \ B = \{1, 3, 5\}, \ C = \{2, 4, 6\}, \ D = \{4, 5, 6\}, \ E = \{1, 3, 4, 6\},$$

is an event (these are not the only events since they are not the only subsets of S). These are all distinct (different) events because no two of these subsets are equal. If we actually were to perform the experiment (roll the die) and we got a 1, then events A, B, and E are said to have occurred since each of these has 1 as an element. Events C and D did not occur since $1 \notin C$ and $1 \notin D$. If we got a 4 when the die was rolled, then we would say that events C, D, and E occurred since 4 is an element of each of them. Notice that no matter which outcome we observe when the experiment is performed, *many* different events have each occurred (as we shall see, exactly half the possible events occur for any particular outcome).

In Example 2.1.3, which consists of selecting one student at random from the student body of UCLA, the sample space consists of the roster of students at the school. Assuming there are thirty-five thousand students at UCLA, we can represent the sample space by

$$S = \{x : x = 1, 2, 3, \ldots, 35{,}000\}.$$

The sets

$$A_1 = \{1, 50\}, \qquad B = \{2, 76, 140, 64\},$$

$$C = \{10\}, \qquad D = \{x : x = 101, 102, \ldots, 960\},$$

etc., are all events since each is a subset of S. If, when we performed the experiment, we happened to select the student numbered 176, then D occurred since $176 \in D$, but none of the other three events listed occurred. If we happened to select the student numbered 9999, then none of the four events listed occurred since this number doesn't belong to any of them.

We shall frequently want to take a word description of an event and translate it into a subset of the sample space. For example, suppose that in Example 2.1.2 we had numbered the 52 cards in the following way: A, 2, 3, ..., K of hearts are numbered 1 through 13 respectively; A, 2, 3, ..., K of diamonds are numbered 14 through 26, respectively; A, 2, 3, ..., K of spades are numbered 27 through 39, respectively; A, 2, 3, ..., K of clubs are numbered 40 through 52, respectively. Then we might define the following events in words:

> A: a red card is drawn
> B: a spade is drawn
> C: an ace is drawn
> D: a face card is drawn (aces are not counted as face cards).

Then, since an event occurs if any of its elements occurs, we have

$$A = \{1, 2, \ldots, 26\}$$
$$B = \{27, 28, \ldots, 39\}$$
$$C = \{1, 14, 27, 40\}$$
$$D = \{11, 12, 13, 24, 25, 26, 37, 38, 39, 50, 51, 52\}.$$

In Example 2.1.4 a pair of dice is rolled one time. The sample space S is the set of 2-tuples

$$S = \{(x_1, x_2): x_1 = 1, 2, \ldots, 6; \; x_2 = 1, 2, \ldots, 6\}.$$

The events:

A: the sum of the two dice is 3
B: the sum of the two dice is 7
C: the two dice show the same number

are, as subsets,

$A = \{(1, 2), (2, 1)\}$
$B = \{(1, 6), (2, 5), (3, 4), (4, 3), (5, 2), (6, 1)\}$
$C = \{(1, 1), (2, 2), (3, 3), (4, 4), (5, 5), (6, 6)\}.$

Suppose that in Example 2.1.5, Doug, Joe, and Hugh are playing a game called "odd man loses." That is, if two of the coins' faces match and the third person's does not, the third person loses. The sample space is

$$S = \{(x_1, x_2, x_3): x_1 = H \text{ or } T, x_2 = H \text{ or } T, x_3 = H \text{ or } T\},$$

where the first position corresponds to Doug's coin, the second to Joe's, the third to Hugh's. Define the events:

A: Doug loses
B: Doug doesn't lose
C: Joe loses
D: No one loses.

Then, written as subsets, we have

$A = \{(H, T, T), (T \; H, H)\}$
$B = \{(H, H, T), (H, T, H), (H, H, H), (T, T, H), (T, H, T), (T, T, T)\}$
$\;\;\; = \bar{A}$
$C = \{(T, H, T), (H, T, H)\}$
$D = \{(H, H, H), (T, T, T)\}.$

EXAMPLE 2.1.7. Suppose our experiment consists of a hundred-yard dash involving four college-age sprinters. It is clear that there are many facets of the experiment that we might be interested in, such as the name of the winner, the winning time, the order in which the four cross the finish line, the time of the second-place man, or the time of the third-place man. Which facets were of interest would determine the sample space to be used. For example, if we were going to refer only to the winning time we could use

$$S_1 = \{t: 0 \leq t \leq 15\}$$

(measuring time in seconds). Or, if we were interested in the times of both the first- and second-place men, we could use

$$S_2 = \{(t_1, t_2): 0 \leq t_1 \leq 15, t_1 < t_2 \leq 20\}.$$

In this latter case t_1 is the time of the winner and t_2 is the time of the second-place man; thus the requirement $t_2 > t_1$. For either of these sample spaces we could define the event A: winning time is between 9.45 and 9.65 seconds; using S_1 we would have

$$A = \{t: 9.45 \leq t \leq 9.65\}.$$

And using S_2 we would have

$$A = \{(t_1, t_2): 9.45 \leq t_1 \leq 9.65, t_1 < t_2 \leq 20\}.$$

Again, then, the sample space is not unique. If A were the only event of interest, we would undoubtedly decide to use S_1 rather than S_2 since this would require keeping track of only the single time of the first-place finisher.

EXERCISE 2.1

1. Specify a sample space for the experiment which consists of drawing 1 ball from an urn containing 10 balls of which 4 are white and 6 are red. (Assume that the balls are numbered 1 through 10.)

2. Specify a sample space for the experiment which consists of drawing 2 balls with replacement from the urn containing 10 balls (that is, the first ball removed is replaced in the urn before the second is drawn out). Again assume that they are numbered.

3. Specify a sample space for the experiment which consists of drawing 2 balls without replacement from the urn containing 10 balls (that is, the first ball removed is not replaced in the urn before the second is drawn out). Assume that they are numbered.

4. For the sample space given in problem 1, define the events (as subsets):

A: a white ball is drawn
B: a red ball is drawn.

5. For the sample space given in problem 2, define the events (as subsets):

C: the first ball is white
D: the second ball is white
E: both balls are white.

Does $C \cap D = E$?

6. A cigarette company packs 1 of 5 different slips, labelled a, b, c, d, e, respectively, with each pack it produces. Suppose that you buy 2 packs of cigarettes of this brand. What is a good sample space for the experiment whose outcome is the pair of slips you receive with the 2 packs?

7. Suppose that all of the residents of a particular town are bald or have brown hair or have black hair. Furthermore, each resident has blue eyes or brown eyes. We select one resident at random. Give a sample space S for this experiment and define, as subsets, these events:

   A: the selected resident is bald
   B: the selected resident has blue eyes
   C: the selected resident has brown hair and brown eyes.

8. Three girls, Marie, Sandy, and Tina, enter a beauty contest. Prizes are awarded for first and second place. Specify a sample space for the experiment which consists of the choice of the two winners. Define as subsets, the events:

   A: Marie wins
   B: Marie gets second prize
   C: Tina and Sandy get the prizes.

9. Three cards are selected at random without replacement from a deck which contains 3 red, 3 blue, 3 green, and 3 black cards. Give a sample space for this experiment and define the events:

   A: all the selected cards are red
   B: 1 card is red, 1 green, and 1 blue
   C: 3 different colors occur
   D: all 4 colors occur.

10. A small town contains 3 grocery stores (call them 1, 2, 3). Four ladies living in this town each randomly and independently pick a store in which to shop (in this town). Give a sample space for the experiment which consists of the selection of stores by the ladies and define the events:

    A: all the ladies choose store 1
    B: half the ladies choose store 1 and half choose store 2
    C: all the stores are chosen (by at least one lady).

11. Nine horses are entered in a race (numbered 1 through 9). Prizes are given for win, place, and show (first, second, and third place, respectively). Give a reasonable sample space for this experiment and, as subsets, define the events.

    A: horse number 5 wins
    B: horse number 5 shows
    C: horse number 5 does not finish in the money.

12. Two light bulbs are placed on test until both fail. Assume that each will burn no more than 1600 hours. Define a reasonable sample space for this experiment and describe, as subsets, the events:

    A: Both bulbs fail in less than 1000 hours
    B: Neither bulb fails in less than 1000 hours
    C: The shortest time to failure (of the two) is 1000 hours
    D: The longest time to failure (of the two) is 1000 hours.

13. Each day a newspaper boy sells papers on the same corner; he is supplied with 30 papers. On any particular day, the number he will sell is not known. Define a reasonable sample space for the experiment which consists of the number of sales he will make on a given day. Define the events:

> A: He sells at least 5 papers
> B: He sells at most 5 papers
> C: He sells exactly 5 papers.

14. (Refer to Exercise 13.) Consider the experiment which consists of the number of sales this newspaper boy will make on two successive days. Give a reasonable sample space for the experiment and define the events:

> A: He sells at least 5 papers the first day
> B: He sells at least 5 papers the second day
> C: He sells at least 5 papers both days.

15. A large retail radio and television store receives a shipment of 50 radios, all of the same make and model. Each radio either works correctly or it does not. A receiving clerk must verify that each radio accepted does work correctly (or not, in which case it is returned). What is a reasonable sample space for this experiment? Define the events:

> A: All work correctly
> B: None work correctly
> C: The first two do not work correctly, but the rest do.

2.2. Probability Axioms

As we shall see in this section, the theory of probability is concerned with consistent ways of assigning numbers to events (subsets of the sample space S) which are called the *probabilities of occurrence* of these events. It is because we want to have the ability to compute the probability of occurrence of any subset of the sample space that we permit every subset to be called an event. In almost every problem, then, we shall be aware that there are many events which we could define and whose probabilities we could compute in addition to the particular few events that we shall have interest in. It seems much more satisfying to have the ability to compute probabilities for any conceivable event (most of which are not of interest) than suddenly to come across a problem for which we cannot derive an answer because the quantity of interest is not an event.

Most people with an intuitive feeling for what probability should be give a relative frequency interpretation to numbers called probabilities. For example, most would be quick to agree that the probability of a head occurring if we flip a fair coin is one-half, meaning that if the coin is fair then half the time we should observe a head. There are two immediate consequences to a relative frequency motivation for probability. First, the relative frequency of occurrence of something we are sure will occur should be 1; thus its probability should be 1. For example, if we flip a coin one time the relative frequency of the event of observing a head or a tail should be 1, thus the probability of this event should be 1. Second, a relative frequency can never be negative; thus the probability of any event should be nonnegative. These two rules are the first two axioms of an abstract probability theory in order that assignments of probabilities will have these intuitive properties. We shall insist that the probabilities of occurrence of any events satisfy these two requirements.

There is only one other rule or axiom which we shall insist is always satisfied—the additivity property of probability. If we refer to the experiment which consists of one roll of one die, it is only reasonable to expect that the probability that we observe a 1 or a 2 should be equal to the sum of the probability of observing a 1 plus that of observing a 2 since relative frequencies of occurrence have this property. More generally, we would expect the probability of any event A to be equal to the sum of the probabilities of any two nonoverlapping events which together constitute the event A (their union is A). This is the additivity requirement.

Probability can be called a *measure* applied to the events that can occur when an experiment is performed. To the extent that the assumptions that generate the numerical values of the measure are "correct" for the given problem, the probabilities coincide with the relative frequency notions just discussed. To ensure that this is the case, the probability measures must satisfy the three axioms given below.

Formally, a probability function is a real-valued set function defined on the class of all subsets of the sample space S; the value that is associated with a subset A is denoted by $P(A)$. The assignment of probabilities must satisfy the following three rules (in order that the set function may be called a probability function):

1. $P(S) = 1$
2. $P(A) \geq 0$ for all $A \subset S$
3. $P(A_1 \cup A_2 \cup A_3 \cup \cdots) = P(A_1) + P(A_2) + P(A_3) + \cdots$
   $$\text{if } A_i \cap A_j = \varnothing \quad \text{for all } i \neq j.$$

Note that these are the three axioms just discussed. For any experiment, the sample space S plays the role of the universal set; thus any complements which we refer to are taken with respect to S.

Many consequences or theorems can be derived for any probability function. Let us take a look at some of these now. It will be recalled that $\varnothing \subset S$. Thus our probability measure must assign some number to this event. The number that must always be assigned is 0, as is proved in Theorem 2.2.1.

Theorem 2.2.1. $P(\varnothing) = 0$ for any S.

Proof: $S \cup \varnothing = S$ and thus $P(S \cup \varnothing) = P(S) = 1$ by axiom 1. But $S \cap \varnothing = \varnothing$ so that $P(S \cup \varnothing) = P(S) + P(\varnothing) = 1 + P(\varnothing)$ by axiom 3. Thus $1 + P(\varnothing) = 1$; that is, $P(\varnothing) = 0$.   ◀

A second consequence of the assumed axioms is given as Theorem 2.2.2. We shall see many instances in which it saves a great deal of effort in computing probabilities.

Theorem 2.2.2. $P(\bar{A}) = 1 - P(A)$, where $\bar{A}$ is the complement of A with respect to S.

Proof: $A \cup \bar{A} = S$ so $P(A \cup \bar{A}) = P(S) = 1$ by axiom 1. But $A \cap \bar{A} = \varnothing$ and thus $P(A \cup \bar{A}) = P(A) + P(\bar{A})$ by axiom 3. Thus we have established that $P(A) + P(\bar{A}) = 1$, from which the result follows immediately.   ◀

Axiom 3 tells us that if two events A and B have no elements in common, then the probability of their union is the sum of their individual probabilities. Theorem 2.2.3 derives a preliminary result which is used to establish Theorem 2.2.4, regarding $P(A \cup B)$ when A and B have elements in common.

Theorem 2.2.3. $P(\bar{A} \cap B) = P(B) - P(A \cap B)$.

Proof: By referring to Figure 2.2.1 it can be seen that

$$B = (\bar{A} \cap B) \cup (A \cap B).$$

Then $P(B) = P((\bar{A} \cap B) \cup (A \cap B))$. Furthermore, $(\bar{A} \cap B) \cap (A \cap B) = \varnothing$; thus, $P((\bar{A} \cap B) \cup (A \cap B)) = P(\bar{A} \cap B) + P(A \cap B)$, and we have established that

$$P(B) = P(\bar{A} \cap B) + P(A \cap B)$$

from which the desired result follows immediately.   ◀

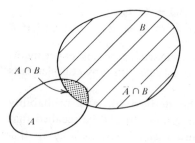

FIGURE 2.2.1   Decomposition of set B.

Theorem 2.2.4. $P(A \cup B) = P(A) + P(B) - P(A \cap B)$.

Proof: By referring to Figure 2.2.1, it can be seen that we can write

$$A \cup B = A \cup (\bar{A} \cap B),$$

and thus $P(A \cup B) = P(A \cup (\bar{A} \cap B))$.   Furthermore, $A \cap (\bar{A} \cap B) = \varnothing$ so that

$$P(A \cup (\bar{A} \cap B)) = P(A) + P(\bar{A} \cap B)$$
$$= P(A) + P(B) - P(A \cap B),$$

by Theorem 2.2.3.   Thus

$$P(A \cup B) = P(A) + P(B) - P(A \cap B). \qquad \blacktriangleleft$$

Notice that the probability function we have discussed in this section is a special sort of (real-valued) set function. For every experiment that can be performed we assume that we can define the sample space S which has as elements all possible outcomes; the class of all subsets of S can then be defined and probabilities are the values assumed by a set function defined on the class of all subsets. The set function whose values are called probabilities is distinguished by the fact that the three axioms just discussed must be satisfied.

It is important to realize that these axioms will not give a unique assignment of probabilities to events; rather the axioms simply clarify relationships between probabilities that we assign so that we shall be consistent with our intuitive notions of probability. For example, if a rocket has been designed to take a man to the moon, then the experiment which consists of firing the rocket and the man at the moon can be thought of as having two outcomes: success and failure. Success would be the safe arrival of the man on the moon, failure anything else that might occur. Then the axioms do not imply that the probability of the event {success} must be $\frac{1}{2}$ or $\frac{3}{4}$ or .99 or any other particular value. If we denote this probability by p, they do imply that $0 \leq p \leq 1$ and that the probability of the event {failure} must be $1 - p$. Beyond this, p is still unspecified.

Actual specification of the value of p must come from analytical considerations of the experiment performed and the mechanism behind it. For the rocket example just mentioned, this would consist of detailed examination of the rocket design and conditions under which it is to be fired, in addition to any prior test firings or performance data available. Generally, considerations of prior data and their implications regarding the value of p fall into the realm of statistics, the topic for the latter half of this volume.

EXERCISE 2.2

1. Given $S = \{1, 2, 3\}$, $A = \{1\}$, $B = \{3\}$, $C = \{2\}$, $P(A) = \frac{1}{3}$, $P(B) = \frac{1}{3}$, find:

    (a) $P(C)$ (d) $P(\bar{A} \cap \bar{B})$

    (b) $P(A \cup B)$ (e) $P(\bar{A} \cup \bar{B})$

    (c) $P(\bar{A})$ (f) $P(B \cup C)$.

2. Let S, A, B, C be defined as in problem 1, but now let $P(A) = \frac{1}{2}$, $P(B) = \frac{1}{5}$. Compute the probabilities asked for in (a) through (f).

3. Let S, A, B, C be defined as in problem 1 and let $P(A) = 1$. Compute the probabilities asked for in (a) through (f). Could we let $P(A)$ be 2?

4. Define $S = \{a, b\}$, $B = \{b\}$. Give three different assignments of probabilities to the subsets of S.

5. Prove, from the axioms, that probabilities are monotonic; that is, $P(A) \leq P(B)$ if $A \subset B$.

6. Prove, from the axioms, that $P(A) \leq 1$ for all A.

7. Given an experiment such that $P(A) = \frac{1}{2}$, $P(B) = \frac{1}{2}$, $P(A \cup B) = \frac{2}{3}$, compute:

    (a) $P(\bar{A})$ (e) $P(\bar{A} \cup \bar{B})$

    (b) $P(\bar{B})$ (f) $P(A \cap \bar{B})$

    (c) $P(A \cap B)$ (g) $P(\bar{A} \cap B)$

    (d) $P(\bar{A} \cap \bar{B})$ (h) $P(\bar{A} \cup B)$.

8. Given an experiment such that $P(A) = \frac{1}{2}$, $P(B) = \frac{1}{3}$, $P(A \cap B) = \frac{1}{4}$, compute:

    (a) $P(A \cup B)$ (d) $P(A \cap \bar{B})$

    (b) $P(\bar{A} \cup B)$ (e) $P(\bar{A} \cap \bar{B})$

    (c) $P(\bar{A} \cap B)$ (f) $P(\bar{A} \cup \bar{B})$.

9. Is it possible to have an assignment of probabilities such that $P(A) = \frac{1}{2}$, $P(A \cap B) = \frac{1}{3}$, $P(B) = \frac{1}{4}$?

10. If we know that $P(A \cup B) = \frac{2}{3}$ and $P(A \cap B) = \frac{1}{3}$, can we determine $P(A)$ and $P(B)$?

11. Suppose we have a simple experiment with sample space $S_1 = \{s, f\}$. Thus only two possible outcomes could occur when the experiment is performed. Assume

    $$P_1(\{s\}) = p, \qquad P_1(\{f\}) = 1 - p$$

    where $0 \leq p \leq 1$. If we have a *compound* experiment consisting of two repetitions of the simple experiment, a reasonable sample space for the compound experiment is

    $$S = S_1 \times S_1$$
    $$= \{(s, s), (s, f), (f, s), (f, f)\}$$

    Under certain conditions, to be explored in some detail in Section 2.7, it is reasonable to assign probabilities to single-element subsets of S by using the product of probabilities of observing the appropriate outcomes from the sample space S_1 of the simple experiment. Thus

$$P(\{s, s\}) = P_1(\{s\}) \cdot P_1(\{s\}) = p^2$$

$$P(\{s, f\}) = P_1(\{s\}) \cdot P_1(\{f\}) = p(1 - p), \text{ etc.}$$

Show that this rule for computing probabilities of events in the compound experiment satisfies the three axioms.

12. A fair coin is to be flipped two times. Letting s represent heads and f represent tails, this could be looked at as being a compound experiment, as discussed above in Exercise 11.
     (a) What would you say is a reasonable value for p?
     (b) Using the value assigned above to p, compute the probability that the same face occurs on both flips.

13. For the situation described in Exercise 12, assume that the coin used is two-headed and answer (a) and (b).

14. The probability that both coins show the same face, as discussed in Exercises 11 and 12, is $p^2 + (1 - p)^2$. Show that this probability is minimized with $p = 1/2$.
     (Hint: Look at the first and second derivatives with respect to p.)

15. In the game "matching coins" two people each flip a penny (or some other coin). If the two coins show the same face, person 1 wins and gets both coins; if they don't, person 2 wins and gets both coins. Can you see any application of the result mentioned in Exercise 14, in playing this game (would you prefer to be person 1 or person 2)?

2.3. Single-Element Events and the Equally Likely Case

When we are dealing with an experiment that has a finite number of possible outcomes (thus S is a finite set), the concept of a single-element event becomes rather important; as we shall see, the specification of the values of the probability function P for the single-element events then completely specifies the values of P for all events. First, the definition of a single-element event is as follows.

> DEFINITION 2.3.1. A *single-element event* A is a subset of S which has only one element of S belonging to it; that is, if there exists only one $x \in S$ such that $x \in A \subset S$, then A is a single-element event.

For the sample space $\{H, T\}$, the single-element events are $\{H\}$, $\{T\}$. For the sample space $\{1, 2, 3, 4, 5, 6\}$, the single-element events are $\{1\}$, $\{2\}$, $\{3\}$, $\{4\}$, $\{5\}$, $\{6\}$. For the sample space $\{(0, 0), (0, 1), (1, 0), (1, 1)\}$, the single-element events are $\{(0, 0)\}$, $\{(0, 1)\}$, $\{(1, 0)\}$, $\{(1, 1)\}$. As can easily be seen, if S has n elements, then there are exactly n distinct single-element events. The following theorem shows how the probabilities of occurrence of the single-element events imply the probabilities of occurrence of any event, no matter how many elements belong to it.

Theorem 2.3.1. Given a sample space S and any event $A \subset S$,

$$P(A) = P(A_1) + P(A_2) + \cdots + P(A_k)$$

where $A_1, A_2, \ldots, A_k$ are distinct single-element events and

$$A = A_1 \cup A_2 \cup \cdots \cup A_k.$$

Proof: Suppose that $A_1, A_2, \ldots, A_k$ are distinct single-element events. Then $A_i \cap A_j = \varnothing$ for all $i \neq j$ and, by axiom 3, $P(A_1 \cup A_2 \cup \cdots \cup A_k) = \sum_{i=1}^{k} P(A_i)$. Thus, if $A = A_1 \cup A_2 \cup \cdots \cup A_k$, we have

$$P(A) = \sum_{i=1}^{k} P(A_i). \qquad \blacktriangleleft$$

EXAMPLE 2.3.1. Suppose we roll a six-sided die. Then $S = \{1, 2, 3, 4, 5, 6\}$. The single-element events are $\{1\}, \{2\}, \{3\}, \{4\}, \{5\}, \{6\}$. Theorem 2.3.1 tells us that if we know the probability of occurrence for each of these six events, we can use this information to compute the probability of occurrence of any other event of interest. For example,

$$P(\{1, 4, 6\}) = P(\{1\}) + P(\{4\}) + P(\{6\}).$$

In many experiments it is quite reasonable to assume that each single-element event is as likely to occur as any other. For example, if our experiment consists of one flip of a fair coin, it is reasonable to assume that the single-element events of $S = \{H, T\}$ are equally likely to occur. Or, if our experiment consists of one roll of a fair die, it is reasonable to assume that the single-element events of $S = \{1, 2, 3, 4, 5, 6\}$ are equally likely to occur. If we make the assumption that all single-element events of S are equally likely, there is a very simple way in which we can assign probabilities to all the subsets of S and still be consistent with the three axioms.

Assume that we have an experiment with k possible outcomes, and from the description of the experiment we are willing to assume that each of the single-element events is as likely to occur as any other. Then, since the total probability of the whole sample space is to be 1, the common value of the probability for each of the single-element events must be $1/k$. Furthermore, since any event is the union of single-element events (as we just saw in Theorem 2.3.1), the probability of any event $A \subset S$ then is given by the ratio of the number of elements in A (the number of single-element events whose union is A) to the number of elements in S. That is, we use the rule

$$P(A) = \frac{n(A)}{n(S)} \quad \text{for } A \subset S$$

where $n(A)$ is the number of elements in A (see Example 1.3.2 in Chapter 1). That this rule will satisfy the three axioms given in Section 2.2 is proved in the next theorem.

Theorem 2.3.2.   If S has k elements the rule

$$P(A) = \frac{n(A)}{n(S)}$$

satisfies the three axioms for a probability function.

Proof: If S has k elements, then $n(S) = k$.

$$P(S) = \frac{n(S)}{n(S)} = \frac{k}{k} = 1$$

so axiom 1 is satisfied. If A is any subset of S, it contains a nonnegative number of elements; i.e., $n(A) \geq 0$ for all $A \subset S$. Then

$$\frac{n(A)}{k} = \frac{n(A)}{n(S)} = P(A) \geq 0 \qquad \text{for all } A \subset S$$

and axiom 2 is satisfied. If $A \cap B = \emptyset$, then A and B have no elements in common and we would know that $n(A \cup B) = n(A) + n(B)$. Thus

$$\frac{n(A \cup B)}{n(S)} = \frac{n(A)}{n(S)} + \frac{n(B)}{n(S)};$$

that is,

$$P(A \cup B) = P(A) + P(B).$$

Clearly this line of reasoning is valid for the union of any number of non-overlapping events and thus axiom 3 is also satisfied.   ◄

Thus, for any problem in which we are justified in assuming equally likely single-element events, we now have a rule which enables us to compute the probability of occurrence of any event.

EXAMPLE 2.3.2.   Suppose we roll a fair die one time. What is the probability of getting an even number? What is the probability of getting a number which is greater than 4? Our sample space is $S = \{1, 2, 3, 4, 5, 6\}$. Since the die is fair, we assume the single-element events to be equally likely to occur; each then has probability $\frac{1}{6}$ of occurrence. Let A be the event that an even number occurs and let B be the event that we get a number greater than 4.

$$A = \{2, 4, 6\}, \qquad B = \{5, 6\},$$
$$n(A) = 3, \qquad n(B) = 2, \qquad n(S) = 6,$$

and we get $P(A) = \frac{3}{6}$, $P(B) = \frac{2}{6}$.

EXAMPLE 2.3.3. We roll a pair of fair dice one time. What is the probability that the sum of the two numbers is 2? That it is 7? That it is 11?

Our sample space is $S = \{(x_1, x_2): x_1 = 1, 2, \ldots, 6; x_2 = 1, 2, \ldots, 6\}$. Since the first die can have a number from 1 through 6 on it and, quite independently, the second die can also have any number 1 through 6 on it, we reason that there are $6 \cdot 6 = 36$ elements belonging to S (or we simply list them all and count them). Thus $n(S) = 36$. Let A be the event that the sum is 2, B the event that the sum is 7, C the event that the sum is 11. Then

$$A = \{(1, 1)\}$$
$$B = \{(1, 6), (2, 5), (3, 4), (4, 3), (5, 2), (6, 1)\}$$
$$C = \{(5, 6), (6, 5)\}$$

and we see that $n(A) = 1$, $n(B) = 6$, $n(C) = 2$. Then since we are assuming that the dice are fair, $P(A) = \frac{1}{36}$, $P(B) = \frac{6}{36}$, $P(C) = \frac{2}{36}$.

EXAMPLE 2.3.4. A certain class has 20 students. Of the 20, 7 are blue-eyed blond girls, 4 are blue-eyed brunette girls, 5 are blue-eyed blond boys, and the remaining 4 are brown-eyed brunette boys. We select 1 student at random. What is the probability that the selected student is a girl? Has blue eyes? Is a brunette? Is a brown-eyed blond? For ease of notation, we assume that the 20 students are numbered in some specified order. Then our sample space is

$$S = \{1, 2, \ldots, 20\}$$

and the single-element events are equally likely to occur (since the student is selected at random). Define the following events:

A: the selected student is a girl
B: the selected student has blue eyes
C: the selected student is a brunette
D: the selected student is a brown-eyed blond.

Then $N(S) = 20$, $N(A) = 11$, $N(B) = 16$, $N(C) = 8$, $N(D) = 0$, and we have

$$P(A) = \tfrac{11}{20}, \qquad P(B) = \tfrac{16}{20}, \qquad P(C) = \tfrac{8}{20}, \qquad P(D) = 0.$$

EXAMPLE 2.3.5. Assume four people enter a tournament; the game to be played is one which cannot end in a tie (such as tennis, chess, cribbage, dominoes, etc.). For concreteness, assume the four people are named Doug, Joe, Hugh and Ray. The tournament will be single elimination; this means that Doug and Joe will play each other (game 1) and the loser will be eliminated. Simultaneously or subsequently Hugh and Ray will play each other (game 2) and the loser will be eliminated. The two winners then play each other (game 3); the winner of this game wins the tournament. What is the probability that Ray wins the tournament? Let us first find a reasonable sample space for this experiment. Since three games will be played, and the winners of the games are the important quantities to record, let us adopt a sample space whose elements are 3-tuples. The first position records the winner of game 1, the second position records the winner of game 2, and the third position records the winner of game 3 (the winner of the tournament). Thus we will use

$$S = \{(\text{Doug, Hugh, Doug}), (\text{Doug, Hugh, Hugh}), (\text{Joe, Hugh, Joe}),$$
$$(\text{Joe, Hugh, Hugh}), (\text{Doug, Ray, Doug}), (\text{Doug, Ray, Ray}),$$
$$(\text{Joe, Ray, Joe}), (\text{Joe, Ray, Ray})\}.$$

To answer questions about who will win the tournament, we must assign values to the probabilities of occurrence of the single-element events. *If the players are all of equal ability*, each single-element event would be assigned probability $\frac{1}{8}$. Under this assumption

$$P(\text{Ray wins}) = P(\{(\text{Doug, Ray, Ray}), (\text{Joe, Ray, Ray})\})$$
$$= P(\{(\text{Doug, Ray, Ray})\}) + P(\{(\text{Joe, Ray, Ray})\})$$
$$= \tfrac{2}{8} = \tfrac{1}{4}.$$

Not surprisingly, with this assumption, we also have

$$P(\text{Doug wins}) = P(\text{Joe wins}) = P(\text{Hugh wins}) = \tfrac{1}{4}.$$

Many important problems have equally likely single-element events. As we have seen, computations of probabilities in these cases reduce to counting the number of elements in S and the number of elements in the events of interest. Probabilities are then given by the ratio of these quantities. Since counting the number of elements belonging to an event plays a role in many practical problems, the next two sections are devoted to counting techniques.

EXERCISE 2.3

1. If two fair coins are flipped, what is the probability that the two faces are alike?

2. If we draw 1 card at random from a standard deck of 52, what is the probability that it is red? That it is a diamond? That it is an ace? That it is the ace of diamonds?

3. Five different colored rubber bowls with the same identical dog food in them are laid out in a row. If a dog chooses a bowl at random from which to eat, what is the probability that he selects the blue one? If a second dog is used, what is the probability that he selects the blue one? What is the probability that both choose the blue one?

4. A pair of fair dice is rolled once. Compute the probability that the sum is equal to each of the integers 2 through 12.

5. A one is painted on the head side and a two is painted on the tail side of each of 3 fifty-cent pieces. If the 3 coins are all tossed once (together), compute the probability that the sum of the three numbers occurring is each of the integers 3 through 6.

6. Forty people are riding on the same railroad car. Of this number, 5 are Irish ladies with blue coats, 2 are Irish men with green coats, 1 is an Irish man with

a black coat, 7 are Norwegian ladies with brown coats, 2 are Norwegian ladies with blue coats, 6 are Norwegian men with black coats, 4 are German men with green coats, 3 are German ladies with black coats, 5 are German ladies with blue coats, and 5 are German men with black coats. If we select one person at random from this car, what is the probability that the selected person is a man? Is wearing a green coat? Is wearing a brown coat? Is a Norwegian? Is a German? Is a German wearing a green coat?

7. Suppose that a die has been loaded in such a way that the probability of a particular number occurring is proportional to that number. Compute the probabilities of all the single-element events and use these to compute the probability of occurrence of (a) an even number and (b) a number greater than 4.

8. Suppose that a die has been loaded in such a way that the particular number occurring is inversely proportional to that number. Compute the probabilities of all the single-element events and use these to compute the probability of occurrence of (a) an even number and (b) a number greater than 4.

9. For the tournament example (Example 2.3.5), keep the elements of the sample space in the order listed. Instead of assuming they are all equally likely, assume the probabilities assigned to the single-element events are, respectively, 4, 8, 1, 5, 4, 2, 1, 2, each divided by 27. Compute the probabilities that each of the four contestants will win the tournament.

10. Using the assignment of probabilities in Exercise 9, compute the probability that each of the four contestants loses his first game.

11. Assume eight players of equal ability enter a single-elimination tournament (no ties allowed). What is the probability that each will be the winner of the tournament? What is the probability that player 1 wins his first two games and loses the final?

12. Assume we have a sample space S with n distinct elements, numbered 1 through n. The single-element event whose element is 1 is assigned probability 1.
    (a) What value must be assigned the probabilities of occurrence of the other single-element events?
    (b) What are the practical implications of such an assignment?

2.4. Counting Techniques

As was mentioned in the last section, the solutions of many probability problems depend on being able to count the numbers of elements belonging to particular sets. A number of techniques are invaluable aids in counting the number of elements belonging to certain sets and thus they are of use in solving probability problems in which these sets occur. Unfortunately, there is no general technique that is universally applicable to all counting problems (other than making a complete list and counting the number of items in it); thus it is necessary to attempt to tailor-make counting methods to the particular counting problem in hand, a process that is always frustrating and frequently fruitless. Be that as it may, almost all counting problems seem to yield eventually to a closely reasoned analysis.

The first technique of some generality that we shall discuss is a very simple one frequently referred to as the *multiplication principle*. This is defined as follows.

DEFINITION 2.4.1. If a first operation can be performed in any of n_1 ways and a second operation can then be performed in any of n_2 ways, both operations can be performed (the second immediately following the first) in $n_1 \cdot n_2$ ways.

For example, if we can travel from town A to town B in 3 ways and from town B to town C in 4 ways, then we can travel from A to C in a total of $3 \cdot 4 = 12$ ways. Or, if the operation of tossing a die gives rise to 1 of 6 possible outcomes and the operation of tossing a second die gives rise to 1 of 6 possible outcomes, then the operation of tossing a pair of dice gives rise to $6 \cdot 6 = 36$ possible outcomes.

EXAMPLE 2.4.1. Suppose that a set A has n_1 elements and a second set B has n_2 elements. Then the Cartesian product $A \times B$ (see Definition 1.2.6) has $n_1 n_2$ elements. The Cartesian product $A \times A$ has n_1^2 elements, $B \times B$ has n_2^2 elements.

This definition can immediately be extended to any number of operations. The operation of tossing three coins gives rise to $2^3 = 8$ possible outcomes. The operation of rolling 5 six-sided dice gives rise to $6^5 = 7776$ possible outcomes. If the set A_i has n_i elements, $i = 1, 2, \ldots, k$, the Cartesian product $A_1 \times A_2 \times \cdots \times A_k$ has $n_1 n_2 \cdots n_k$ different elements.

DEFINITION 2.4.2. An arrangement of n symbols in a definite order is called a *permutation* of the n symbols.

We shall frequently want to know how many different n-tuples can be made using n different symbols each only once (that is, how many permutations are possible). The multiplication principle will immediately give us the answer. We can count the number of n-tuples by reasoning as follows. In listing all the possible n-tuples, we would perform n natural operations. First we must fill the leftmost position of the n-tuple. Then we must fill the second leftmost position, the third leftmost position, etc. Since we could put any of the n elements available into the leftmost position, this operation can be performed in n ways. After we fill the leftmost position, we can use any of the remaining $n - 1$ elements to fill the second leftmost position; we can use any of the remaining $n - 2$ elements to fill the third leftmost position and so on until we finally arrive at the rightmost position and want to count the number of ways it can be filled. We have used $n - 1$ elements at this point to fill the first $n - 1$ positions and thus have left only one element which must be used to fill the final position. Then the total number of ways we can perform all n operations (which is also the total number of n-tuples we could make with the given n symbols) is given by the product of the numbers of ways of doing the individual operations. Thus the number of different n-tuples is

$$n(n - 1)(n - 2) \cdots 2 \cdot 1$$

which we write $n!$ (read n-factorial).

EXAMPLE 2.4.2.   Suppose that the same 5 people park their cars on the same side of the street in the same block every night. How many different orderings of the 5 cars parked on the street are possible? The different orderings of the cars parked on the street could be represented by 5-tuples with 5 distinct elements; then if we can count the number of different 5-tuples that are possible, we also know the number of different orderings on the street. The number of 5-tuples possible is of course $5! = 5 \cdot 4 \cdot 3 \cdot 2 \cdot 1 = 120$; thus these 5 people could park their cars on the street in a different order every night for 4 months without repeating an ordering they had already used.

DEFINITION 2.4.3.   The number of r-tuples we can make $(r \leq n)$, using n different symbols (each only once), is called the *number of permutations of n things r at a time* and is denoted by $_nP_r$.

How might we compute the value of $_nP_r$? Each r-tuple has exactly r positions. The leftmost position could be filled by any of the n symbols; the second leftmost position could then be filled by any of the remaining $n - 1$ symbols, etc. By the time we are ready to fill the rth position, we have used $(r - 1)$ symbols already and any of the remaining $n - (r - 1)$ symbols could be used in the rth position. Thus the total number of r-tuples we could construct is $n(n - 1) \cdots (n - r + 1)$ and we have

$$_nP_r = n(n - 1) \cdots (n - r + 1).$$

If we multiply this number by $(n - r)!/(n - r)!$, we certainly do not change its value; but in so doing we get a form of $_nP_r$ that is much easier to remember:

$$_nP_r = n(n - 1)(n - 2) \cdots (n - r + 1)\frac{(n - r)!}{(n - r)!}$$

$$= \frac{n(n - 1)(n - 2) \cdots (n - r + 1)(n - r)(n - r - 1) \cdots 2 \cdot 1}{(n - r)!}$$

$$= \frac{n!}{(n - r)!}$$

EXAMPLE 2.4.3.   (a) Fifteen cars enter a race. In how many different ways could trophies for first, second and third place be awarded? This answer is simply

$_{15}P_3 = 15!/12! = 2730$ since the question is equivalent to asking how many permutations are there of 15 objects, 3 at a time.

(b) How many of the 3-tuples just counted have car number 15 in the first position? This can be answered in two ways. First we might reason that there are $_{14}P_2 = 14!/12! = 182$ ways in which the last two positions could be filled, having already put 15 into the first position of the 3-tuple. Alternatively, there must obviously be equal numbers of 3-tuples (in the totality of all possible) having car 15 in first place as there are having car 14 in first place, 13 in first place, etc. Thus if we divide the total number of 3-tuples by 15, we should get the number that have car 15 in first place; this gives $2730/15 = 182$—the same answer derived above.

EXAMPLE 2.4.4. (a) How many three-letter "words" can we make using the letters w, i, n, t, e, r (allowing no repetition)? (A "word" is any arrangement of letters, regardless of whether it is in actual fact a word listed in the dictionary for some language.) This is, of course, just $_6P_3 = 6!/3! = 120$. The number of four-letter words is $_6P_4 = 360$, etc.

(b) Suppose that repetition of a letter is allowed in making three-letter words using w, i, n, t, e, r. How many three-letter words can we make? The answer is $6 \cdot 6 \cdot 6 = 6^3 = 216$, since we now would be able to fill each position with six letters. The number of four-letter words we could make, allowing repetition, then is $6 \cdot 6 \cdot 6 \cdot 6 = 6^4 = 1296$.

(c) How many three-letter words are there with one or more repeated letters? How many four-letter words are there with one or more repeated letters? We know that there are 216 three-letter words if repetitions are allowed and 120 three-letter words if repetition is not allowed. Thus, there are $216 - 120 = 96$ three-letter words with one or more repeated letters. Analogously, there are $1296 - 360 = 936$ four-letter words with one or more repeated letters.

DEFINITION 2.4.4. The number of subsets, each of size r, that a set with n elements has is called the number of *combinations* of n things r at a time and is denoted by $\binom{n}{r}$.

Remember that sets are not ordered. If we knew $\binom{n}{r}$ and multiplied by $r!$, we would have to get $_nP_r$, the number of permutations of n things r at a time, since each distinct subset of r elements would give rise to $r!$ different r-tuples. Thus we get

$$\binom{n}{r} r! = {}_nP_r = \frac{n!}{(n-r)!}$$

which, dividing through by $r!$, gives us

$$\binom{n}{r} = \frac{n!}{r!\,(n-r)!}.$$

The most difficult part of many counting problems is deciding whether ordering should be of importance. If ordering does not matter we want combinations; if ordering is of importance we want permutations.

EXAMPLE 2.4.5.   (a) How many distinct 5-card hands can be dealt from a standard 52-card deck? Since the 5-card hand remains unchanged if you received the same 5 cards, but in a different order, the answer is

$$\binom{52}{5} = \frac{52!}{5!\,47!} = 2,598,960.$$

(b) How many distinct 13-card hands can be dealt from a standard deck?

$$\binom{52}{13} = 6.35 \times 10^{11}.$$

(c) Suppose that 10 boys go out for basketball at a particular school. How many different teams could be fielded from this school?

$$\binom{10}{5} = 252.$$

This is the number of different collections or subsets of 5 boys that could be chosen from the 10. If different allocations of boys to the positions are to be counted as different teams, then there are

$$5!\,(252) = 30,240 = {}_{10}P_5$$

different teams.

(d) One of the 10 boys out for basketball is named Joe. How many of the 252 teams include Joe as a member? If we want to count only those teams which include Joe, then we need only count how many ways might we select 4 additional individuals to be on the team. This is

$$\binom{9}{4} = 126.$$

$\binom{n}{k}$ is frequently called a combinatorial coefficient. Tables are available giving the value of $\binom{n}{k}$ for varying values of n and k. Let us note a few facts regarding these coefficients. First, $\binom{n}{0} = 1$ for all n as can be seen by simply evaluating the factorials involved. As is also easily seen, $\binom{n}{1} = n$ for all n. One further result which will be useful is the identity $\binom{n}{k} = \binom{n}{n-k}$ for any n and k. By writing these two coefficients out in factorial form, it is immediately evident that they are always equal. Recalling that $\binom{n}{k}$ is the number of subsets of size k that we can construct for a set having n elements, we realize that every time we write down a subset of k items, we leave behind a subset of size $n - k$. Thus it is not surprising that there are always equal numbers of subsets of these two sizes.

For fixed n, $\binom{n}{k}$ is an increasing function of k for $k \leq (n + 1)/2$, and a decreasing function of k, for $k > (n + 1)/2$. This can be seen as follows: First,

$$\frac{\binom{n}{k}}{\binom{n}{k+1}} = \frac{n!/k!\,(n-k)!}{n!/(k+1)!\,(n-k-1)!}$$

$$= \frac{k+1}{n-k}$$

Note then that

$$\binom{n}{k} \leq \binom{n}{k+1}$$

so long as

$$\frac{k+1}{n-k} \leq 1,$$

i.e., so long as

$$k \leq \frac{n-1}{2}.$$

Thus the largest value occurs at $[(n - 1)/2] + 1 = (n + 1)/2$ as claimed. Then, if n is odd, $\binom{n}{k}$ increases to a maximum with $k = (n + 1)/2$ and since $\binom{n}{k} = \binom{n}{n-k}$, notice that

$$\binom{n}{n - \frac{n+1}{2}} = \binom{n}{\frac{n-1}{2}}$$

so the two middle terms are equal. If n is even the (unique) maximum value of $\binom{n}{k}$ occurs with $k = n/2$ (since $(n + 1)/2$ is not an integer in this case).

Let us now state and prove the *binomial theorem*, which may already be familiar from high school algebra, at least for some values of n. This result is useful in a surprising number of different problems, not the least of which, for our purposes, will be in describing the binomial random variable.

Theorem 2.4.1.   If x and y are any two real numbers and n is a positive integer, then

$$(x + y)^n = \sum_{i=0}^{n} \binom{n}{i} x^i y^{n-i}, \quad \text{where} \quad \binom{n}{i} = \frac{n!}{(n-i)!\, i!}.$$

Proof: We note first of all that

$$(x + y)^1 = x + y = \binom{1}{0} x^0 y^{1-0} + \binom{1}{1} x^1 y^{1-1} = \sum_{i=0}^{1} \binom{1}{i} x^i y^{1-i}.$$

Now we assume that

$$(x + y)^k = \sum_{i=0}^{k} \binom{k}{i} x^i y^{k-i}$$

and wish to use only this assumption to establish that

$$(x + y)^{k+1} = \sum_{i=0}^{k+1} \binom{k+1}{i} x^i y^{k+1-i}.$$

Now

$$(x + y)^{k+1} = (x + y)(x + y)^k$$

$$= (x + y) \sum_{i=0}^{k} \binom{k}{i} x^i y^{k-i}$$

$$= \sum_{i=0}^{k} \binom{k}{i} x^{i+1} y^{k-i} + \sum_{i=0}^{k} \binom{k}{i} x^i y^{k+1-i}.$$

Combining like terms in these two sums, we have

$$(x + y)^{k+1} = y^{k+1} + \sum_{i=1}^{k} \left[\binom{k}{i-1} + \binom{k}{i} \right] x^i y^{k+1-i} + x^{k+1}.$$

Now

$$\binom{k}{i-1} + \binom{k}{i} = \frac{k!}{(k+1-i)!\,(i-1)!} + \frac{k!}{(k-i)!\, i!}$$

$$= \frac{i \cdot k! + (k+1-i)k!}{(k+1-i)!\, i!}$$

$$= \frac{(k+1)k!}{(k+1-i)!\, i!} = \binom{k+1}{i}.$$

Since $\binom{k+1}{0} = \binom{k+1}{k+1} = 1$, we have

$$(x + y)^{k+1} = \sum_{i=0}^{k+1} \binom{k+1}{i} x^i y^{k+1-i}$$

and we thus have established that

$$(x + y)^n = \sum_{i=0}^{n} \binom{n}{i} x^i y^{n-i},$$

for $n = 2, 3, 4, 5, \ldots$ ◄

The following example shows one particular use for this theorem.

EXAMPLE 2.4.6.   (a) How many subsets does a set S with n elements have? We are able to compute the answer by using the binomial theorem and the combinatorial coefficients. Clearly, if we add together the number of subsets of S having $0, 1, 2, \ldots, n$ elements, respectively, this sum would be the total number of subsets of S. $\binom{n}{k}$ is the number of subsets of S, each having k elements, for $k = 0, 1, 2, \ldots, n$. Thus, the total number of subsets is

$$\sum_{k=0}^{n} \binom{n}{k} = \sum_{k=0}^{n} \binom{n}{k} 1^k 1^{n-k} = (1 + 1)^n = 2^n.$$

A set with 2 elements has $2^2 = 4$ subsets; a set with 5 elements has $2^5 = 32$ subsets; a set with 100 elements has $2^{100} = 1.27 \times 10^{30}$ subsets.

(b) If $S = \{1, 2, \ldots, n\}$, how many subsets of S have 1 as an element? Clearly, only one single-element event has 1 as an element. The number of subsets of size 2 having 1 as an element is $\binom{n-1}{1}$ since the second element in the subset could be any one of the remaining $n - 1$. The number of subsets of size 3 having 1 as an element is $\binom{n-1}{2}$; the number of subsets of size r having 1 as an element is $\binom{n-1}{r-1}$ where $r = 1, 2, \ldots, n$. Thus, the total number of subsets having 1 as an element is

$$1 + \binom{n-1}{1} + \binom{n-1}{2} + \cdots + \binom{n-1}{n-1};$$

but, by the binomial theorem, this sum equals 2^{n-1}. Obviously, no matter which single element belonging to S we consider, it belongs to 2^{n-1} events. Since there are 2^n events in total, then exactly half of all possible events occurs, no matter which element belonging to S is the outcome we observe.

EXERCISE 2.4

1. How many ways can 3 different books be arranged side by side on a shelf?
2. If an item sold by a vending machine costs 40 cents, and the money deposited

into the machine must consist of a quarter, a dime, and a nickel, in how many different orders could the money be inserted into the machine?

3. Six people are about to enter a cave in single file. In how many ways could they arrange themselves in a row to go through the entrance?

4. A bag contains 1 red, 1 black and 1 green marble. I randomly select 1 of the marbles and record its color. I then replace it in the bag, shake the bag, and randomly select a second marble, again recording its color. The second marble is then replaced and a third marble is randomly selected and its color recorded. How many different samples of 3 colors could occur?

5. An ant farm contains both red and black ants. A particular passage in the farm is so narrow that only 1 ant can get through at a time. If 4 ants follow each other through the passage, how many different color patterns (having 4 elements) could be produced (assuming that red ants are indistinguishable from one another, as are black ants)?

6. A particular city is going to give 3 awards to outstanding residents. If 4 people are eligible to receive them, in how many different ways could they be distributed among the 4 people (assuming that no person may receive more than 1 award)?

7. If a set has 3 elements, how many subsets does it have?

8. Could we define a set that has exactly 9 subsets?

9. How many selections of 5 dominoes can be made from a regular 28-domino double-6 set?

10. In how many ways could 2 teams be chosen from an 8-team league?

11. How many committees of 3 people could be chosen from a group of 10?

12. How many 5-man squads could be chosen from a company of 20 men?

13. Given a set of 15 points in a plane, how many lines would be necessary to connect all possible pairs of points?

14. A "complete graph of order 3" is given by connecting 3 points in all possible ways. If 15 points are joined in all possible ways, how many complete graphs of order 3 would be included? Of order $k = 4, 5, 6, \ldots, 15$?

15. Given a box with 2 25-watt, 3 40-watt, and 4 100-watt bulbs, in how many ways could 3 bulbs be selected from the box?

16. Referring to the light bulbs in Exercise 15, how many of these selections of three bulbs would include both 25-watt bulbs? How many would include no 25-watt bulbs?

17. How many bulb selections defined in Exercise 15 would include exactly 1 of each of the 3 different wattages?

18. Many states use automobile license plates which have three letters followed by three digits.
   (a) Assuming all possible combinations can be used, how many different plates can be made?

(b) If 196 three-letter combinations have a bad connotation, and will not be used, how many plates can be made?

(c) Assume that each license plate again is to have a total of six positions, three of which must be letters and three of which must be digits. How many different plates could be made (ignore the restriction in (b))?

19. A small town has 100 registered voters, 60 of whom favor a school bond issue. The town treasurer is going to take a poll of 10 registered voters.

(a) How many different selections of 10 people can be made from the 100 registered voters?

(b) How many of these will include 6 or more people who favor the bond issue?

(c) How many will not include 6 or more people who favor the bond issue?

20. If n is any even number, show that

$$\binom{n}{1} + \binom{n}{3} + \cdots + \binom{n}{n-1} = \binom{n}{0} + \binom{n}{2} + \cdots + \binom{n}{n}.$$

(Hint: Consider the binomial expansion of $(1-1)^n$.)

21. A ship has a flag pole which has three different positions in which it can fly a flag. Assume that the ship carries four different flags (one of each kind) for making signals.

(a) How many different one-flag signals could be displayed? (Assume that the same flag, flown in a different position, is a different signal.)

(b) How many different two-flag signals could be displayed? (Again, the two positions used are of importance.)

(c) How many different three-flag signals could be displayed?

(d) How many different signals can be displayed? (Don't forget the one with no flags.)

22. Permutations of things that are not all different can be handled in much the same way as the cases covered in the text. Suppose we have n objects, n_1 of which are alike, n_2 of which are alike, but different from all others, etc., out to n_k are of a k-th kind. Note then that $\Sigma \, n_i = n$. Suppose we want to know the number of different permutations, or arrangements in a row, that are possible with these n objects. The answer is no longer $n!$, since those items that are alike are indistinguishable and could be permuted among themselves without changing the arrangement. Thus the answer must be less than $n!$. We can choose any n_1 of the n positions to put the first kind of object into. This operation can be performed in $\binom{n}{n_1}$ ways. Once the objects of the first type have been assigned their positions, any n_2 of the remaining $n - n_1$ positions may be chosen for the second type of object; this operation can be performed in $\binom{n - n_1}{n_2}$ ways; similarly the third type of object may be assigned its positions in $\binom{n - n_1 - n_2}{n_3}$ different ways. Thus, we can see that the total

number of permutations is

$$
\binom{n}{n_1}\binom{n-n_1}{n_2}\binom{n-n_1-n_2}{n_3}\cdots\binom{n-n_1-n_2-\cdots-n_{k-1}}{n_k}
$$

$$
= \frac{n!}{n_1!\,(n-n_1)!} \cdot \frac{(n-n_1)!}{n_2!\,(n-n_1-n_2)!} \cdot \frac{(n-n_1-n_2)!}{n_3!\,(n-n_1-n_2-n_3)!} \cdots
$$

$$
\frac{\left(n-\sum_1^{k-1} n_i\right)!}{n_k!\left(n-\sum_1^k n_i\right)!}
$$

$$
= \frac{n!}{n_1!\,n_2!\,n_3!\cdots n_k!}
$$

because of the cancellation that occurs from one term to the next (and recall that $n = \sum_1^k n_i$). This quantity is frequently called a multinomial coefficient and is denoted $\binom{n}{n_1\,n_2\,n_3\,\cdots\,n_k}$. That is

$$
\binom{n}{n_1\,n_2\,\cdots\,n_k} = \frac{n!}{\displaystyle\prod_{i=1}^k n_i!}.
$$

(a) How many different 10-letter sequences can be made using the letters in statistics?

(b) How many different 11-letter sequences can be made using the letters in Mississippi?

23. Twenty automobiles enter a race. Eight are made by manufacturer A, seven by manufacturer B, and the rest by manufacturer C. Keeping track of only the manufacturer of the cars, in how many ways can the cars cross the finish line? (Refer to Exercise 22.)

24. (a) How many of the arrangements listed in Exercise 23 have one of A's cars in first position?

(b) How many of the arrangements listed in Exercise 23 have A's cars in both of the first two positions?

25. In a pharmaceutical study, two drugs (A and B) and a placebo are to be compared for effects. Fifty people are to be used in the study, of whom 20 are to receive drug A, 20 receive drug B, and the remainder receive the placebo. Assuming the people are numbered 1 to 50, how many different ways could the assignment of the drugs and placebo be made?

26. In Exercise 21, assume that of the four flags, two are solid red and two are solid blue. Answer the questions posed there.

2.5. Some Particular Probability Problems

In this section we shall take up a few problems that should help acquaint the reader with counting techniques and their applications to probability problems.

EXAMPLE 2.5.1.   A bag contains 4 red and 2 white marbles. If these are randomly laid out in a row, what is the probability that the 2 end marbles are white? That they are not white? That the 2 white marbles are side by side? For convenience we assume that the white marbles are numbered 1 and 2 and that the red marbles are numbered 3 through 6. Then we might adopt as our sample space S the collection of $6! = 720$ permutations of 6 things; i.e.,

$$S = \{(x_1, x_2, \ldots, x_6): x_i = 1, 2, 3, \ldots, 6, \text{ for all } i \text{ and } x_i \neq x_j \text{ for } i \neq j\}.$$

If the marbles are randomly laid out in a row, then each of these 6-tuples is equally likely to occur and we can use our equally likely formula for computing probabilities. Define A to be the event that the first and last marbles are white (the collection of 6-tuples with marbles 1 and 2 on the ends) and B to be the event that marbles 1 and 2 are side by side. Then the number of elements belonging to A is

$$n(A) = 2 \cdot 4! = 48$$

(the white marbles could be on the ends in 2 ways and, for either of these, the red marbles could be arranged between in 4! ways). We also find that

$$n(B) = 5 \cdot 2 \cdot 4! = 240.$$

(There are 5 side-by-side positions for the white marbles to occupy, namely 12, 23, 34, 45, 56; whichever of these is the one to occur, the white marbles can occupy the selected pair of positions in 2 ways and the red balls can be permuted in the remaining positions in 4! ways.) As we noted above,

$$n(S) = 6! = 720$$

and we have

$$P(A) = \tfrac{48}{720} = \tfrac{1}{15}$$

and

$$P(B) = \tfrac{240}{720} = \tfrac{1}{3}.$$

$\bar{A}$ is the event that the white marbles are not on the 2 end positions; we immediately have

$$P(\bar{A}) = 1 - P(A) = \tfrac{14}{15}.$$

An alternative sample space for this problem can be derived as follows. If we pretend that the marbles are going to be put into numbered spots, then all possible outcomes of the experiment can be recorded by keeping track of just the 2 positions that the white marbles occupy; all remaining positions are, of course, filled with the red marbles. All possible pairs of positions are equally likely to occur if the marbles are put down randomly. Thus,

$$S = \{(x_1, x_2): x_1 = 1, 2, \ldots, 5;\ x_2 = 2, \ldots, 6;\ x_1 < x_2\}.$$

Note that S does list every possible pair of position numbers that we could select for the white marbles and that it lists each one only once. The number of elements belonging to S is equal to the number of subsets of size 2 that a set with 6 elements has, i.e.,

$$n(S) = \binom{6}{2} = 15.$$

If we define A and B as above, exactly 1 of these subsets consists of the largest and the smallest elements of S and exactly 5 of them consist of consecutive pairs. Thus,

$$n(A) = 1, \qquad n(B) = 5$$

and, as above,

$$P(A) = \tfrac{1}{15}, \qquad P(\bar{A}) = \tfrac{14}{15}, \qquad P(B) = \tfrac{1}{3}.$$

In many examples, more than one equally likely sample space is possible; used correctly any one of them will give answers to problems of interest.

EXAMPLE 2.5.2. Suppose that we select a whole number at random between 100 and 999, inclusive. What is the probability that it has at least one 1 in it? What is the probability that it has exactly two 3's in it? For a sample space, we choose

$$S = \{x: x = 100, 101, \ldots, 999\}.$$

Then $n(S) = 900$ and, since the number is chosen at random, we assume that all single-element events are equally likely to occur. Define the events:

A: the selected number has at least one 1 in it
B: the selected number has exactly two 3's in it.

We shall find it easy to compute $n(\bar{A})$, then use this to get $P(\bar{A})$, and finally use Theorem 2.2.2 to compute $P(A) = 1 - P(\bar{A})$. We shall compute $n(B)$ directly. The event $\bar{A}$ would be the collection of 3-digit numbers, each of which contains no 1's. The first position of any such number can be filled in 8 ways (since the first digit can be neither 1 nor 0) and each of the succeeding two positions can be filled in 9 ways (since 1 cannot occur in either). Thus,

$$n(\bar{A}) = 8 \cdot 9 \cdot 9 = 648$$

and

$$P(\bar{A}) = \frac{n(\bar{A})}{n(S)} = .72$$

$$P(A) = 1 - P(\bar{A}) = .28.$$

To compute $n(B)$ we can reason as follows: If the first digit is a 3, then one of the succeeding digits must be a 3 and the other can be any of the remaining 9 digits. These two succeeding digits can occur in either of 2 orders, so there are $9 \cdot 2 = 18$ 3-digit numbers having a 3 in the first position and each containing exactly two 3's.

If the first digit is not a 3, then this position can be filled in 8 ways (neither 0 nor 3 can be used). The last two positions must then both be filled with 3's. Thus,

$$n(B) = 18 + 8 = 26$$

and

$$P(B) = \frac{n(B)}{n(S)} \doteq .029.$$

EXAMPLE 2.5.3.   Suppose that n people are in a room. If we make a list of their birthdates (month and day of the month), what is the probability that there will be one or more repetitions in the list? (What is the probability that two or more people have the same birthday?) We shall make the assumption that there are only 365 days available for each birthday (ignoring February 29) and that each of these days is equally likely to occur for any individual's birthday. (It can, in fact, be shown that this is the worst possible assumption we might make relative to this probability; that is, if days in March or some other month are more likely to occur as birthdays, then the probability of one or more repeated birthdays is larger than if all days are equally likely.) Our sample space is the collection of all possible n-tuples that could occur for the birthdays, numbering the days of the year sequentially from 1 to 365. Thus,

$$S = \{(x_1, x_2, \ldots, x_n): x_i = 1, 2, \ldots, 365; \; i = 1, 2, \ldots, n\}.$$

The first position in each n-tuple gives the first person's birthday; the second position gives the second person's birthday, etc. Assuming that all days of the year are equally likely for each person's birthday implies that each of these n-tuples is equally likely to occur. By using the counting techniques presented in Section 2.4, we see that

$$n(S) = 365^n.$$

Define A to be the event that there is one or more repetitions of the same number in the n-tuple that occurs. Then $\bar{A}$ is the collection of n-tuples which have no repetitions; we can see rather easily that

$$n(\bar{A}) = 365 \cdot 364 \cdot 363 \cdots (365 - n + 1)$$

which gives us

$$P(\bar{A}) = \frac{n(\bar{A})}{n(S)} = \frac{365 \cdot 364 \cdot 363 \cdots (365 - n + 1)}{365^n},$$

Again from Theorem 2.2.2,

$$P(A) = 1 - P(\bar{A}).$$

Table 2.1 gives the values of $P(\bar{A})$ and $P(A)$ for various values of n. It is somewhat surprising that the probability of a repetition exceeds $\frac{1}{2}$ for as few as 23 people in the room and that for 60 people it is a virtual certainty.

EXAMPLE 2.5.4.   Suppose that Mrs. Riley claims to be a clairvoyant. Specifically she claims that if she is presented with 8 cards, 4 of which are red and 4 of which are black, she will correctly identify the color of at least 6 of the cards

TABLE 2.1

n	$P(\bar{A})$	$P(A)$
10	.871	.129
20	.589	.411
21	.556	.444
22	.524	.476
23	.493	.507
24	.462	.538
25	.431	.569
30	.294	.706
40	.109	.891
50	.030	.970
60	.006	.994

without being able to see their colors. If she is guessing and has no special ability, what is the probability that she would correctly identify at least 6 out of the 8 cards? (She will identify 4 of the cards as red and 4 of the cards as black.) We arbitrarily decide to present her with the 4 red cards first, one by one, and then present her with the 4 black cards. The sample space for the experiment is the set of all possible guesses she might give for the colors of the cards; that is,

$$S = \{(x_1, x_2, \ldots, x_8): x_i = R \text{ or } B, \text{ for all } i; \text{ exactly 4 } x_i\text{'s are } R\}.$$

If she is guessing, then each of the single-element events is equally likely to occur. Since each 8-tuple belonging to S contains exactly 4 R's and exactly 4 B's, we can compute $n(S)$ by counting how many ways we might select 4 positions from the 8 in which to place the R's; thus,

$$n(S) = \binom{8}{4} = 70.$$

Define

A: she identifies at least 6 cards correctly
B: she identifies exactly 6 cards correctly
C: she identifies all 8 cards correctly.

Since she will call 4 cards red and 4 cards black, it is not possible for her to be correct on exactly 7 cards; thus,

$$A = B \cup C$$

and, since

$$B \cap C = \varnothing,$$

$$P(A) = P(B) + P(C).$$

Clearly, $n(C) = 1$ so $P(C) = \frac{1}{70}$. If B is to occur, she must identify exactly 3 of the 4 red cards correctly and exactly 3 of the 4 black cards correctly. The 1 red card that she is wrong on could be any of the 4 and the 1 black card she is wrong on could be any one of the 4. Thus, the number of 8-tuples having 1 B in the first 4 positions and 1 R in the last 4 positions is

$$n(B) = 4 \cdot 4 = 16$$

and we have

$$P(B) = \tfrac{16}{70}.$$

Thus

$$P(A) = \tfrac{1}{70} + \tfrac{16}{70} = \tfrac{17}{70} \doteq .243;$$

if she only guesses, there is slightly less than 1 chance in 4 of her doing as well as she claims.

EXERCISE 2.5

1. In how many ways could a dozen oranges be chosen from a table holding 30 oranges? (How many distinct collections of 12 oranges could be made?)

2. How many arrangements could be made of 5 red balls and 1 orange ball?

3. In how many of the arrangements counted in problem 2 are the red balls all together?

4. A certain market uses red boxes and green boxes for displays at Christmas time. In how many ways could the market arrange 20 boxes in a row if 15 of them are red and 5 are green? If there are 10 boxes of each color?

5. Ten people in total are nominated for a slate of 3 offices. If every group of 3 people has the same probability of winning, what is the probability that a particular person will be on the winning slate? That a particular pair of people will be on the winning slate?

6. Two people are to be selected at random to be set free from a prison with a population of 100. What is the probability that the oldest prisoner is 1 of the 2 selected? That the oldest and youngest are the pair selected?

7. In a certain national election year, governors were to be elected in 30 states. Assume that in every state there were only 2 candidates (called the Republican and Democratic candidates, respectively). What is the probability that the Republicans carried all 30 states, assuming that each state was equally likely to elect either party? What is the probability that the same party carried all the states?

8. A, B, and C are going to race. What is the probability that A will finish ahead of C, given that all are of equal ability (and no ties can occur)? What is the probability that A will finish ahead of both B and C?

9. Each of 5 people is asked to distinguish between vanilla ice cream and French vanilla custard (each is given a small sample of both and asked to identify which is ice cream). If all 5 people are guessing, what is the probability that all will correctly identify the ice cream? If all 5 are guessing, what is the probability that at least 4 will identify the ice cream correctly?

10. Compute the probability that a group of 5 cards drawn at random from a 52-card deck will contain
    (a) exactly 2 pair
    (b) a full house (3 of one denomination and 2 of another)
    (c) a flush (all 5 from the same suit)
    (d) a straight (5 in sequence, beginning with ace or deuce or trey, ... , or ten).

11. n people are in a room. Compute the probability that at least 2 have the same birth month. Evaluate this probability for $n = 3, 4, 5, 6$.

12. A student is given a true-false examination with 10 questions. If he gets 8 or more correct, he passes. If he is guessing, what is his probability of passing the examination?

13. Eight black and 2 red balls are randomly laid out in a row. What is the probability that the 2 red balls are side by side? That the 2 red balls are occupying the end positions?

14. A person is to be presented with 3 red and 3 white cards in a random sequence. He knows that there will be 3 of each color; thus he will identify 3 cards as being of each color. If he is guessing, what is his probability of correctly identifying all 6 cards? Of identifying exactly 5 correct? Exactly 4?

15. In Example 2.5.4, what is the probability that Mrs. Riley identifies exactly 4 cards correctly?

16. In your psychology course you are given an examination with 10 multiple choice questions. To pass the exam you must get seven or more correct. Assuming you are guessing at the answer to every question, what is the probability that you will pass the exam if
    (a) each question has 3 alternative answers?
    (b) each question has 4 alternative answers?
    (c) the first five questions have 3 alternative answers and the last five all have 4 alternative answers?

17. (See Example 2.3.5.) Assume that 16 teams enter a single-elimination tournament. If they are all of equal ability, what is the probability your favorite will win the tournament?

18. A cereal manufacturer is considering changing the design of the box for one of his products. Each of 6 persons is individually shown the old design and the new design and asked which one he prefers. Assuming that each person really has no preference for one over the other, what is the probability that at least 5 of the 6 people express a preference for the new design?

19. The research division of a light bulb manufacturer has developed a new coating for the filaments of their bulbs which they feel extends the lifetime of the bulbs. To compare the lifetime of the new bulbs with the old, ten bulbs are made with the new coating and ten regular bulbs are selected; these are paired together, one new with one old. These 10 pairs are then put on test and, for each pair, it is noted whether the old or the new bulb failed first. Assuming that the new process really doesn't change the bulb lifetime, what is the probability that the old bulb fails first, in at least 9 of the 10 pairs?

Suppose that the experiment is performed and it is found that the old bulb did in fact fail first in 9 of the pairs. Would you think the new process should be adopted?

2.6. Conditional Probability

In some applications we shall be given the information that an event A occurred and will be asked the probability that another event B also occurred. For example, we might be given that the card we selected from a regular 52-card bridge deck was red and then might want to know the probability that the card selected was the ace of hearts. Or, when running an opinion poll, we might be given the fact that the person we have selected is a Republican and then might also want to know the probability that he favors our current actions in Viet Nam. Or, when conducting a medical research experiment, we might be given that a randomly selected person has a family history of diabetes and then ask what is the probability that this person also has diabetes. In each of these examples we are given that an event A has occurred and want to know the probability that a second event B also has occurred.

If we are given that A has already occurred, then in effect we now have the event A as our sample space since we know that any element $x \in \bar{A}$ did not occur. Thus it would seem reasonable to measure the probability that B also has occurred by the relative proportion of the time that A and B occur together (relative to the total probability of A occurring). This is in fact the definition of the conditional probability of B occurring, given that A has occurred, as we see in the following definition.

DEFINITION 2.6.1. The *conditional probability* of B occurring, given that A has occurred (written $P(B \mid A)$) is $P(B \mid A) = P(B \cap A)/P(A)$, if $P(A) > 0$.

EXAMPLE 2.6.1. We roll a pair of fair dice 1 time and are given that the 2 numbers that occur are not the same. Compute the probability that the sum is 7 or that the sum is 4 or that the sum is 12. Define the event

A: the two numbers that occur are different.

Then we are given that A has occurred. Also define the events

B: the sum is 7
C: the sum is 4
D: the sum is 12.

Then, assuming equally likely single-element events, we find that

$$P(A) = \tfrac{5}{6}, \qquad P(B) = \tfrac{1}{6}, \qquad P(C) = \tfrac{1}{12}, \qquad P(D) = \tfrac{1}{36},$$
$$P(A \cap B) = \tfrac{1}{6}, \qquad P(A \cap C) = \tfrac{1}{18}, \qquad P(A \cap D) = 0$$

and thus we have

$$P(B \mid A) = \frac{\frac{1}{6}}{\frac{5}{6}} = \frac{1}{5}$$

$$P(C \mid A) = \frac{\frac{1}{18}}{\frac{5}{6}} = \frac{1}{15}$$

$$P(D \mid A) = \frac{0}{\frac{5}{6}} = 0.$$

EXAMPLE 2.6.2.   From past experience with the illnesses of his patients, a doctor has gathered the following information: 5 % feel that they have cancer and do have cancer, 45 % feel that they have cancer and don't have cancer, 10 % do not feel that they have cancer and do have it, and finally 40 % feel that they do not have cancer and really do not have it. These percentage figures then imply the following probabilities for a randomly selected patient from this doctor's practice. Define the events

A: the patient feels he has cancer
B: the patient has cancer.

Then

$$P(A \cap B) = .05, \qquad P(A \cap \bar{B}) = .45,$$
$$P(\bar{A} \cap B) = .1, \qquad P(\bar{A} \cap \bar{B}) = .4.$$

Then we have

$$P(A) = P(A \cap B) + P(A \cap \bar{B}) = .5$$
$$P(B) = P(A \cap B) + P(\bar{A} \cap B) = .15.$$

The probability that a patient has cancer then, given that he feels he has it, is

$$P(B \mid A) = \frac{.05}{.5} = .1.$$

The probability that he has cancer given that he does not feel that he has it is

$$P(B \mid \bar{A}) = \frac{.1}{.5} = .2$$

The probability he feels he has cancer given that he does not have it is

$$P(A \mid \bar{B}) = \frac{.45}{.85} = \frac{9}{17}.$$

The probability he feels he has cancer given that he does have it is

$$P(A \mid B) = \frac{.05}{.15} = \frac{1}{3}.$$

One of the major uses of conditional probability is to allow easy computation of the probabilities of intersections of events for certain experiments. The equation

$$P(B \mid A) = \frac{P(B \cap A)}{P(A)}$$

implies that

$$P(B \cap A) = P(A \cap B) = P(A)P(B \mid A). \tag{2.1}$$

The equation

$$P(A \mid B) = \frac{P(A \cap B)}{P(B)}$$

implies that

$$P(A \cap B) = P(B)P(A \mid B). \tag{2.2}$$

One or the other of these formulas for $P(A \cap B)$ is useful in many problems.

EXAMPLE 2.6.3.   We select 2 balls at random without replacement from an urn which contains 4 white and 8 black balls. (a) Compute the probability that both balls are white. (b) Compute the probability that the second ball is white.
   (a) Define

A: the first ball is white
B: the second ball is white
C: both balls are white.

Then

$$A \cap B = C \quad \text{and} \quad P(C) = P(A \cap B) = P(A)P(B \mid A) = \tfrac{4}{12} \cdot \tfrac{3}{11} = \tfrac{1}{11}.$$

   (b) Clearly $B = (A \cap B) \cup (\bar{A} \cap B)$ and

$$(A \cap B) \cap (\bar{A} \cap B) = \varnothing.$$

Then

$$\begin{aligned}
P(B) &= P(A \cap B) + P(\bar{A} \cap B) \\
&= P(A)P(B \mid A) + P(\bar{A})P(B \mid \bar{A}) \\
&= \tfrac{4}{12} \cdot \tfrac{3}{11} + \tfrac{8}{12} \cdot \tfrac{4}{11} \\
&= \tfrac{1}{3}.
\end{aligned}$$

Notice that the probability of drawing a white ball the second draw (even though the first one is not replaced) is $\tfrac{1}{3}$, the same as the probability that the first ball drawn is white. This can be shown to be the case generally. $P(B)$ is an unconditional probability and, as indicated by the way it was computed above, it is an average of the two conditional probabilities $P(B \mid A)$ and $P(B \mid \bar{A})$.

EXAMPLE 2.6.4.   Box 1 contains 4 defective and 16 nondefective light bulbs. Box 2 contains 1 defective and 1 nondefective light bulb. We roll a fair die 1 time. If we get a 1 or a 2, then we select a bulb at random from box 1. Otherwise we select a bulb from box 2. What is the probability that the selected bulb will be defective? Define

A: we select a bulb from box 1
B: the selected bulb is defective.

Then $P(A) = \tfrac{1}{3}$, $P(\bar{A}) = \tfrac{2}{3}$, $P(B \mid A) = \tfrac{1}{5}$, and $P(B \mid \bar{A}) = \tfrac{1}{2}$. Since

$$B = (B \cap A) \cup (B \cap \bar{A})$$

and

$$(B \cap A) \cap (B \cap \bar{A}) = \varnothing,$$

we have

$$P(B) = P(B \cap A) + P(B \cap \bar{A})$$
$$= P(A)P(B \mid A) + P(\bar{A})P(B \mid \bar{A})$$
$$= \tfrac{1}{3} \cdot \tfrac{1}{5} + \tfrac{2}{3} \cdot \tfrac{1}{2} = \tfrac{6}{15} = \tfrac{2}{5}.$$

The following result, frequently called Bayes' theorem or Bayes' formula, is useful in many applied problems. It is named after the Reverend Thomas Bayes, one of the early writers on probability theory; it has recently been applied in many different sorts of problems and plays an important role in many branches of applied statistics. We shall, in Chapter 9, study some Bayesian methods of statistical inference. These are all based on Bayes' theorem.

Theorem 2.6.1.   (Bayes) Suppose that we are given k events A_1, $A_2, \ldots, A_k$ such that

1. $A_1 \cup A_2 \cup \cdots \cup A_k = S$
2. $A_i \cap A_j = \varnothing$,   for all $i \neq j$

(these events form a *partition* of S); then for any event $E \subset S$,

$$P(A_j \mid E) = \frac{P(A_j)P(E \mid A_j)}{\sum\limits_{i=1}^{k} P(A_i)P(E \mid A_i)}, \qquad j = 1, 2, \ldots, k.$$

Proof: Because of properties 1 and 2 listed above, we know that for any event $E \subset S$ we can write

$$E = (E \cap A_1) \cup (E \cap A_2) \cup \cdots \cup (E \cap A_k)$$

where

$$(E \cap A_i) \cap (E \cap A_j) = \varnothing, \qquad \text{for all } i \neq j.$$

Thus

$$P(E) = P(E \cap A_1) + P(E \cap A_2) + \cdots + P(E \cap A_k)$$
$$= P(A_1)P(E \mid A_1) + P(A_2)P(E \mid A_2) + \cdots + P(A_k)P(E \mid A_k)$$

by Equation 2.1. By definition

$$P(A_j \mid E) = \frac{P(A_j \cap E)}{P(E)}, \qquad j = 1, 2, \ldots, k.$$

And by using Equation 2.1 and the result above for $P(E)$, we have

$$P(A_j \mid E) = \frac{P(A_j)P(E \mid A_j)}{\sum\limits_{i=1}^{k} P(A_i)P(E \mid A_i)}$$

which is the desired result.   ◀

EXAMPLE 2.6.5. Assume that the probability is .95 that the jury selected to try a criminal case will arrive at the appropriate verdict. That is, given a guilty defendant on trial, the probability is .95 that the jury will find him guilty and, conversely, given an innocent man on trial, the probability is .95 that the jury will find him innocent. Suppose that the local police force is quite diligent in its duties and that 99% of the people brought before the court are actually guilty. Let us compute the probability that a defendant is innocent, given that the jury finds him innocent. Let G be the event that the defendant is guilty and let J be the event that the jury finds him guilty. Then, we are given that $P(J \mid G) = P(\bar{J} \mid \bar{G}) = .95$, $P(G) = .99$, and we want to compute $P(\bar{G} \mid \bar{J})$. G and $\bar{G}$ form a partition of the sample space, so from Bayes' Theorem

$$P(\bar{G} \mid \bar{J}) = \frac{P(\bar{J} \mid \bar{G})P(\bar{G})}{P(\bar{J} \mid \bar{G})P(\bar{G}) + P(\bar{J} \mid G)P(G)}$$

$$= \frac{(.95)(.01)}{(.95)(.01) + (.05)(.99)} = .161.$$

Thus, there is about 1 chance in 6 he really is innocent, if found innocent, and about 5 chances in 6 he is guilty, although found innocent. Similarly, the probability is .0005 that he is innocent, if found guilty, and is .9995 that he is in fact guilty if found guilty.

EXERCISE 2.6

1. An urn contains 4 balls numbered 1, 2, 3, 4, respectively. Two balls are drawn without replacement. Let A be the event that the sum is 5 and let B_i be the event that the first ball drawn has an i on it, $i = 1, 2, 3, 4$. Compute $P(A \mid B_i)$, $i = 1, 2, 3, 4$, and $P(B_i \mid A)$, $i = 1, 2, 3, 4$.

2. Suppose that the two balls of problem 1 are drawn with replacement. Let A and B_i be defined as above and compute $P(A \mid B_i)$ and $P(B_i \mid A)$, $i = 1, 2, 3, 4$.

3. A fair coin is flipped 4 times. What is the probability that the fourth flip is a head, given that each of the first 3 flips resulted in heads?

4. A fair coin is flipped 4 times. What is the probability that the fourth flip is a head, given that 3 heads occurred in the 4 flips? Given that 2 heads occurred in the 4 flips?

5. Urn 1 contains 2 red and 4 blue balls, urn 2 contains 10 red and 2 blue balls. If an urn is chosen at random and a ball is removed from the chosen urn, what is the probability that the selected ball is blue? That it is red?

6. Suppose that in problem 5, instead of selecting an urn at random we roll a die and select the ball from urn 1 if a 1 occurs on the die and otherwise select the ball from urn 2. What is the probability that the selected ball is blue? That it is red?

7. Five cards are selected at random without replacement from a 52-card deck. What is the probability that they are all red? That they are all diamonds?

8. An urn contains 2 red, 2 white, and 2 blue balls. Two balls are selected at random without replacement from the urn. Compute the probability that the second ball drawn is red.

9. An urn contains 2 black and 5 brown balls. A ball is selected at random. If the ball drawn is brown, it is replaced and two additional brown balls are also put into the urn. If the ball drawn is black, it is not replaced in the urn and no additional balls are added. A ball is then drawn from the urn the second time. What is the probability that it is brown?

10. The two-stage experiment described in problem 9 was performed and we are given that the ball selected at the second stage was brown. What is the probability that the ball selected at the first stage was also brown?

11. Suppose that medical science has a cancer-diagnostic test that is 95 % accurate both on those that do have cancer and on those that do not have cancer. If .005 of the population actually does have cancer, compute the probability that a particular individual has cancer, given that the test says he has cancer.

12. In a large midwestern school, 1 % of the student body participates in the inter-collegiate athletic program; 10 % of these people have a grade point of 3 or more (out of 4) whereas 20 % of the remainder of the student body have a grade point of 3 or more. What proportion of the total student body has a grade point of 3 or more? Suppose we select 1 student at random from this student body and find that he has a grade point of 3.12. What is the probability that he participates in the intercollegiate athletic program?

13. Two different suppliers, A and B, provide a manufacturer with the same part. All supplies of this part are kept in a large bin. In the past, 5 % of the parts supplied by A and 9 % of the parts supplied by B have been defective. A supplies four times as many parts as B. Suppose you reach into the bin and select a part, and find it is nondefective. What is the probability that it was supplied by A?

14. Two fair dice are rolled one time. Given that the sum of the two numbers that occurred was at least 7, compute the probability that it was equal to i, for $i = 7, 8, 9, 10, 11, 12$.

15. Assume that, among families with two children, there are equal numbers of families with (boy, boy), (boy, girl), (girl, boy) and (girl, girl), where the ordering in the 2-tuple indicates the order of birth. We select a family with two children at random.
    (a) What is the probability that the family has 2 boys, given that it has at least one boy?
    (b) Suppose you are married and currently have one child, a boy. You are expecting your second child. Clearly, once you have two children, you will have at least one boy. Is the answer to (a) the probability that your second child will be a boy?

16. Sixteen teams enter a single-elimination tournament, playing a game which cannot end in a tie. Thus, a total of four rounds will be required to determine the winner. Assume that the probability your team wins its first game is .9

and that the probabilities it will win its succeeding games are .8, .7, .6, respectively, given that it won its preceding games.

(a) What is the probability that your team will win the tournament?

(b) What is the probability that your team is eliminated in the third round? The second round? The first round?

17. You are given an exam which contains eight true-false questions. To pass the exam you must get at least six questions correct.

   (a) Given that you are guessing in answering each one, what is the probability that you pass the exam?

   (b) Assume you know the answer to the first question and thus must guess on only the last seven. Given that you get the first question right, what is the probability that you pass the exam?

   (c) Given that you know the answers to both of the first two questions, what is the probability that you pass the exam?

18. Every 25th person to enter an amusement park is given a prize.

   (a) What is the probability that you receive a prize when you enter?

   (b) Given that you have waited, and observed that none of the fifteen people to enter before you received a prize, what is the probability that you receive a prize?

2.7. Independent Events

It is possible to define and be interested in events A and B such that if we know that A has occurred, then we are also certain that B has occurred (take any example where $A \subset B$). In such a case there is certainly a degree of dependence between A and B. As we shall see in this section, it is also possible to have two events A and B such that the knowledge that A has occurred gives no information on whether or not B also has occurred. Two such events will be called independent (also referred to as being statistically independent). The formal definition of independent events is as follows.

DEFINITION 2.7.1. Two events, A and B, are *independent* if and only if

$$P(A \cap B) = P(A)P(B).$$

Before looking at some examples of independent events, let us note that if A and B are independent, then

$$P(A \mid B) = \frac{P(A \cap B)}{P(B)} = \frac{P(A)P(B)}{P(B)} = P(A), \qquad \text{granted } P(B) \neq 0.$$

Thus, the conditional probability of A occurring is the same as the unconditional probability of A. The knowledge that B occurred didn't change the probability of A's occurrence. Similarly, it is easily seen that $P(B \mid A) = P(B)$ if A and B are independent.

EXAMPLE 2.7.1. Assume that the numbers given in the cells of Table 2.2 give the probabilities of a randomly selected individual falling into the given cell.

TABLE 2.2

	Gets cancer	Does not get cancer
Smoker	.50	.20
Nonsmoker	.10	.20

That is, if we let A be the event that the selected individual is a smoker and let B be the event that the selected individual gets cancer, then

$$P(A \cap B) = .5, \qquad P(A \cap \bar{B}) = .2, \qquad P(\bar{A} \cap B) = .1$$

and

$$P(\bar{A} \cap \bar{B}) = .2.$$

Since

$$P(A) = P(A \cap B) + P(A \cap \bar{B}) = .7$$
$$P(B) = P(A \cap B) + P(\bar{A} \cap B) = .6,$$

we see that $P(A \cap B) = .5 \neq (.7)(.6)$, so A and B are not independent.

EXAMPLE 2.7.2. If 2 fair dice are rolled 1 time, show that the 2 events

A: the sum of the 2 dice is 7
B: the 2 dice have the same number

are not independent. As we have seen before, $P(A) = P(B) = \frac{1}{6}$; $A \cap B = \varnothing$ so $P(A \cap B) = 0$ which is not $\frac{1}{6} \cdot \frac{1}{6}$. Thus, the 2 events are not independent.

Two events which cannot happen simultaneously are said to be mutually exclusive, as is given in the next definition.

DEFINITION 2.7.2. A and B are *mutually exclusive* if and only if

$$A \cap B = \varnothing.$$

The definitions of independence and mutually exclusive are frequently confused; this is probably caused by the fact that in common English usage the word independent is frequently used to mean "having nothing to do with." The phrase "having nothing to do with" could be interpreted to mean that two events could not happen together, which we have just defined as being mutually exclusive. The two definitions (2.7.1 and 2.7.2) are certainly not the same in content, as we can see from the following theorem.

Theorem 2.7.1. Assume that $P(A) \neq 0$ and $P(B) \neq 0$. Then A and B independent implies that they are not mutually exclusive and A and B mutually exclusive implies that they are not independent.

Proof: Suppose that A and B are independent. Then $P(A \cap B) = P(A)P(B) \neq 0$, since $P(A) \neq 0$ and $P(B) \neq 0$. Thus they are not mutually exclusive. Now suppose A and B are mutually exclusive. Then $A \cap B = \varnothing$ and $P(A \cap B) = 0$. But since $P(A) \neq 0$ and $P(B) \neq 0$, $P(A)P(B) \neq 0$ and they are then not independent.   ◄

Independence of three events is defined as follows.

DEFINITION 2.7.3. A, B, and C are independent if and only if:

1. $P(A \cap B) = P(A)P(B)$
2. $P(A \cap C) = P(A)P(C)$
3. $P(B \cap C) = P(B)P(C)$
4. $P(A \cap B \cap C) = P(A)P(B)P(C)$.

Many examples can be given which show that the first three of these conditions do not imply the fourth and vice versa. The following is an example in which the first three equations hold but the fourth does not.

EXAMPLE 2.7.3.   Assume a fair coin is flipped two times and define the three events

A: Head on first flip,
B: Head on second flip,
C: Same face on both flips.

Then we can easily see that $P(A) = P(B) = P(C) = \frac{1}{2}$ and that $P(A \cap B) = P(A \cap C) = P(B \cap C) = \frac{1}{4}$, so equations 1, 2, and 3 of Definition 2.7.3 are satisfied; A, B, and C are called *pairwise independent events*. But, since

$$P(A \cap B \cap C) = \tfrac{1}{4} \neq (\tfrac{1}{2})^3,$$

they do not satisfy equation 4 above and thus are not independent events.

Many practical experiments consist of independent trials; by independent trials we mean that a particular outcome in one trial has no effect on the outcome observed in another trial. This idea is made more exact in the following definition.

DEFINITION 2.7.4.   An experiment consists of n *independent trials* if and only if: (1) S is the Cartesian product of n sets $S_1, S_2, \ldots, S_n$, and (2) the probability of occurrence of a single-element event $A \subset S$ is the product of the probabilities of occurrence of appropriate single-element events $A_i \subset S_i$, $i = 1, 2, \ldots, n$; that is,

$$P(A) = P_1(A_1)P_2(A_2) \cdots P_n(A_n)$$

where $A \subset S$, $A_i \subset S_i$, $i = 1, 2, \ldots, n$ and $A, A_1, \ldots, A_n$ are each single-element events.

Note immediately then that an experiment that consists of n independent trials has n-tuples as elements of its sample space. Furthermore, probabilities of single-element events are assigned in a special way; this special way in fact gives an easy method of computing many probabilities.

EXAMPLE 2.7.4.   A television-radio retail sales outlet receives a shipment of five radios from a manufacturer. In the past, 90% of the radios received from this manufacturer work correctly when first tried; 10% have required some adjustment or repair before working correctly. Let us discuss the experiment which consists of checking which of the five radios just received work correctly. We assume that each radio works correctly (nondefective, labeled n) or does not work correctly (defective, labeled d). Thus, the simple experiments which consist of checking individual radios could all use the sample space

$$S_i = \{n, d\}, \qquad i = 1, 2, 3, 4, 5.$$

Based on the historical information given, we should also assume

$$P_i(\{n\}) = .9, \qquad P_i(\{d\}) = .1, \qquad i = 1, 2, 3, 4, 5.$$

The experiment which consists of checking all 5 radios could have the sample space

$$S = S_1 \times S_2 \times S_3 \times S_4 \times S_5.$$

Furthermore, since the individual radios are separate physical entities, it seems reasonable that the occurrence or non-occurrence of defects would be independent. That is, we have an experiment which consists of independent trials. Then, using the rule mentioned in Definition 2.7.4, we would assign probabilities to the single-element events of S by multiplying together the probabilities of observing the appropriate results in the individual trials. For example,

$$P(\{(n, n, n, n, n)\}) = (.9)(.9)(.9)(.9)(.9) = (.9)^5$$
$$P(\{(d, d, d, d, d)\}) = (.1)(.1)(.1)(.1)(.1) = (.1)^5$$
$$P(\{(n, d, n, d, n)\}) = (.9)(.1)(.9)(.1)(.9) = (.1)^2(.9)^3$$
$$P(\{(d, d, n, n, n)\}) = (.1)(.1)(.9)(.9)(.9) = (.1)^2(.9)^3$$

and so on. Once the probabilities of the single-element events are known, of course, we can compute the probabilities of any other events of interest.

EXAMPLE 2.7.5.   The Iowa State University football team plays 11 different teams on successive Saturdays. If we assume that their performances from one Saturday to the next are independent, then the experiment which consists of their full schedule for a year is made up of 11 independent trials. Define

$$S_i = \{W, L, T\}, \qquad i = 1, 2, \ldots, 11,$$

where W stands for win, L for loss, and T for tie, and the sample space S for the year's games can be written

$$S = S_1 \times S_2 \times \cdots \times S_{11}.$$

Then the probability of occurrence of any single-element event belonging to S is defined to be the product of the probabilities of the single-element events for the appropriate trials; that is

$$P(\{(x_1, x_2, \ldots, x_{11})\}) = P_1(\{x_1\})P_2(\{x_2\}) \cdots P_{11}(\{x_{11}\})$$

where

$$(x_1, x_2, \ldots, x_{11}) \in S, \qquad x_1 \in S_1, x_2 \in S_2, \ldots, x_{11} \in S_{11}.$$

The differing subscript of P for the various trials is meant to indicate that the probability of Iowa State winning (or losing or tying) a game is not necessarily the same for all weeks. The probability that they win all their games is then

$$\prod_{i=1}^{11} P_i(\{W\});$$

the probability that they lose all their games is

$$\prod_{i=1}^{11} P_i(\{L\}).$$

EXERCISE 2.7

1. One fair coin is flipped 2 times. Are the 2 events

    A: a head occurs on the first flip
    B: a head occurs on the second flip

    independent?

2. A fair coin is flipped 2 times. Let A be the event that a head occurs on the first flip and let B be the event that the same face does not occur on both flips. Are A and B independent?

3. An urn contains 4 balls numbered 1, 2, 3, 4, respectively. Two balls are drawn without replacement. Let A be the event that the first ball drawn has a 1 on it and let B be the event that the second ball has a 1 on it. Are A and B independent?

4. If the drawing is done with replacement in problem 3, are A and B independent?

5. A pair of dice is rolled 1 time. Let A be the event that the first die has a 1 on it, B the event that the second die has a 6 on it, and C the event that the sum is 7. Are A, B, and C independent?

6. A fair coin is flipped 3 times. Let A be the event that a head occurs on the first flip, let B be the event that at least 2 tails occur, and let C be the event that we get exactly 1 head or that we get tail, head, head in that order. Show that these 3 events satisfy equation 4 of Definition 2.7.3, but not equations 1, 2, or 3.

7. Prove that if A and B are independent, so are $\bar{A}$ and $\bar{B}$.

8. The probability that a certain basketball player scores on a free throw is .7. If in a game he gets 15 free throws, compute the probability that he makes

them all. Compute the probability that he makes 14 of them. What assumptions have you made in deriving your answer?

9. Three teams, A, B, and C, enter a round-robin tournament. (Each team plays 2 games, 1 against each of the possible opponents. The winner of the tournament, if there is a winner, is the team winning both its games.) Assume that the game played is one in which a tie is not allowed. We assume the following probabilities:

$$P(A \text{ beats } B) = .7$$
$$P(B \text{ beats } C) = .8$$
$$P(C \text{ beats } A) = .9.$$

Compute the probability that

    (a) A wins.
    (b) B wins.
    (c) No one wins.

10. In Example 2.7.4, assume that 10 radios are received.
   (a) What would you use as a sample space for the experiment?
   (b) What is the probability that all 10 work correctly?
   (c) What is the probability that at least 9 work correctly?

11. A department store has a sales display of 1000 light-bulbs, 10 of which are defective.
   (a) If you buy 20 of these bulbs, what is the probability you get no defectives?
   (b) If you buy 20 of these bulbs, what is the probability you get exactly one defective bulb?

12. In one morning a Fuller brush salesman can call at 16 homes. If the probability of his making a sale is .1, at each home, what is the probability he makes at least one sale, during his morning?

13. Assume that a student is carrying 4 courses in a given quarter, and that his probabilities of getting A's are .2, .5, .1, .7, respectively, in these courses.
   (a) What is the probability that he gets all A's?
   (b) What is the probability that he gets exactly 3 A's?
   Can you find anything to criticize in the way you've answered this question?

14. n repeated independent trials are performed, each resulting in the occurrence of a success or failure. The probability a success occurs is p for each trial.
   (a) Show that the probability exactly x successes occur is

$$\binom{n}{x} p^x (1 - p)^{n-x}, \qquad x = 0, 1, 2, \ldots, n.$$

   (b) Show that the sum of the probabilities given in (a) is 1, i.e.,

$$\sum_{x=0}^{n} \binom{n}{x} p^x (1 - p)^{n-x} = 1$$

15. Two independent trials are performed, each resulting in the occurrence of a success or failure. Assume that the probability a success occurs on trial i is

p_i, $i = 1, 2$. What is the probability of observing exactly x successes, $x = 0, 1, 2$?

16. An urn contains 15 balls, of which 5 are white. Five balls are drawn at random from the urn, without replacement. What is the probability that x white balls are selected, for $x = 0, 1, 2, 3, 4, 5$?

17. Assume the sampling in Exercise 16 is done with replacement. What is the probability that x white balls are selected, for $x = 0, 1, 2, 3, 4, 5$?

2.8. Discrete and Continuous Sample Spaces

The reader may have noted that the sample spaces used in all the preceding examples were finite sets. Such sample spaces are special cases of what are called discrete sample spaces. Discrete sample spaces are defined as follows.

DEFINITION 2.8.1. A *discrete* sample space is one which has a finite or a countably infinite number of elements.

Generally, discrete sample spaces are those for which it is meaningful to consider single-element events. That is, knowledge of the probabilities of occurrence of the single-element events is sufficient for computing the probabilities of any event. All of the finite sample spaces we have seen thus far have been of this sort. The following example gives a problem in which the sample space is countably infinite, yet we still are able to compute probabilities from knowledge of the single-element events.

Before looking at this example, let us note what our third axiom is saying when we have an infinite sample space. In an infinite sample space it is possible to have an infinite number of nonoverlapping events. (Consider, for example, the single-element events.) Axiom 3 says then that the probability assigned to the union of an infinite number of nonoverlapping events must be equal to the value to which an infinite series converges (the sum of the individual probabilities assigned to the events).

EXAMPLE 2.8.1. Suppose we flip a fair coin until we get a head. Compute the probability that it takes less than 4 flips to conclude the experiment. Compute the probability that it takes an even number of flips to conclude the experiment. Our sample space is

$$S = \{H, TH, TTH, \dots\}.$$

H signifies that we get a head on the first flip, TH that we *first* got a head on the second flip, TTH that we first got a head on the third flip, etc. Clearly we can set up a one-to-one correspondence between the elements of S and the positive integers; thus S has a countably infinite number of elements and is a discrete sample space.

Assuming that the coin is fair and that the flips are independent, we can easily assign probabilities to the single-element events as follows.

$$P(\{H\}) = \tfrac{1}{2}, \qquad P(\{TH\}) = \tfrac{1}{2} \cdot \tfrac{1}{2} = \tfrac{1}{4},$$

$$P(\{TTH\}) = \tfrac{1}{2} \cdot \tfrac{1}{2} \cdot \tfrac{1}{2} = \tfrac{1}{8}, \qquad P(\{TTTH\}) = \tfrac{1}{2} \cdot \tfrac{1}{2} \cdot \tfrac{1}{2} \cdot \tfrac{1}{2} = \tfrac{1}{16},$$

etc. First of all we should make sure that this assignment satisfies axiom 1. Clearly

$$
\begin{aligned}
P(S) &= \tfrac{1}{2} + \tfrac{1}{4} + \tfrac{1}{8} + \tfrac{1}{16} + \cdots \\
&= \tfrac{1}{2}(1 + \tfrac{1}{2} + \tfrac{1}{4} + \tfrac{1}{8} + \cdots) \\
&= \tfrac{1}{2} \cdot \frac{1}{1 - \tfrac{1}{2}} = 1,
\end{aligned}
$$

since the parenthetic expression is a geometric progression with $r = \tfrac{1}{2}$. Thus we do satisfy axiom 1. Define

  A: it takes less than 4 flips to conclude the experiment
  B: it takes an even number of flips to conclude the experiment.

That is,

$$A = \{H, TH, TTH\}$$
$$B = \{TH, TTTH, TTTTTH, \ldots\}.$$

Then

$$
\begin{aligned}
P(A) &= \tfrac{1}{2} + \tfrac{1}{4} + \tfrac{1}{8} = \tfrac{7}{8} \\
P(B) &= \tfrac{1}{4} + \tfrac{1}{16} + \tfrac{1}{64} + \tfrac{1}{256} + \cdots \\
&= \tfrac{1}{4}(1 + \tfrac{1}{4} + \tfrac{1}{16} + \tfrac{1}{64} + \cdots) \\
&= \tfrac{1}{4} \cdot \frac{1}{1 - \tfrac{1}{4}} = \tfrac{1}{3}.
\end{aligned}
$$

   EXAMPLE 2.8.2.   Suppose that we select a number at random from the positive integers. What is the probability that it is even? The sample space for this experiment is

$$S = \{1, 2, 3, \ldots\},$$

so S is discrete. However, it is not possible to assign the same probability to each of the single-element events. No matter how small the value of this common probability, the sum of the probabilities of the single-element events is infinite. That is, suppose we say

$$P(\{x\}) = p > 0 \qquad \text{for } x = 1, 2, 3, \ldots$$

Then

$$P(S) = \sum_{x=1}^{\infty} P(\{x\}) = \sum_{x=1}^{\infty} p$$

and this sum diverges, no matter how close to 0 p is. Thus it would appear that there is no way we could satisfactorily describe such an experiment. This conclusion is correct because, after a little reflection, we would have to conclude that there is absolutely no way in which we could perform the experiment. It is impossible to select an integer at random from the set of positive integers. The experiment itself makes no sense; this is why we cannot describe it. In spite of the foregoing it would seem reasonable to give $\frac{1}{2}$ as the probability of selecting an even number since every other integer is even. This conclusion is actually based on reasoning such as the following. If we were to select randomly an integer from the set $\{1, 2, 3, \ldots, M\}$ where M is very large, the probability that the number is even is $\frac{1}{2}$ if M is even and $\frac{1}{2} - (1/2M)$ if M is odd. In either case, it is about $\frac{1}{2}$ and, as M gets bigger and bigger, it gets closer and closer to $\frac{1}{2}$ (for odd M; for even M it is $\frac{1}{2}$).

Many examples give rise to continuous sample spaces. These are defined as follows:

DEFINITION 2.8.2. A *continuous* (one-dimensional) sample space is one which has as elements all of the points in some interval on the real line.

Thus the sets

$$\{x : 0 < x < 1\}, \qquad \{x : 10 \leq x \leq 20\}, \qquad \{x : x > 0\}$$

are all examples of continuous sample spaces. Generally, if the experiment consists of observing something which could lie anywhere along a continuous line, we shall want to use a continuous sample space. Again, subsets of the continuous interval are events; the single-element events would be the sets of single points in the interval. In the case of continuous sample spaces, however, we shall generally have to conclude that the probabilities of occurrence of these individual points must be zero. Otherwise, since there are a noncountable infinity of them in any interval, we would not be able to satisfy our first axiom. Thus, in continuous sample spaces, we cannot compute the probabilities of occurrence of any event from knowledge of the probabilities of the single-element events.

This situation is analogous to that encountered in geometry. The reader will recall that an individual point has zero length (and height). Surely a line that is, say, one inch long comprises a large number of individual points; yet the length of the line is not the summation of the lengths of the individual points. The measure *length* is a function whose nontrivial domain is a set of intervals and which takes on the value zero at each point. Similarly, when using continuous sample spaces a probability function's nontrivial domain is a set of intervals; it must assign the value zero to individual points. (This statement is actually too strong, since we could assign $\frac{1}{3}$, for example, as the probability of occurrence of some single point, leaving a probability of $\frac{2}{3}$ to be assigned to intervals; we shall have no need in our study for this generality.)

Another difficulty is also encountered. In discrete sample spaces, all subsets are called events and probabilities can consistently be assigned to them. It can be shown that in a continuous sample space S, not all subsets are probabilizable; that is, certain examples of subsets can be constructed such that any assignment of probabilities to them is inconsistent with our three axioms. Such subsets are not called events and thus are not assigned probabilities. The reader should rest assured, however, that in any practical problem we shall be able to call any outcomes of interest events. In more advanced courses it is shown that the subsets of S belonging to the class of Borel sets are all probabilizable; furthermore, the class of Borel sets includes all of the continuous-interval subsets of S, as well as unions and intersections of such continuous-interval subsets. Without going further into the matter, we merely acknowledge the fact that not all subsets of S are called events if S consists of all the points in some continuous interval; those that are events are actually Borel sets.

Continuous sample spaces, in quite general examples, are most easily studied after we are acquainted with random variables (to be introduced in Chapter 3). Let us content ourselves here with looking at a particular type of continuous sample space and the way in which we assign probabilities to events. The particular type of continuous sample space we shall investigate now is one which has equally likely elements. As we have seen, each of the single-element events (single points) must then have probability 0. Therefore, what we really mean by saying that the elements of S are equally likely to occur is that the probability of the occurring point lying in a continuous subinterval of S is proportional to the length of the subinterval. If, for example, A and B are both continuous subintervals of S and if A is twice as long as B, then the probability of A containing the outcome is twice as big as the probability of B containing the outcome.

It can be shown that for continuous sample spaces we must specify a rule for assigning probabilities to continuous subintervals rather than specifying probabilities for all single-element events. In this section, as has been noted, we shall be concerned only with the particular continuous sample space which has equally likely outcomes. As mentioned above, this implies that the probability of the outcome lying in a continuous subinterval should be proportional to the length of the subinterval. Accordingly, suppose we are given an experiment whose outcome is equally likely to lie anywhere in the real interval from a to b inclusive. Then

$$S = \{x: a \leq x \leq b\}.$$

Suppose A is a continuous subinterval of S; i.e.,

$$A = \{x: c \leq x \leq d\},$$

where $c > a$, $d < b$. (See Figure 2.8.1.) For any continuous subinterval A,

FIGURE 2.8.1   Continuous sample space.

define the set function

$$L(A) = \text{length of } A.$$

Then, for the particular subset defined above we have

$$L(A) = d - c,$$

and for S we have

$$L(S) = b - a.$$

One rule for assigning probabilities then is

$$P(A) = \frac{L(A)}{L(S)} = \frac{d - c}{b - a}$$

if A is any continuous subinterval of S. For any subset B of S which is a union of nonoverlapping continuous subintervals of S, we define $L(B)$ to be the sum of the lengths of the nonoverlapping subintervals belonging to B. That is, if

$$B = C_1 \cup C_2 \cup \cdots, \qquad \text{where } C_i \cap C_j = \varnothing, \qquad \text{for } i \neq j,$$

then

$$L(B) = L(C_1) + L(C_2) + \cdots$$

and we have

$$P(B) = \frac{L(B)}{L(S)} = \frac{L(C_1)}{L(S)} + \frac{L(C_2)}{L(S)} + \cdots = P(C_1) + P(C_2) + \cdots.$$

Theorem 2.8.1.   The rule

$$P(A) = \frac{L(A)}{L(S)} \qquad \text{for } A \subset S$$

satisfies the three axioms.

Proof: $P(S) = L(S)/L(S) = 1$ so axiom 1 is satisfied. $L(A) \geq 0$ for any $A \subset S$ so $P(A) = L(A)/L(S) \geq 0$ and axiom 2 is satisfied.   By the way in which lengths of unions of intervals are defined, we have

$$P(C_1 \cup C_2 \cup \cdots) = \frac{L(C_1 \cup C_2 \cdots)}{L(S)}$$

$$= \frac{L(C_1)}{L(S)} + \frac{L(C_2)}{L(S)} + \cdots = P(C_1) + P(C_2) + \cdots$$

if $C_i \cap C_j = \varnothing$ for all $i \neq j$.   ◄

Note that if A is a single-element event then $L(A) = 0$ and we have $P(A) = 0$, as discussed above.

Let us now consider some specific examples.

EXAMPLE 2.8.3.   Doug is a 2-year-old boy. From his family history it seems plausible to assume that his adult height is equally likely to lie between 5 feet 9 inches and 6 feet 2 inches. Making this assumption, what is the probability that he will stand at least 6 feet high as an adult? What is the probability that his adult height will lie between 5 feet 10 inches and 5 feet 11 inches?

We use as our sample space

$$S = \{x: 69 \le x \le 74\}$$

where we are recording his achieved adult height in inches. Define

$$A = \{x: 72 \le x \le 74\}$$
$$B = \{x: 70 \le x \le 71\}.$$

Then

$$L(S) = 5, \qquad L(A) = 2, \qquad L(B) = 1$$

and we have

$$P(A) = \tfrac{2}{5}, \qquad P(B) = \tfrac{1}{5}.$$

EXAMPLE 2.8.4.   Assume that you daily ride a commuter train from your home in Connecticut into Manhattan. The station you leave from has trains leaving for Manhattan at 7 a.m., 7:13 a.m., 7:20 a.m., 7:25 a.m. 7:32 a.m., 7:45 a.m., and 7:55 a.m. It is your practice to take the first train that leaves after your arrival at the station. Due to the vagaries of your rising time and the traffic you encounter driving to the station, you are equally likely to arrive at the station at any instant between 7:15 a.m. and 7:45 a.m. On a particular day, what is the probability that you have to wait less than 5 minutes at the station? Less than 10 minutes? Suppose the 7:25 a.m. and 7:45 a.m. trains are expresses. What is the probability that you catch an express on a given day?

Let us, for convenience, take as our sample space

$$S = \{x: 0 \le x \le 30\}$$

where the elements of S are actually meant to represent minutes after 7:15 a.m. that might occur until your arrival time. (Refer to Figure 2.8.2.) Define the events

A: you wait less than 5 minutes
B: you wait less than 10 minutes
C: you catch an express.

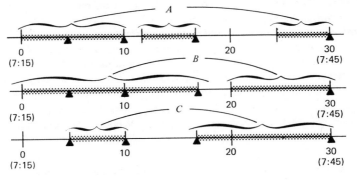

FIGURE 2.8.2   Events A, B, and C.

Then

$$A = \{x : 0 \le x < 10 \text{ or } 12 \le x < 17 \text{ or } 25 \le x < 30\}$$
$$B = \{x : 0 \le x < 17 \text{ or } 20 \le x < 30\}$$
$$C = \{x : 5 \le x < 10 \text{ or } 17 \le x < 30\}$$

and

$$L(S) = 30, \qquad L(A) = 20, \qquad L(B) = 27, \qquad L(C) = 18.$$

Thus

$$P(A) = \tfrac{2}{3}, \qquad P(B) = \tfrac{9}{10}, \qquad P(C) = \tfrac{3}{5}.$$

EXERCISE 2.8

1. A fair die is rolled until a 1 occurs. Compute the probability that:
    (a) 10 rolls are needed
    (b) less than 4 rolls are needed
    (c) an odd number of rolls is needed.

2. A fair pair of dice is rolled until a 7 occurs (as the sum of the 2 numbers on the dice). Compute the probability that
    (a) 2 rolls are needed
    (b) an even number of rolls is needed.

3. You fire a rifle at a target until you hit it. Assume the probability that you hit it is .9 for each shot and that the shots are independent. Compute the probability that:
    (a) it takes more than 2 shots.
    (b) the number of shots required is a multiple of 3.

4. Hugh takes a written driver's license test repeatedly until he passes it. Assume the probability that he passes it any given time is .1 and that the tests are independent. Compute the probability that:
    (a) it takes him more than 4 attempts
    (b) it takes him more than 10 attempts.

5. A traffic light on a route you travel every day turns red every 4 minutes, stays red 1 minute and then turns green again (thus it is green 3 minutes, red 1, etc.), with the red part of the signal starting on the hour, every hour.

    (a) If you arrive at the light at a random instant between 7:55 a.m. and 8:05 a.m., what is the probability that you have to stop at the light?

    (b) If you arrive at the light at a random instant between 7:54 a.m. and 8:04 a.m. what is the probability that you have to stop for the light?

6. The plug on an electric clock with a sweep second hand is pulled at a random instant of time within a certain minute. What is the probability that the second hand is between the 4 and 5? Between the 1 and the 2? Between the 1 and the 6?

7. A point is chosen at random between 0 and 1 on the x-axis in the (x, y) plane. A circle centered at the origin is then drawn in the plane, with radius determined by the chosen point. Compute the probability that the area of the circle is less than $\pi/2$.

8. A 12-inch ruler is broken into 2 pieces at a random point along its length. What is the probability that the longer piece is at least twice the length of the shorter piece?

9. The game of odd man out is played with 3 people, each flipping a single coin. All 3 flip their coins simultaneously; if one face is different than the other two, its owner is the odd man and he loses.

    (a) What is the probability there is an odd man on a given turn, assuming all three coins are fair?

    (b) If there is no odd man on the first turn, the coins are all flipped again, until the odd man is determined. What is the probability that an even number of turns is required to determine the loser?

10. Answer Exercise 9 assuming that four people play (here 3 must match and one is different).

11. Answer Exercise 9 assuming that n people play ($n - 1$ must match). Does this seem like a feasible game as n gets large?

12. The assignment of probabilities to continuous intervals can be done in many different ways. It is possible to think of the assignment as being made by the area between the x-axis and a continuous function $f(x)$ defined over the sample space S. That is, if $S = \{x : a \le x \le b\}$ then the domain of $f(x)$ is S and the probability that x lies in an interval from c to d, say, where $c > a$ and $d < b$, is defined to be the area between $f(x)$ and the x-axis, over the interval from c to d. Then the area between a and b must be 1.

    (a) Show that the rule discussed in Theorem 2.8.1 implies that $f(x)$ must be a constant over the interval from a to b, if $S = \{x : a \le x \le b\}$.

    (b) What must the constant value of $f(x)$ be equal to?

13. For the continuous sample space discussed in Example 2.8.3, assume that probabilities are assigned to intervals by the area between $f(x)$ and the x-axis, where $f(x)$ is as pictured below. Answer the questions posed in the text for this case.

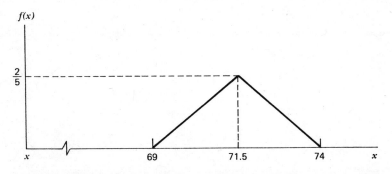

14. For the continuous sample space discussed in Example 2.8.4, assume that probabilities are assigned to intervals by the area between $f(x)$ and the x-axis, where $f(x)$ is as pictured below. Answer the questions posed in the text for this case.

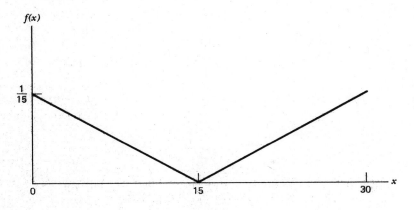

3

Random Variables and Distribution Functions

Probability theory itself is an interesting mathematical discipline. It can be used in many different contexts; one of its most important applications is in the description of random variables.

Much headway was made in science by assuming an essentially deterministic world—one in which it was assumed that the identical operation or experiment, performed repeatedly under "identical" conditions, would always give rise to the same result. However, as time passed and technology created more and more precise methods and equipment for observation, it became apparent that the same experiment, performed under "identical" conditions, does not give exactly the same result each time. It is this variability of observed results that is ideally described with probability theory and random variables. Since the outcomes or results of experiments or observations are usually summarized by numbers, this variability of the results of experiments is represented by a variability of the numbers observed. As we shall see, we can think of random variables as results of an experiment that is affected by some chance mechanism; thus the description of random variables gives us a method of describing the variability of outcomes for the experiment.

3.1. Random Variables

Let us first mention some specific examples of random variables. A man who bets 1 dollar in a gambling game is interested in the net change in his capital as a result of his bet. The man firing a rifle at a target is interested in the distance between the point at which his bullet strikes the target and the bull's-eye of the target. A jockey is interested in the total elapsed time from when his horse leaves the starting gates until it crosses the finish line. A variable whose value is determined by the outcome of an experiment is called a random variable. Formally, the definition of a random variable is as follows.

DEFINITION 3.1.1.   A *random variable* X is a real-valued function of the elements of a sample space S.

Note then that the domain of definition of a random variable is a sample space and that its range is a set of real numbers.

Especially now, when we are first studying random variables, we shall use a functional notation to stress the fact that a random variable is a special sort of mathematical function. The Greek letter ω will be used to represent a generic element of the sample space (regardless of whether the elements are 1-tuples or n-tuples) and $X(\omega)$ will be the functional representation of the random variable (the rule) X. We shall reserve capital letters from the end of the alphabet (X, Y, Z, U, V, W, etc.) to represent random variables and lower case letters (x, y, z, u, v, w, etc.) to stand for particular values in the range of the random variable.

We shall concern ourselves with two different sorts of random variables, called discrete and continuous random variables (actually, a mixture of these two types could also be defined but it occurs so rarely in practice that we shall not do so). The difference between these two is essentially the same as the difference between discrete and continuous sample spaces, as discussed in Section 2.8. Since every random variable $X(\omega)$ is a real-valued function of the elements of S, the range of $X(\omega)$ is always a set of real numbers. It is a basic difference between the ranges of discrete and continuous random variables that forces us to differentiate between them now. The formal definition follows.

DEFINITION 3.1.2.   A random variable X is called *discrete* if its range R_X is a discrete set of real numbers. A random variable X is called *continuous* if its range R_X is an interval or union of intervals on the real line and if it has probability zero of equalling any single value in R_X.

Let us consider some specific examples.

EXAMPLE 3.1.1.   We roll a pair of fair dice 1 time. Let X be the sum of the 2 numbers that occur. Then the sample space is

$$S = \{(x_1, x_2): x_1 = 1, 2, \ldots, 6; \quad x_2 = 1, 2, \ldots, 6\}$$

and we have defined

$$X(\omega) = x_1 + x_2 \quad \text{for } \omega = (x_1, x_2) \in S.$$

The range of X is $R_X = \{2, 3, \ldots, 12\}$ so X is a discrete random variable.

EXAMPLE 3.1.2. A random sample of 3 people is selected from the list of registered voters in the county of Monterey, California. Let Y be the number of Republicans that occur in the sample. For convenience, we use as our sample space

$$S = \{(x_1, x_2, x_3): x_1 = 0, 1; \; x_2 = 0, 1; \; x_3 = 0, 1\}$$

where the 3 elements in the 3-tuples represent the people selected in the sample as follows: x_1 is 1 if the first person is a Republican and 0 otherwise; x_2 is 1 if the second person selected is a Republican and 0 otherwise; x_3 is defined similarly for the third person. With our 3-tuples defined in this way, we can express Y as

$$Y(\omega) = x_1 + x_2 + x_3, \quad \text{for } \omega = (x_1, x_2, x_3) \in S.$$

The range of Y is $R_Y = \{0, 1, 2, 3\}$, so Y is a discrete random variable.

EXAMPLE 3.1.3. Suppose the instant at which you leave your residence for your day's activities is equally likely to fall anywhere in the interval from 7:30 a.m. to 8:00 a.m. inclusive. Let Z be the length of time between the instant you leave your residence and 8:00 a.m. A reasonable sample space can be defined by calling 7:30 a.m. zero on our time scale and using a minute for our unit. Thus we have

$$S = \{x: 0 \leq x \leq 30\}.$$

Using this sample space the random variable Z can be written in functional notation as

$$Z(\omega) = 30 - \omega, \quad \text{for } \omega \in S.$$

Then the range of Z is $R_Z = \{x: 0 \leq x \leq 30\}$ so Z may be a continuous random variable.

EXAMPLE 3.1.4. We select 1 student at random from those registered at the University of Washington. Let U be the weight of the selected student. We assume that there are 25,000 students registered and that they are numbered from 1 through 25,000. A reasonable sample space then would be

$$S = \{x: x = 1, 2, 3, \ldots, 25{,}000\}.$$

The random variable U then is defined to be

$$U(\omega) = \text{weight of student } \omega, \quad \text{for } \omega \in S.$$

In actual fact, no more than 25,000 distinct values of U could occur. We might, however, find it convenient to think of U as being a continuous random variable and to use continuous methods in describing its behavior. If we thought of using such a continuous approach, we might assume that no student at Washington weighs less than 50 pounds or more than 500 pounds, in which case the range of U would be $R_U = \{x: 50 \leq x \leq 500\}$ and U then may be a continuous random variable.

If X is a discrete random variable, we can use the probability measure defined on the subsets of S to define the probability function for X. This can be used to evaluate probability statements about X and is defined as follows:

DEFINITION 3.1.3. The *probability function* for X is a function (denoted by $p_X(x)$) of a real variable x and is defined to be

$$p_X(x) = P(X(\omega) = x), \qquad \text{for all real } x.$$

If we have a given experiment and a sample space S for that experiment and have defined the discrete random variable X on the elements of S, then we can find $p_X(x)$ as follows. Let

$$A(x) = \{\omega : X(\omega) = x\}.$$

Then $p_X(x) = P(A(x))$ for all real x. Note immediately that $A(x)$ may equal $\varnothing$ for many x and thus $p_X(x)$ is 0 for any such x.

EXAMPLE 3.1.5. Consider the discrete random variable X defined in Example 3.1.1. (X is the sum of the 2 numbers that occur when a pair of fair dice is rolled.) The range of X is the set $\{2, 3, \ldots, 12\}$. Then

$$A(2) = \{(1, 1)\}$$
$$A(3) = \{(1, 2), (2, 1)\}$$
$$A(4) = \{(1, 3), (2, 2), (3, 1)\}$$

$$\begin{array}{cc} \cdot & \cdot \\ \cdot & \cdot \\ \cdot & \cdot \end{array}$$

$$A(11) = \{(5, 6), (6, 5)\}$$
$$A(12) = \{(6, 6)\}$$
$$A(x) = \varnothing, \quad \text{for all other } x.$$

Then we have

$$\begin{aligned} p_X(x) &= \tfrac{1}{36}, & x &= 2 \\ &= \tfrac{2}{36}, & x &= 3 \\ &= \tfrac{3}{36}, & x &= 4 \\ & \quad \cdot & & \quad \cdot \\ & \quad \cdot & & \quad \cdot \\ & \quad \cdot & & \quad \cdot \\ &= \tfrac{2}{36}, & x &= 11 \\ &= \tfrac{1}{36}, & x &= 12 \\ &= 0, & & \text{otherwise.} \end{aligned}$$

We can, of course, use $p_X(x)$ to compute any probabilities of interest regarding X when X is a discrete random variable. For example, the probability that X is an odd number is

$$p_X(3) + p_X(5) + p_X(7) + p_X(9) + p_X(11) = \tfrac{1}{2}.$$

We compute the probability that X lies in any interval by summing the probability function over the interval (strictly speaking, we are summing only over those points in the interval that have positive probability). Thus,

$$P(4 \leq X(\omega) \leq 8) = \sum_{x=4}^{8} p_X(x) = \tfrac{23}{36}$$

$$P(0 \leq X(\omega) \leq 3) = \sum_{x=2}^{3} p_X(x) = \tfrac{1}{12}.$$

EXAMPLE 3.1.6. In Example 3.1.2. we selected 3 people at random from the list of registered voters in Monterey county. We assume that the proportion of registered voters who are Republicans is .4 and that the number of registered voters is large enough to consider the selection of the 3 as being independent trials with probability .4 of selecting a Republican. That is, the probabilities of the single-element events can be computed as though we have independent trials. Thus,

$$P(\{(0, 0, 0)\}) = (.6)^3 = .216$$
$$P(\{(1, 0, 0)\}) = (.4)(.6)^2 = .144$$
$$P(\{(0, 1, 0)\}) = (.6)(.4)(.6) = .144,$$

etc. The range of Y, the number of Republicans who appear in the sample, is $\{0, 1, 2, 3\}$. We define

$$A(0) = \{(0, 0, 0)\}$$
$$A(1) = \{(1, 0, 0), (0, 1, 0), (0, 0, 1)\}$$
$$A(2) = \{(1, 1, 0), (1, 0, 1), (0, 1, 1)\}$$
$$A(3) = \{(1, 1, 1)\}.$$

The probability function for Y then is

$$
\begin{aligned}
p_Y(y) &= .216, &&\text{for } y = 0 \\
&= .432, &&\text{for } y = 1 \\
&= .288, &&\text{for } y = 2 \\
&= .064, &&\text{for } y = 3.
\end{aligned}
$$

EXAMPLE 3.1.7. The game of Chuck-a-Luck is played as follows. Three fair dice are rolled. You as the bettor are allowed to bet 1 dollar (or some other amount) on the occurrence of one of the integers 1, 2, 3, 4, 5, 6. Suppose you bet on the occurrence of a 5. Then if one 5 occurs (on the 3 dice) you win 1 dollar, if two 5's occur you win 2 dollars, and if three 5's occur you win 3 dollars. If no 5's occur you lose your 1 dollar. Let V be the net amount you win in one play of this game. The range of V is $R_V = \{-1, 1, 2, 3\}$ so V is a discrete random variable. We choose as our sample space

$$S = \{(x_1, x_2, x_3): x_i = 1, 2, \ldots, 6; \; i = 1, 2, 3\}.$$

Since we are assuming the dice to be fair, the 3 dice constitute 3 independent trials, each with probabilitity $\frac{1}{6}$ of giving a 5. We immediately then are able to assign probabilities to all of the single-element events (see Section 2.7) and can use these to find the probability function for V:

$$
\begin{aligned}
p_V(v) &= \tfrac{125}{216}, && \text{for } v = -1 \\
&= \tfrac{75}{216}, && \text{for } v = 1 \\
&= \tfrac{15}{216}, && \text{for } v = 2 \\
&= \tfrac{1}{216}, && \text{for } v = 3.
\end{aligned}
$$

As you might gather, it is quite significant that $p_V(-1) > \frac{1}{2}$. We shall soon see exactly why this should play a role in Las Vegas casinos' offers to play this game.

Since the probability function for a discrete random variable gives the probabilities of the random variable being equal to various values in its range (and since a random variable assigns a real number to every $\omega \in S$), we know that

$$
\sum_{R_X} p_X(x) = 1, \qquad \text{for any discrete random variable } X. \tag{3.1}
$$

Furthermore, since probabilities cannot be negative,

$$
p_X(x) \geq 0, \qquad \text{for all } x. \tag{3.2}
$$

Any function of x that satisfies Equations 3.1 and 3.2 could in fact be looked at as being the probability function for some random variable X. We have purposely not mentioned a probability function for a continuous random variable, because it would be zero for all x. We shall have an alternative function to evaluate probabilities about continuous random variables.

A word about future notation. From now on we shall write $P(X = x)$ for $P(X(\omega) = x)$. That is, the functional dependence of X upon the elements $\omega \in S$ will be suppressed. Unless necessary, we shall not even need to refer explicitly to the sample space S. We can always in theory make such a reference, but as we gain skill and practice in deriving probability functions it represents a superfluous step which will be deleted.

EXERCISE 3.1

1. An urn contains 4 balls numbered $1, 2, 3, 4$, respectively. Let Y be the number that occurs if 1 ball is drawn at random from the urn. What is the probability function for Y?

2. Consider the urn defined in problem 1. Two balls are drawn from the urn without replacement. Let Z be the sum of the 2 numbers that occur. Derive the probability function for Z.

3. Assume that the sampling in problem 2 is done with replacement and define Z in the same way. Derive the new probability function for Z.

4. Two balls are drawn with replacement from the urn mentioned in problem 1. Let X be the sum of the squares of the 2 numbers drawn and derive the probability function for X.

5. A class in statistics contains 10 students, 3 of whom are 19, 4 are 20, 1 is 21, 1 is 24 and 1 is 26. Two students are selected at random without replacement from this class. Let X be the average age of the 2 selected students and derive the probability function for X.

6. A man has four keys in his pocket and, since it is dark, cannot see which is his door key. He will try each key in turn until he finds the right one. Let X be the number of keys tried (including the right one) to open the door. What is the probability function for X?

7. Five cards are dealt from a standard 52-card deck. Let Y be the number of red cards that are dealt. What is the probability function for Y?

8. For the experiment defined in 7, let Z be the number of spades dealt. What is the probability function for Z?

9. Let X equal the number of times per week that you enter a supermarket. What would you use as a probability function for X?

10. Let V be the (whole) number of glasses of milk you drink per day. What would you use as a probability function for V?

3.2. Distribution Functions and Density Functions

We saw several particular examples in the last section of discrete random variables and the probability function for discrete random variables. The probability function can be derived very naturally from the probability measure on the subsets of the sample space and can be used to evaluate probability statements about the random variable. The distribution function for the random variable X can also be used to evaluate probability statements about X; the advantage of the distribution function is that it is appropriate for either discrete or continuous random variables. The distribution function is defined as follows.

DEFINITION 3.2.1. The *distribution function* for a random variable X (denoted by $F_X(t)$) is a function of a real variable t such that

(1) the domain of definition of F_X is the whole real line
(2) for any real t $F_X(t) = P(X \leq t)$.

Note, then, that the distribution function associates with each real t the probability that X is less than or equal to t; it is expressing the probability that X will equal a value in the interval from $-\infty$ to t. Thus, as t increases, it is not possible for F_X to decrease. As we shall see, the distribution function provides a powerful tool for many types of application.

We can immediately derive the distribution function of a discrete random variable X if we know the probability function of X. Clearly,

$$F_X(t) = \sum_{x \le t} p_X(x), \tag{3.3}$$

where the summation is over all x from the range of X that satisfy $x \le t$.

EXAMPLE 3.2.1.   In Example 3.1.6 we found that

$$\begin{aligned} p_Y(y) &= .216, && \text{for } y = 0 \\ &= .432, && \text{for } y = 1 \\ &= .288, && \text{for } y = 2 \\ &= .064, && \text{for } y = 3 \\ &= 0, && \text{otherwise.} \end{aligned}$$

The distribution function for Y then is

$$\begin{aligned} F_Y(t) &= 0, && t < 0 \\ &= .216, && 0 \le t < 1 \\ &= .648, && 1 \le t < 2 \\ &= .936, && 2 \le t < 3 \\ &= 1, && t \ge 3. \end{aligned}$$

(See Figure 3.2.1)

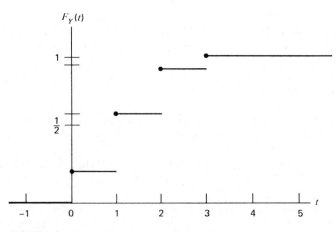

FIGURE 3.2.1   A discrete distribution function.

EXAMPLE 3.2.2.   In Example 3.1.7 we found

$$p_V(v) = \tfrac{125}{216}, \qquad \text{for } v = -1$$
$$= \tfrac{75}{216}, \qquad \text{for } v = 1$$
$$= \tfrac{15}{216}, \qquad \text{for } v = 2$$
$$= \tfrac{1}{216}, \qquad \text{for } v = 3.$$

Then the distribution function for V is

$$F_V(t) = 0, \qquad t < -1$$
$$= \tfrac{125}{216}, \qquad -1 \le t < 1$$
$$= \tfrac{200}{216}, \qquad 1 \le t < 2$$
$$= \tfrac{215}{216}, \qquad 2 \le t < 3$$
$$= 1, \qquad t \ge 3.$$

(See Figure 3.2.2.)

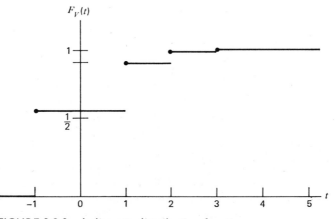

FIGURE 3.2.2   A discrete distribution function.

The distribution function F_X can be used to evaluate probability state-ments of any kind regarding X; in particular, if we define the two events

$$A: X \le a$$
$$B: a < X \le b$$

where a and b are any two real numbers such that $a < b$, we obviously have

$$A \cap B = \varnothing$$

and thus

$$P(A \cup B) = P(A) + P(B).$$

That is,

$$P(X \le b) = P(X \le a) + P(a < X \le b),$$

or

$$F_X(b) = F_X(a) + P(a < X \leq b).$$

Thus

$$P(a < X \leq b) = F_X(b) - F_X(a), \tag{3.4}$$

which shows us how to evaluate the probability that X lies in a particular interval, given F_X.

EXAMPLE 3.2.3.   Given the random variable V discussed in Example 3.2.2 (refer to Figure 3.2.2), use F_V to compute

$$P(0 < V \leq 3), \qquad P(V \leq 0), \qquad P(-1 < V \leq 0).$$

Using Equation 3.4 (and Definition 3.2.1),

$$P(0 < V \leq 3) = F_V(3) - F_V(0)$$
$$= 1 - \tfrac{125}{216} = \tfrac{91}{216}$$
$$P(V \leq 0) = F_V(0) = \tfrac{125}{216},$$
$$P(-1 < V \leq 0) = F_V(0) - F_V(-1)$$
$$= \tfrac{125}{216} - \tfrac{125}{216} = 0.$$

This last probability is zero since the distribution function F_V is constant or level on the interval $-1 < t \leq 0$. This means that there are no values of t in the interval $-1 < t \leq 0$ which V will equal with positive probability. This is the case in general; if the distribution function of a random variable X is constant over any interval, then the probability of X taking on values in that interval is zero.

We have seen how we might derive the distribution function for a discrete random variable X from the probability function for X. We might logically then ask how we could get the probability function for X from the distribution function for X. To answer this, consider the real numbers t in the interval $b - h < t \leq b$ where h is some positive number. As h tends to zero, this interval clearly tends to the single point $t = b$; that is, the limit of any such interval as $h \to 0$ is the single point $t = b$ since $b - h \to b$. Then we might expect, as is indeed the case, that

$$\lim_{h \to 0} P(b - h < X \leq b) = P(X = b).$$

But

$$\lim_{h \to 0} P(b - h < X \leq b) = \lim_{h \to 0} [F_X(b) - F_X(b - h)]$$
$$= F_X(b) - \lim_{h \to 0} F_X(b - h)$$
$$= F_X(b) - F_X(b-),$$

FIGURE 3.2.3   Distribution function and probability function.

where we have written $F_X(b-)$ for $\lim_{h \to 0} F_X(b - h)$. Thus, if b is a point of discontinuity of F_X, then b is a value that X takes on with positive probability; the probability that $X = b$ is the size of the jump at $F_X(b)$. (Refer to Figure 3.2.3.)

EXAMPLE 3.2.4.   Suppose that we are given a random variable U with distribution function

$$
\begin{aligned}
F_U(t) &= 0, & t &< 0 \\
&= \tfrac{1}{2}, & 0 &\le t < 2 \\
&= \tfrac{5}{6}, & 2 &\le t < 3 \\
&= 1, & t &\ge 3.
\end{aligned}
$$

Then we see that there are three points at which F_U is discontinuous, namely $t = 0, 2, 3$. These then are the values which U will equal with positive probability, and we have

$$
\begin{aligned}
p_U(u) &= F_U(0) - F_U(0-) = \tfrac{1}{2}, & \text{at } u &= 0 \\
&= F_U(2) - F_U(2-) = \tfrac{1}{3}, & \text{at } u &= 2 \\
&= F_U(3) - F_U(3-) = \tfrac{1}{6}, & \text{at } u &= 3.
\end{aligned}
$$

Since there are no other points of discontinuity of F_U, $p_U(u)$ is 0, otherwise.

The distribution function for any random variable X is always a function of a real variable. We might inquire whether there are definite rules which a function must satisfy in order that it could be the distribution function of some random variable X. The answer is yes, of course. A function $H(t)$

of a real variable must satisfy the following four requirements in order to be a distribution function. First,

$$0 \le H(t) \le 1, \qquad \text{for all } t. \tag{3.5}$$

This is necessary since $H(t)$ must be a probability for any real t and probabilities are bounded by 0 and 1, as we know. Second,

$$\lim_{t \to -\infty} H(t) = 0, \qquad \lim_{t \to \infty} H(t) = 1. \tag{3.6}$$

This requirement is analogous to axiom 1; it is equivalent to saying that the total probability of value 1 for the sample space is accounted for if we look at all real t. The third requirement is

$$H(a) \le H(b), \qquad \text{for all } a \le b, \tag{3.7}$$

which is necessary because

$$P(a < X \le b) = H(b) - H(a).$$

If we did not insist that Equation 3.7 be satisfied, then we would be allowing a negative probability of the event

$$a < X \le b.$$

Finally, we require that

$$\lim_{h \to 0} H(b + h) = H(b), \qquad \text{for all } b, \text{ where } h > 0; \tag{3.8}$$

that is, that H is continuous from the right at all points. If this were not required and we had, for example,

$$\lim_{h \to 0} H(b + h) = H(b) + \tfrac{1}{4},$$

then we would not be able to attribute the probability of $\tfrac{1}{4}$ that has been added to $H(b)$ to either specific values that X might take on or to specific intervals it might lie in. To summarize, any function $H(t)$ which satisfies Equations 3.5 through 3.8 can be thought of as the distribution function of some random variable; conversely, it is easily shown that any distribution function F_X will satisfy all four statements.

We noted in Definition 3.1.2 that the probability was zero for a continuous random variable X being equal to any value in its range. Since,

$$P(X = b) = F_X(b) - F_X(b-),$$

we see that, if X is continuous, then

$$F_X(b) - F_X(b-) = 0, \qquad \text{for all } b.$$

That is,

$$F_X(b) = F_X(b-)$$

and we see that the distribution function of X has no jumps in it. If X is a continuous random variable, then $F_X(t)$ must be a continuous function of t for all t.

EXAMPLE 3.2.5.   The function

$$H(t) = 0, \quad t < 0$$
$$= t, \quad 0 \le t \le 1$$
$$= 1, \quad t > 1$$

rather obviously satisfies Equation 3.5 through 3.8. Thus it may be thought of as the distribution function for some random variable X. Since $H(t)$ is a continuous function of t, the corresponding random variable X is continuous.

EXAMPLE 3.2.6.   The function

$$H(t) = 0, \quad t < 0$$
$$= 1 - e^{-t}, \quad t \ge 0$$

satisfies Equations 3.5 through 3.8 and thus is the distribution function for some random variable Y. Since $H(t)$ is a continuous function of t, the corresponding random variable Y is also continuous.

It has been pointed out that the probability function, $P(Y = b)$, must be zero for all b if Y is a continuous random variable. Thus, the probability function cannot be used (as an alternative to the distribution function) to compute the probability that $a \le Y \le b$ if Y is continuous. Suppose, though, that we can find a function $f_Y(y)$ such that

$$F_Y(t) = \int_{-\infty}^{t} f_Y(y)\, dy$$

for all values t. Notice, then, that we would have

$$P(a < Y \le b) = F_Y(b) - F_Y(a)$$
$$= \int_{-\infty}^{b} f_Y(y)\, dy - \int_{-\infty}^{a} f_Y(y)\, dy$$
$$= \int_{a}^{b} f_Y(y)\, dy$$

so $f_Y(y)$ could be used to evaluate the probability that Y takes on a value in any specified interval. For all continuous random variables with which we shall be concerned, such a function f_Y does always exist and is called the probability density function for the continuous random variable Y. We formally define the probability density function for a continuous random variable Y as follows.

DEFINITION 3.2.2. The *probability density function* for a continuous random variable Y (denoted by $f_Y(y)$) is a function of a real variable y such that: (1) the domain of f_Y is the whole real line and (2) for any real number t

$$F_Y(t) = \int_{-\infty}^{t} f_Y(y)\, dy.$$

The *range* of Y is $R_Y = \{y : f_Y(y) > 0\}$.

Let us note some general facts about density functions before examining specific examples. First, we find that

$$f_Y(t) = \frac{d}{dt} \int_{-\infty}^{t} f_Y(y)\, dy = \frac{d}{dt} F_Y(t),$$

so we may derive the density function simply by taking the derivative of the distribution function with respect to t. (In the cases covered in the text, it may happen that this derivative will not exist at (at most) a countable number of points $t_1, t_2, \ldots$. Since the probability that Y equals these points, or any others, is zero, we shall encounter no difficulty if we arbitrarily define the density to equal any convenient value at such points.) Second, since

$$f_Y(t) = \frac{d}{dt} F_Y(t)$$

note that the value of the density function f_Y at t is equal to the slope of the distribution function at t; since $F_Y(t)$ gives the total accumulation of probability on the real line up to and including t, f_Y at t gives the relative rate at which the probability is accumulating in a neighborhood of t.

Third, since the probability function for a continuous random variable is equal to zero at all points, the following four probabilities must all be equal:

$$P(a \leq Y \leq b) = P(a \leq Y < b)$$
$$= P(a < Y < b)$$
$$= P(a < Y \leq b)$$
$$= \int_{a}^{b} f_Y(y)\, dy;$$

that is, it is immaterial whether the endpoints are included in computing the probability that Y will take on a value in the interval from a to b. Note that the same result is true for the measure length.

Finally, since definite integrals can be thought of as giving the area between the function being integrated and the horizontal axis, between the definite limits, we can graphically think of the probability that $a < Y < b$ as being the area under f_Y between a and b.

EXAMPLE 3.2.7.   Consider the distribution function given in Example 3.2.5:

$$F_X(t) = 0, \qquad t < 0$$
$$= t, \qquad 0 \le t \le 1$$
$$= 1, \qquad t > 1.$$

The derivative of F_X is

$$\frac{d}{dt} F_X(t) = 0, \qquad \text{for } t < 0 \text{ and } t > 1$$
$$= 1, \qquad \text{for } 0 < t < 1.$$

The derivative doesn't exist at $t = 0$ and $t = 1$, but we can conveniently define

$$f_X(x) = 1, \qquad 0 < x < 1$$
$$= 0, \qquad \text{otherwise,}$$

and use f_X to compute

$$P(\tfrac{1}{4} < X < \tfrac{3}{4}) = \int_{1/4}^{3/4} 1 \, dx = \tfrac{3}{4} - \tfrac{1}{4} = \tfrac{1}{2}$$

$$P(-1 < X < \tfrac{1}{2}) = \int_{-1}^{1/2} f_X(x) \, dx$$

$$= \int_{-1}^{0} f_X(x) \, dx + \int_{0}^{1/2} f_X(x) \, dx$$

$$= \int_{-1}^{0} 0 \, dx + \int_{0}^{1/2} 1 \, dx = \tfrac{1}{2}.$$

(Refer to Figure 3.2.4.)

EXAMPLE 3.2.8.   The continuous distribution function discussed in Example 3.2.6 was

$$F_Y(t) = 0, \qquad t < 0$$
$$= 1 - e^{-t}, \qquad t \ge 0.$$

The derivative of F_Y is

$$\frac{d}{dt} F_Y(t) = 0, \qquad t < 0$$
$$= e^{-t}, \qquad t > 0.$$

We then can define

$$f_Y(y) = e^{-y}, \qquad y > 0$$
$$= 0, \qquad \text{otherwise,}$$

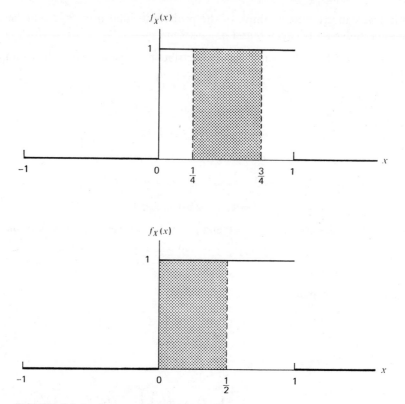

FIGURE 3.2.4   Areas under a density function.

and use f_Y to compute

$$P(1 < Y < 3) = \int_1^3 e^{-y}\, dy = e^{-1} - e^{-3}$$

$$P(Y > 4) = \int_4^\infty e^{-y}\, dy = e^{-4}$$

$$P(Y < 2) = \int_{-\infty}^2 f_Y(y)\, dy = \int_0^2 e^{-y}\, dy = 1 - e^{-2}.$$

Let us conclude this section by commenting that any function $h(t)$ can be considered the probability density function of some continuous random variable if

$$h(t) \geq 0, \qquad \text{for all } t, \tag{3.9}$$

and

$$\int_{-\infty}^\infty h(t)\, dt = 1. \tag{3.10}$$

EXERCISE 3.2

1. Verify that

$$F_X(t) = 0, \qquad t < -3$$
$$= \tfrac{1}{3}, \qquad -3 \le t < -1$$
$$= \tfrac{2}{3}, \qquad -1 \le t < 0$$
$$= 1, \qquad t \le 0$$

is a distribution function and derive the probability function for X.

2. Verify that

$$F_Z(t) = 0, \qquad t < -2$$
$$= \tfrac{1}{2}, \qquad -2 \le t < 0$$
$$= 1, \qquad t \ge 0$$

is a distribution function and specify the probability function for Z. Use it to compute $P(-1 \le Z \le 1)$.

3. Verify that

$$F_W(t) = 0, \qquad t < 3$$
$$= \tfrac{1}{3}, \qquad 3 \le t < 4$$
$$= \tfrac{1}{2}, \qquad 4 \le t < 5$$
$$= \tfrac{2}{3}, \qquad 5 \le t < 6$$
$$= 1, \qquad t \ge 6$$

is a distribution function and specify the probability function for W. Use it to compute $P(3 < W \le 5)$.

4. Verify that

$$F_Y(t) = 0, \qquad t < 0$$
$$= \tfrac{1}{4}, \qquad 0 \le t < 5$$
$$= \tfrac{1}{3}, \qquad 5 \le t < 7$$
$$= \tfrac{1}{2}, \qquad 7 \le t < 100$$
$$= \tfrac{5}{6}, \qquad 100 \le t < 102$$
$$= 1, \qquad t \ge 102$$

is a distribution function and specify the probability function for Y. Use it to compute $P(Y \le 100)$.

5. The random variable Z has the probability function

$$p_Z(x) = \tfrac{1}{3}, \qquad \text{for } x = 0, 1, 2$$
$$= 0, \qquad \text{otherwise.}$$

What is the distribution function for Z?

6. The random variable U has the probability function

$$p_U(-3) = \tfrac{1}{3}$$
$$p_U(0) = \tfrac{1}{6}$$
$$p_U(4) = \tfrac{1}{3}.$$

What is the distribution function for U?

7. Verify that

$$F_X(t) = 0, \qquad t < -1$$
$$= \frac{t+1}{2}, \qquad -1 \le t \le 1$$
$$= 1, \qquad t > 1$$

is a distribution function and specify the probability density function for X. Use it to compute $P(-\tfrac{1}{2} \le X \le \tfrac{1}{2})$.

8. Verify that

$$F_Y(t) = 0, \qquad t < 0$$
$$= \sqrt{t}, \qquad 0 \le t \le 1$$
$$= 1, \qquad t > 1$$

is a distribution function and specify the probability density function for Y. Use it to compute $P(\tfrac{1}{4} < Y < \tfrac{3}{4})$.

9. Verify that

$$F_Z(t) = 0, \qquad t < 0$$
$$= t^2, \qquad 0 \le t < \tfrac{1}{2}$$
$$= 1 - 3(1 - t)^2, \qquad \tfrac{1}{2} \le t < 1$$
$$= 1, \qquad t \ge 1$$

is a distribution function and derive the density function for Z.

10. Is

$$H(t) = 0, \qquad t < -1$$
$$= 1 - t^2, \qquad -1 \le t < \tfrac{1}{2}$$
$$= \tfrac{1}{2} + t^2, \qquad \tfrac{1}{2} \le t < 1$$
$$= 1, \qquad t \ge 1$$

a distribution function?

11. Verify that

$$H(t) = 0, \qquad t < 0$$
$$= \tfrac{1}{2}, \qquad 0 \le t < \tfrac{1}{2}$$
$$= t, \qquad \tfrac{1}{2} \le t \le 1$$
$$= 1, \qquad t > 1$$

is a distribution function. This is the distribution function of what is called a *mixture random variable*. It has one point of discontinuity (at $t = 0$) and thus there is a positive probability ($\frac{1}{2}$) of the random variable equalling zero. Note that the distribution function is continuous and increasing for $\frac{1}{2} \leq t \leq 1$. Thus the random variable takes on particular values in this interval with probability zero; we could define a pseudo density function for the random variable lying in subintervals of this interval.

12. Given X has probability density function

$$f_X(x) = 1, \qquad 99 < x < 100$$
$$= 0, \qquad \text{otherwise,}$$

derive $F_X(t)$.

13. Y is a continuous random variable with

$$f_Y(y) = 2(1 - y), \qquad 0 < y < 1$$
$$= 0, \qquad \text{otherwise.}$$

Derive $F_Y(t)$.

14. Z is a continuous random variable with probability density function

$$f_Z(z) = 10e^{-10z}, \qquad z > 0$$
$$= 0, \qquad \text{otherwise.}$$

Derive $F_Z(t)$.

15. Given the sample space $S = \{\omega : a \leq \omega \leq b\}$, Theorem 2.8.1 showed that the rule

$$P(A) = \frac{L(A)}{L(S)} \quad \text{for } A \subset S$$

where $L(A)$ is the length of A, satisfies the three axioms for a probability measure. Let X be the *identity* random variable defined on this sample space, i.e. let $X(\omega) = \omega$ for $\omega \in S$ and derive the density function and distribution function for X.

16. Let Y be the identity random variable for the sample space given in Example 2.8.4 and derive the density and distribution function for Y. Use F_Y to answer the questions posed in the example.

17. Assume that probabilities are assigned to intervals by the rule described in Exercise 2.8.13 and let X be the identity random variable.
    (a) Find f_X and F_X.
    (b) Use F_X to answer the questions posed there.

18. Assume that probabilities are assigned to intervals by the rule described in Exercise 2.8.14. Let Y be the identity random variable defined on that sample space. Find F_Y and use it to answer the questions posed there.

3.3. Expected Values and Summary Measures

We have seen that if X is a discrete random variable, we can use either the distribution function F_X or the probability function p_X to evaluate probability statements about X. If X is a continuous random variable, we can use the distribution function F_X or the density function f_X to evaluate probability statements about X. Frequently, problems are phrased that require the notion of the average or expected value of X, not merely a statement of the probability that X will lie in a certain interval. For example, if we bet 1 dollar on the occurrence of a certain outcome in a gambling game, we might ask how much, on the average, we could expect to get in return. Or, if a doctor recommends that we go on a special reducing diet, we might inquire how much weight we could expect to lose. If we stock a stream with 10,000 small trout, how many might we expect to survive to adult size? Each of these questions refers to an average value of a random variable.

Rather than restrict our discussion to simply the average value of a random variable, we can easily refer to the average value of a function of a random variable $H(X)$. This includes as a special case the identity function $H(X) = X$; that is, if we discuss the average of $H(X)$, this discussion includes the average value of X itself. For concreteness, let us assume that the range of X consists of three numbers, x_1, x_2, x_3. We can interpret $p_X(x_i)$, $i = 1, 2, 3$, as giving us the relative proportion of the time that we would observe $X = x_i$ if we were to repeatedly perform the experiment for which X is defined. The values $H(X)$ assumes when $X = x_1$ or x_2 or x_3 are $H(x_i)$, $i = 1, 2, 3$; thus we would expect $H(X)$ to equal $H(x_i)$ the proportion $p_X(x_i)$ of the time. The average value which $H(X)$ would be expected to assume then is

$$H(x_1)p_X(x_1) + H(x_2)p_X(x_2) + H(x_3)p_X(x_3) = \sum_{R_X} H(x)p_X(x).$$

This is in fact the definition given in Definition 3.3.1 below.

If X is a continuous random variable, a very similar line of reasoning is appropriate. To be explicit, suppose X is continuous and $R_X = \{x : 0 \le x \le 1\}$; then f_X is positive on the interval from 0 to 1. We can subdivide the interval from 0 to 1 into n equal length pieces, where n is large. Let x_k be the midpoint of the kth subinterval. (See Figure 3.3.1.) The probability that X equals a value in this subinterval is approximately $f_X(x_k)/n$, the area of the rectangle centered at x_k; thus, at least approximately, we would expect $H(X)$ to equal (approximately) $H(x_k)$ the proportion $f_X(x_k)/n$ of the time, for $k = 1, 2, \ldots, n$. Then, reasoning as above, we would (approximately) expect the average value of $H(X)$, over repeated performances of the experiment, to be

$$\sum_{k=1}^{n} H(x_k) f_X(x_k)/n;$$

The exact expected value of $H(X)$ is given by the limit of this expression as

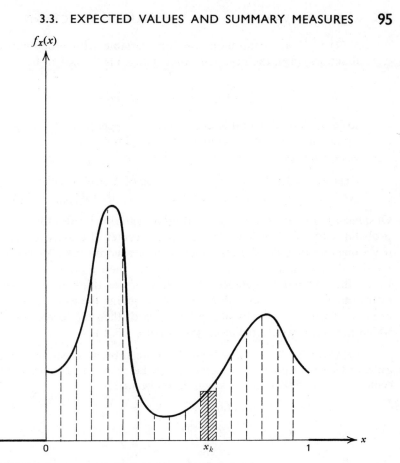

FIGURE 3.3.1   Approximation of areas under $f_X(x)$.

n tends to infinity; but this limit is

$$\lim_{n \to \infty} \sum_{k=1}^{n} H(x_k) f_X(x_k)/n = \int_{R_X} H(x) f_X(x)\, dx,$$

as we see in Definition 3.3.1. Notice that we can just as well extend the limits of integration over the whole real line, rather than just over R_X, since $f_X(x) = 0$ for $x \notin R_X$.

DEFINITION 3.3.1.   (1) If X is a discrete random variable with probability function $p_X(x)$, the *expected value* of $H(X)$ (written $E[H(X)]$) is defined to be

$$E[H(X)] = \sum_{R_X} H(x) p_X(x),$$

so long as the sum is absolutely convergent.

(2) If X is a continuous random variable with probability density function $f_X(x)$, the *expected value* of $H(X)$ is defined to be

$$E[H(X)] = \int_{-\infty}^{\infty} H(x)f_X(x)\,dx,$$

so long as the integral is absolutely convergent. If the integral or the sum is not absolutely convergent, we simply say the expected value does not exist.

DEFINITION 3.3.2.   The expected value of X itself is called the *mean* or *average value* of X and is denoted by μ_X; that is, $\mu_X = E[X]$.

Graphically, the mean value of the random variable locates the middle of its probability function or density function in a center of gravity sense. That is, if we imagine that the density function of a random variable describes the relative distribution of mass in a metal rod, then μ_X locates the point at which the rod would balance. Similarly, if we think of the probability distribution of a discrete random variable as giving the location and amount of mass of various tin cans sitting on a board, then μ_X locates the point at which the board would balance. (See Figure 3.3.2.)

EXAMPLE 3.3.1.   Suppose you bet 1 dollar on the occurrence of a 1 in the game of Chuck-a-Luck (see Example 3.1.7). Let V be the amount you win and compute μ_V and $E[V^2]$. As we saw in Example 3.1.7,

$$\begin{aligned}
p_V(v) &= \tfrac{125}{216}, & v &= -1 \\
&= \tfrac{75}{216}, & v &= 1 \\
&= \tfrac{15}{216}, & v &= 2 \\
&= \tfrac{1}{216}, & v &= 3.
\end{aligned}$$

Then

$$\begin{aligned}
\mu_V = E[V] &= \sum_{R_V} v p_V(v) \\
&= (-1)\tfrac{125}{216} + (1)\tfrac{75}{216} + (2)\tfrac{15}{216} + (3)\tfrac{1}{216} \\
&= -\tfrac{17}{216} = -.079,
\end{aligned}$$

and

$$\begin{aligned}
E[V^2] &= \sum_{R_V} v^2 p_V(v) \\
&= (-1)^2 \tfrac{125}{216} + (1)^2 \tfrac{75}{216} + (2)^2 \tfrac{15}{216} + (3)^2 \tfrac{1}{216} \\
&= \tfrac{269}{216} = 1.245.
\end{aligned}$$

Since $\mu_V = -.079$, we could, on the average, expect to lose about 8 cents every time we bet 1 dollar on this game. A game where $\mu_V = 0$ is called *fair*; thus this is not a fair game. Of course, everything the bettor loses goes to the house, so the casino playing this game can expect to win about 8 cents from every dollar bet on this game, every play. This is part of the reason the game is attractive to a casino.

FIGURE 3.3.2 Balance points for densities and probability functions.

EXAMPLE 3.3.2.   Suppose your doctor recommends that you go on a particular diet for 2 weeks. Considering your build and bone structure, he assumes that the amount of weight you will lose is equally likely to lie between 5 and 10 pounds. What is the average amount you might expect to lose on such a diet? If we let X be the number of pounds you will lose, then

$$f_X(x) = \tfrac{1}{5}, \qquad 5 < x < 10$$

$$= 0, \qquad \text{otherwise.}$$

The amount you might expect to lose is μ_X where

$$\mu_X = E[X] = \int_5^{10} x \cdot \tfrac{1}{5}\, dx = 7\tfrac{1}{2} \text{ pounds.}$$

EXAMPLE 3.3.3.   If, in the previous example, the doctor said that the amount you would lose has probability density function

$$f_X(x) = \tfrac{3}{125}(x - 5)^2, \qquad 5 < x < 10$$

$$= 0, \qquad \text{otherwise,}$$

the expected amount of weight you would lose is

$$\mu_X = E[X] = \int_5^{10} x \cdot \tfrac{3}{125}(x-5)^2 \, dx = 8.75 \text{ pounds.}$$

The mean of a random variable gives a measure of the middle of the probability distribution of the random variable. As we shall see, it is desirable in essentially all applications also to have a measure of the variability of the probability distribution (or of the random variable itself). The most commonly used measure of variability of a random variable is the average of the square of the distance between the random variable and its mean μ_X. If X varies relatively little, it will always be close to its mean value and this average squared distance will be relatively small; if X varies a great deal, then it will differ from its mean value by a relatively large amount some of the time and the average squared distance will be relatively large. This average squared deviation is called the variance of the random variable. Its positive square root is called the standard deviation of the random variable. Both are defined below.

DEFINITION 3.3.3.   The *variance* of a random variable X (denoted by σ_X^2) is defined to be

$$\sigma_X^2 = E[(X - \mu_X)^2];$$

its positive square root is denoted by σ_X and is called the *standard deviation* of X. Thus, $\sigma_X = \sqrt{\sigma_X^2}$.

If there are units attached to the random variable X, the variance would be measured in the squares of those units. The standard deviation is of interest since it is measured in the same units as the original random variable.

EXAMPLE 3.3.4.   Compute the variance and standard deviation of V, defined in Example 3.3.1. By definition,

$$\sigma_V^2 = \sum_{R_V} [v - (-\tfrac{17}{216})]^2 p_V(v)$$

$$= (-\tfrac{199}{216})^2 \tfrac{125}{216} + (\tfrac{233}{216})^2 \tfrac{75}{216} + (\tfrac{449}{216})^2 \tfrac{15}{216} + (\tfrac{665}{216})^2 \tfrac{1}{216}$$

$$= 1.239,$$

and

$$\sigma_V = \sqrt{1.239} = 1.113.$$

EXAMPLE 3.3.5.   Compute the variance and standard deviation of the random variable X, defined in Example 3.3.2. By definition,

$$\sigma_X^2 = \int_5^{10} (x - 7\tfrac{1}{2})^2 \cdot \tfrac{1}{5} \, dx = \tfrac{25}{12} = 2.083.$$

and

$$\sigma_X = \sqrt{2.083} = 1.443.$$

The operation of taking expected values has several important properties, some of which are given below in Theorem 3.3.1. You are asked to prove this result in Exercise 12 below.

Theorem 3.3.1. If X is a random variable, then

1. $E[c] = c,$ where c is a constant
2. $E[cH(X)] = cE[H(X)]$
3. $E[H(X) + G(X)] = E[H(X)] + E[G(X)],$

so long as the expected values involved exist.

Let us immediately use Theorem 3.3.1 to derive an alternative formula for the variance of a random variable, one that is frequently more easily evaluated than is the definition for σ_X^2. We proceed as follows:

$$\sigma_X^2 = E[(X - \mu_X)^2] = E[X^2 - 2X\mu_X + \mu_X^2]$$
$$= E[X^2] - 2\mu_X E[X] + \mu_X^2$$
$$= E[X^2] - \mu_X^2. \tag{3.11}$$

Thus, σ_X^2 can always be calculated as the average of the squares of X minus the square of the average of X.

EXAMPLE 3.3.6. Let us recompute the variances of V and of X as given in Examples 3.3.4 and 3.3.5 to verify that we do in fact get the same answers. We saw in Example 3.3.1 that

$$\mu_V = -\tfrac{17}{216}, \qquad E[V^2] = \tfrac{269}{216}.$$

Then

$$\sigma_V^2 = \frac{269}{216} - \left(-\frac{17}{216}\right)^2 = \frac{58,104 - 289}{46,656} = \frac{57,815}{46,656} = 1.239,$$

as we found in Example 3.3.5. In Example 3.3.2, we found that $\mu_X = 7\tfrac{1}{2}$. Then

$$E[X^2] = \int_5^{10} x^2 \tfrac{1}{5}\, dx = \tfrac{175}{3}$$

and

$$\sigma_X^2 = E[X^2] - \mu_X^2 = \frac{175}{3} - \left(\frac{15}{2}\right)^2 = \frac{700 - 675}{12} = \frac{25}{12} = 2.083,$$

the same result as we had in Example 3.3.5.

Let us discuss an additional measure of the middle of the probability law for a random variable X (in a different sense) and an additional measure of the variability of X (also in a different sense). Both will be special functions of the percentiles of X, which are defined below.

DEFINITION 3.3.4. The $100k$-th *percentile* of X (denoted by t_k) is the smallest possible number such that the probability of not exceeding it is at least k, where $0 < k < 1$; i.e. t_k is the smallest value such that

$$F_X(t_k) \geq k.$$

Note, as in Figure 3.3.3 that the 20-th percentile, $t_{.20}$, is the number such that

FIGURE 3.3.3   Percentiles of X.

$F_X(t_{.20}) = .2$; the 60-th percentile, $t_{.60}$, satisfies $F_X(t_{.60}) = .60$. Thus, t_k is the smallest number such that the probability that $X \leq t_k$ is k. Several of the percentiles are given special names; $t_{.25}$, $t_{.50}$, and $t_{.75}$ are called the *quartiles* of the distribution, because they cut the probability function or density function into four equal probabilities. The 50-th percentile, $t_{.50}$, is also called the *median* of the random variable; it has the property that the random variable is equally likely to lie on either side of it. This is occasionally used as a measure of the middle of the probability law. The difference, $t_{.75} - t_{.25}$, is called the *interquartile range* and is sometimes used as a measure of variability. Note that the interquartile range is the length of an interval which brackets the "middle" 50% of the probability distribution for X.

The inequality is used in Definition 3.3.4 to define the $100k$-th percentile to make sure that the percentiles are defined for all random variables. In particular, if X is a discrete random variable, then its distribution function has jumps in it and there may be no number such that $F_X(t) = k$. By using the given definition, the $100k$-th percentile is defined to be the number t_k at which the distribution function jumped beyond k (see Figure 3.3.4).

EXAMPLE 3.3.7. Suppose that X is a random variable with distribution function

$$F_X(t) = 0, \qquad t < 0$$
$$= \sqrt{t}, \qquad 0 \leq t \leq 1$$
$$= 1, \qquad t > 1.$$

FIGURE 3.3.4 Percentiles and jumps.

Let us find the median of X and the interquartile range for X. Since

$$F_X(t) = \sqrt{t},$$

the median is defined by

$$\sqrt{t_{.50}} = .5$$

and thus $t_{.50} = (.5)^2 = .25$. Similarly,

$$t_{.25} = (.25)^2 = .0625$$

$$t_{.75} = (.75)^2 = .5625.$$

The interquartile range is

$$t_{.75} - t_{.25} = .5.$$

Let us also compute μ_X and σ_X. We find that

$$f_X(t) = \frac{1}{2\sqrt{t}}, \qquad 0 < t < 1$$

$$= 0, \qquad \text{otherwise.}$$

Then

$$\mu_X = \int_0^1 x \frac{1}{2\sqrt{x}}\, dx = \tfrac{1}{3}$$

$$E[X^2] = \int_0^1 x^2 \frac{1}{2\sqrt{x}}\, dx = \tfrac{1}{5}$$

$$\sigma_X^2 = E[X^2] - \mu_X^2 = \tfrac{1}{5} - \tfrac{1}{9} = \tfrac{4}{45},$$

and

$$\sigma_X = \sqrt{\frac{4}{45}} = \frac{2}{3\sqrt{5}}.$$

Note the median and μ_X are not equal in this example. Unless the probability density function is symmetric with respect to some line perpendicular to the x-axis, the point at which the density balances and the point which cuts it into two equal areas will not coincide.

EXAMPLE 3.3.8.   Suppose that we randomly select a number from the telephone book and assume that the last digit in the selected number is equally likely to be $0, 1, 2, \ldots, 9$. Let X be this final digit. Then

$$F_X(t) = 0, \qquad t < 0$$

$$= \frac{j+1}{10}, \qquad \leq t < j + 1, \qquad j = 0, 1, \ldots, 8$$

$$= 1, \qquad t \geq 9.$$

The median for X then is 4 and $t_{.25} = 2$, $t_{.75} = 7$, and the interquartile range is $7 - 2 = 5$.
We also find that $\mu_X = 4.5$, $\sigma_X = 2.87$.

EXERCISE 3.3

1. If

$$p_X(x) = \tfrac{1}{4}, \qquad x = 2, 4, 8, 16$$
$$= 0, \qquad \text{otherwise}$$

   compute:
   (a) $E[X]$ (c) $E[1/X]$ (e) σ_X^2 and σ_X.
   (b) $E[X^2]$ (d) $E[2^{X/2}]$

2. Suppose that $f_X(x) = \tfrac{1}{2}$, $-1 < x < 1$, compute
   (a) $E[X]$ (c) $E[X + 2]$ (e) σ_X^2
   (b) $E[X^2]$ (d) $E[X/4 + 7]$ (f) σ_X.

3. Given

$$f_X(x) = 2(1 - x), \qquad 0 < x < 1$$
$$= 0, \qquad \text{otherwise,}$$

compute:

(a) $E[X]$ (c) $E[(X + 10)^2]$ (e) σ_X^2

(b) $E[X^2]$ (d) $E[1/(1 - X)]$ (f) σ_X.

4. Show that $E[X - \mu_X] = 0$.

5. Given

$$F_X(t) = 0, \qquad t < 0$$
$$= \frac{t}{100}, \qquad 0 \le t \le 100$$
$$= 1, \qquad t > 100$$

find: (a) $t_{.10}$, (b) $t_{.20}$, (c) $t_{.55}$, (d) $t_{.99}$, (e) $t_{.47}$, (f) $t_{.80}$ and (g) the interquartile range.

6. Suppose that

$$F_U(t) = 0, \qquad t < 1$$
$$= \log_e t, \qquad 1 \le t \le e$$
$$= 1, \qquad t > e,$$

find the median, the interquartile range for U, μ_U and σ_U^2.

7. If

$$F_Z(t) = 0, \qquad t < 0$$
$$= 2^t - 1, \qquad 0 \le t \le 1$$
$$= 1, \qquad t > 1,$$

find the median, the interquartile range for Z, μ_Z and σ_Z.

8. If

$$F_X(t) = 0, \qquad t < 0$$
$$= \tfrac{1}{6}, \qquad 0 \le t < 1$$
$$= \tfrac{1}{3}, \qquad 1 \le t < 2$$
$$= \tfrac{1}{2}, \qquad 2 \le t < 3$$
$$= 1, \qquad t \ge 3,$$

find: (a) $t_{.50}$, (b) $t_{.34}$, (c) $t_{.51}$, (d) $t_{.16}$ (e) $t_{.17}$ and (f) the interquartile range.

9. Show that $E[(X - a)^2]$ is minimized if $a = \mu_X$.

10. The adult height of a 3-year-old boy is equally likely to fall in the interval from 5 feet 6 inches to 5 feet 11 inches. What is his expected adult height?

11. A church lottery is going to give away a 3000-dollar car. They sell 10,000 tickets at 1 dollar apiece. If you buy 1 ticket, what is your expected gain? What is your expected gain if you buy 100 tickets? Compute the variance of your gain in these two instances.

12. Prove Theorem 3.3.1.

13. Compute the mean and variance of the sum of the two numbers that occur when a pair of fair dice is rolled one time.

14. According to recent mortality tables, the probability an American citizen will die in his 20-th year is .00178. Assume that a life insurance company will sell a $1000 one-year term policy to a 19-year-old for $5.00. (This means that the company will pay the beneficiaries $1000 if the person dies during his 20-th year; if the person does not die the company retains the $5.00.) What is the expected gain to the insurance company from selling one such policy, ignoring expenses of selling and administration? What is the standard deviation of the gain?

15. Suppose you are allowed to flip a fair coin until the first head appears. Let X be the total number of flips required. You will win 2^X dollars at the occurrence of the first head.
    (a) What is the probability function for X?
    (b) What is the expected value of your winnings?
    (c) If the amount you pay to play a game is equal to the amount you expect to win, the game is called *fair*. How much should you pay to play this game, to make it fair? This game is an example of the *Saint Petersburg paradox*.

16. Suppose, in Exercise 15, the game is truncated at 20 flips. That is, you flip the coin until the first head appears. You win 2^X dollars for $X \leq 19$ and you win 2^{20} dollars $= \$1,048,576$ if $X \geq 20$. How much should you pay to make this a fair game?

3.4. Moments and Generating Functions

We have seen that the mean and the variance of a random variable give some information about the distribution of the random variable. Both of these quantities are defined by the expected value of the random variable raised to the first and to the second power; the expected value of X raised to an integral power is called a moment of the distribution of X. The full collection of moments includes the expected values of all positive integral powers of X. The formal definition of moments of a random variable follows.

DEFINITION 3.4.1. The k-th *moment* of a random variable X (denoted by m_k) is the expected value of X to the k-th power, $k = 1, 2, 3, \ldots$; that is, $m_k = E[X^k]$.

The first moment m_1 is defined to be

$$m_1 = E[X^1] = E[X]$$

and thus $m_1 = \mu_X$. The second moment m_2 is

$$m_2 = E[X^2]$$

and thus $m_2 = \sigma_X^2 + \mu_X^2$. Thus knowledge of m_1 and m_2 for a random variable would immediately allow us to calculate the mean and the variance of the random variable. These two moments give information about the middle of the probability law and the relative variability about that middle value. The higher order moments $m_3, m_4, m_5, \ldots$, can be shown to give information about other facets of the probability law: the relative peakedness, how similar the two tails are on either side of the mean value, etc. Let us look at some particular examples.

EXAMPLE 3.4.1. Suppose that we throw a dart at a target and have probability .9 of hitting the bull's-eye. Let X be 1 if we hit the bull's-eye and 0 if we miss it. Then,

$$\begin{aligned} p_X(x) &= .1, && \text{at } x = 0 \\ &= .9, && \text{at } x = 1 \\ &= 0, && \text{otherwise.} \end{aligned}$$

The k-th moment of X is

$$m_k = E[X^k] = 0^k(.1) + 1^k(.9) = .9 \quad \text{for } k = 1, 2, 3, \ldots.$$

Thus, this random variable X has constant moments of .9 for all k.

EXAMPLE 3.4.2. Suppose that we are going to drive a car to an athletic event. From past experience in making this trip we are willing to assume that our driving time is equally likely to be anywhere from 20 to 30 minutes. If we let X be the number of minutes it will take us to get there, then we have

$$\begin{aligned} f_X(x) &= \tfrac{1}{10}, && 20 < x < 30 \\ &= 0, && \text{otherwise.} \end{aligned}$$

The k-th moment of this random variable is

$$m_k = E[X^k] = \int_{20}^{30} \frac{x^k}{10} \, dx = \frac{30^{k+1} - 20^{k+1}}{10(k + 1)}, \quad k = 1, 2, 3, \ldots.$$

Thus,

$$m_1 = \frac{(30)^2 - (20)^2}{10(2)} = 25$$

$$m_2 = \frac{(30)^3 - (20)^3}{10(3)} = 633\tfrac{1}{3},$$

etc., and we have $\mu_X = 25$, $\sigma_X^2 = 8\tfrac{1}{3}$.

A particular function, which can be used to generate the moments of a random variable X by differentiation, is the moment generating function. This is defined as follows.

DEFINITION 3.4.2.   The *moment generating function* (denoted by $m_X(t)$) for a random variable X is defined to be $m_X(t) = E[e^{tX}]$.

To see why $m_X(t)$ is called a moment generating function, let us look at its derivatives with respect to t, evaluated at $t = 0$. We define

$$m_X^{(k)}(0) = \frac{d^k}{dt^k} m_X(t) \bigg|_{t=0}, \qquad k = 1, 2, 3, \ldots.$$

Then

$$m_X^{(1)}(t) = \frac{d}{dt} E[e^{tX}].$$

Assuming we can interchange the operations of differentiation and expectation, we have

$$m_X^{(1)}(t) = E\left[\frac{d}{dt} e^{tX}\right] = E[Xe^{tX}]$$

and thus

$$m_X^{(1)}(0) = E[Xe^0] = E[X] = m_1.$$

(The interchange of the two operations is permissible if $m_X(t)$ exists and, if the integral or sum defining $m_X(t)$ is improper, if the integral or sum converges uniformly.) Similarly, we have

$$m_X^{(2)}(t) = \frac{d^2}{dt^2} E[e^{tX}]$$

$$= E\left[\frac{d^2}{dt^2} e^{tX}\right]$$

$$= E[X^2 e^{tX}],$$

so that $m_X^{(2)}(0) = E[X^2] = m_2$. Similarly, we see that $m_X^{(k)}(0) = E[X^k] = m_k$. Thus the successive derivatives of $m_X(t)$ evaluated at the origin generate the moments of X. For this reason, $m_X(t)$ is called the moment generating function for X.

Another way to see that the derivatives of m_X, evaluated at the origin, will generate the moments of X is to realize that the Taylor series expansion of e^{tx} about the origin is

$$e^{tx} = 1 + tx + \frac{(tx)^2}{2!} + \frac{(tx)^3}{3!} + \cdots$$

and that this series converges to e^{tx} for all real x. Then

$$m_X(t) = E[e^{tX}]$$

$$= E\left[1 + tX + \frac{(tX)^2}{2!} + \frac{(tX)^3}{3!} + \cdots\right]$$

$$= 1 + tE[X] + \frac{t^2}{2!}E[X^2] + \frac{t^3}{3!}E[X^3] + \cdots$$

$$= 1 + tm_1 + \frac{t^2}{2!}m_2 + \frac{t^3}{3!}m_3 + \cdots;$$

it is easy to see, then, that

$$m_X^{(k)}(0) = m_k.$$

EXAMPLE 3.4.3.   Suppose that the length of time a transistor will work (in a given circuit) is a random variable Y with density function

$$f_Y(y) = 1000e^{-1000y}, \qquad y > 0$$

$$= 0, \qquad\qquad \text{otherwise.}$$

The moment generating function for Y is

$$m_Y(t) = E(e^{tY})$$

$$= \int_0^\infty e^{ty}1000e^{-1000y}\,dy$$

$$= \int_0^\infty 1000e^{-y(1000-t)}\,dy$$

$$= \frac{1000}{1000-t}, \quad \text{for } t < 1000.$$

From this we find that

$$\mu_Y = m_1 = m^{(1)}(0) = \frac{1}{1000}$$

$$m_2 = m^{(2)}(0) = \frac{2}{(1000)^2},$$

so

$$\sigma_Y^2 = m_2 - \mu_Y^2 = \frac{1}{(1000)^2}$$

and

$$\sigma_Y = \frac{1}{1000}.$$

EXAMPLE 3.4.4.   A fair coin is flipped twice. Let Z be the number of heads that occur. Then

$$p_Z(z) = \tfrac{1}{4}, \qquad \text{at } z = 0$$
$$= \tfrac{1}{2}, \qquad \text{at } z = 1$$
$$= \tfrac{1}{4}, \qquad \text{at } z = 2$$
$$= 0, \qquad \text{otherwise.}$$

The moment generating function for Z is

$$m_Z(t) = E[e^{tZ}]$$
$$= \tfrac{1}{4} + \tfrac{1}{2}e^t + \tfrac{1}{4}e^{2t}$$
$$= \tfrac{1}{4}(1 + e^t)^2,$$

from which we find $\mu_Z = m_1 = 1$, $\sigma_Z^2 = m_2 - \mu_Z^2 = \tfrac{1}{2}$.

For some uses we shall want to be able to compute what are called moments about the mean of the random variable. These are defined as follows.

DEFINITION 3.4.3.   The quantity $E[(X - \mu_X)^k] = \mu_k$ is called the k-th *moment about the mean* of X where $k = 1, 2, 3, \ldots$.

As we saw in problem 4 of the last section, $\mu_1 = E[(X - \mu_X)] = 0$ for any random variable X. Furthermore,

$$\mu_2 = E[(X - \mu_X)^2]$$

is the variance σ_X^2 of the random variable. By using the binomial theorem, we can derive the following relationship between the moments of the random variable and the moments about the mean of the random variable.

$$\mu_k = E[(X - \mu_X)^k]$$
$$= E\left[\sum_{i=0}^{k} \binom{k}{i} X^i (-\mu_X)^{k-i}\right]$$
$$= \sum_{i=0}^{k} \binom{k}{i} m_i (-\mu_X)^{k-i}.$$

Thus,

$$\mu_1 = m_1 - \mu_X = 0,$$

as already noted, and

$$\mu_2 = (-\mu_X)^2 + 2m_1(-\mu_X) + m_2$$
$$= m_2 - \mu_X^2$$
$$\mu_3 = (-\mu_X)^3 + 3m_1(-\mu_X)^2 + 3m_2(-\mu_X) + m_3$$
$$= m_3 - 3m_2\mu_X + 2\mu_X^3,$$

etc. Similarly, we can write

$$m_k = E[X^k] = E[(\{X - \mu_X\} + \mu_X)^k]$$

$$= E\left[\sum_{i=0}^{k} \binom{k}{i} \{X - \mu_X\}^i \mu_X^{k-i}\right]$$

$$= \sum_{i=0}^{k} \binom{k}{i} \mu_i \mu_X^{k-i},$$

and thus

$$m_1 = \mu_X + \mu_1 = \mu_X$$

$$m_2 = \mu_X^2 + 2\mu_1\mu_X + \mu_2$$

$$= \mu_X^2 + \mu_2$$

$$m_3 = \mu_X^3 + 3\mu_1\mu_X^2 + 3\mu_2\mu_X + \mu_3$$

$$= \mu_3 + 3\mu_2\mu_X + \mu_X^3,$$

etc., since $\mu_1 = 0$. Thus, as is not surprising, knowledge of the moments of X is equivalent to knowledge of the moments of X about its mean and vice versa.

We can also easily find a generating function for moments about the mean. We define

$$m_{X-\mu_X}(t) = E[e^{t(X-\mu_X)}]$$

and find that $\mu_k = m_{X-\mu_X}^{(k)}(0)$. Furthermore,

$$E[e^{t(X-\mu_X)}] = E[e^{-t\mu_X}e^{tX}]$$

$$= e^{-t\mu_X}E[e^{tX}],$$

so that

$$m_{X-\mu_X}(t) = e^{-t\mu_X}m_X(t).$$

Thus, if we know the moment generating function, we simply multiply it by $e^{-t\mu_X}$ to derive the function that generates moments about μ_X.

EXAMPLE 3.4.5. For the random variable Y defined in Example 3.4.3, we found

$$m_Y(t) = \frac{1000}{1000 - t}, \qquad m_1 = \mu_Y = \frac{1}{1000}.$$

Then

$$m_{Y-(1/1000)}(t) = e^{-t/1000}\frac{1000}{1000 - t}$$

will generate moments about the mean of Y and it is easily verified that

$$\mu_1 = m_{Y-(1/1000)}^{(1)}(0) = 0$$

$$\sigma_Y^2 = m_{Y-(1/1000)}^{(2)}(0) = \frac{1}{(1000)^2}.$$

For the random variable Z defined in Example 3.4.4, we found

$$m_Z(t) = \tfrac{1}{4}(1 + e^t)^2, \qquad m_1 = \mu_Z = 1,$$

and thus

$$m_{Z-1}(t) = \frac{e^{-t}}{4}(1 + e^t)^2$$

will generate moments of Z about its mean. We can also verify for this random variable that

$$\mu_1 = 0, \qquad \mu_2 = \tfrac{1}{2} = \sigma_Z^2.$$

Let us study briefly one additional generating function, one that is frequently called either the factorial moment generating function or the *probability generating function*. Although we shall use the former name, it is commonly referred to as the probability generating function, especially in stochastic process applications of probability. It is defined as follows.

DEFINITION 3.4.4.   The *factorial moment generating function* for a random variable X is defined to be

$$\psi_X(t) = E[t^X].$$

To see first of all why this might be called the factorial moment generating function, we note that:

$$\psi_X^{(1)}(t) = E[Xt^{X-1}]$$
$$\psi_X^{(2)}(t) = E[X(X - 1)t^{X-2}]$$

$$\vdots \qquad\qquad \vdots$$

$$\psi_X^{(k)}(t) = E[X(X - 1) \cdots (X - k + 1)t^{X-k}],$$

and thus

$$\xi_1 = \psi_X^{(1)}(1) = E[X] = \mu_X$$
$$\xi_2 = \psi_X^{(2)}(1) = E[X(X - 1)]$$

$$\vdots \qquad\qquad \vdots$$

$$\xi_k = \psi_X^{(k)}(1) = E[X(X - 1) \cdots (X - k + 1)].$$

The quantities ξ_k are called the factorial moments of X (because of their factorial-like structure). The factorial moments are especially useful for discrete random variables, as we shall see. It can be shown that knowledge of ξ_k, $k = 1, 2, \ldots$, is equivalent to knowledge of the moments of X and vice versa. Thus, if we are easily able to get the factorial moments, then we are also able to evaluate the moments of the random variable, should we desire to do so. In fact, since

$$t^X = e^{X \log_e t},$$

we can see that

$$\psi_X(t) = m_X(\log_e t)$$

and that

$$\psi_X(e^t) = m_X(t).$$

To see why $\psi_X(t)$ is also called a probability generating function, let us assume that X is a discrete random variable and that the range of X is the set of integers $1, 2, \ldots, n$. Then

$$\psi_X(t) = E[t^X]$$

$$= \sum_{i=1}^{n} t^i p_X(i)$$

$$= t p_X(1) + t^2 p_X(2) + \cdots + t^n p_X(n).$$

Then

$$\psi_X^{(1)}(t) = p_X(1) + 2t p_X(2) + \cdots + n t^{n-1} p_X(n)$$

$$\psi_X^{(2)}(t) = 2 p_X(2) + 3 \cdot 2 t p_X(3) + \cdots + n(n-1) t^{n-2} p_X(n)$$

$$\vdots \qquad\qquad \vdots$$

$$\psi_X^{(n-1)}(t) = (n-1)! \, p_X(n-1) + n! \, t p_X(n)$$

$$\psi_X^{(n)}(t) = n! \, p_X(n),$$

and we see that

$$\psi_X^{(k)}(0) = k! \, p_X(k), \qquad k = 1, 2, \ldots, n.$$

That is, the derivatives of $\psi_X(t)$ evaluated at $t = 0$ gives a known constant times the values of the probability function for X. (Note as well that $\psi_X(0) = p_X(0)$.)

EXAMPLE 3.4.6.   For the discrete random variable Z defined in Example 3.4.4, we have

$$\psi_Z(t) = E[t^Z]$$

$$= \tfrac{1}{4} + \tfrac{1}{2}t + \tfrac{1}{4}t^2 = \tfrac{1}{4}(1 + t)^2.$$

Note that

$$\psi_Z(0) = \tfrac{1}{4} = p_Z(0)$$
$$\psi_Z^{(1)}(0) = \tfrac{1}{2} = p_Z(1)$$
$$\psi_Z^{(2)}(0) = \tfrac{1}{2} = 2! \, p_Z(2)$$
$$\psi_Z^{(k)}(t) = 0, \qquad k = 3, 4, 5, \ldots,$$

and that

$$\psi_Z^{(1)}(1) = 1 = \mu_Z$$
$$\psi_Z^{(2)}(1) = \tfrac{1}{2} = E[Z^2] - \mu_Z.$$

EXAMPLE 3.4.7.   Suppose that we flip a fair coin until we get a head (see Example 2.8.1) and let W be the number of flips required to conclude the experiment. Then

$$p_W(k) = \frac{1}{2^k}, \qquad k = 1, 2, 3, \ldots,$$

and

$$\psi_W(t) = E[t^W]$$

$$= \sum_{i=1}^{\infty} t^i \frac{1}{2^i}$$

$$= \sum_{i=1}^{\infty} \left(\frac{t}{2}\right)^i = \frac{1}{1 - t/2} - 1 = \frac{t}{2 - t}, \quad \text{if } |t| < 2.$$

We find that

$$\psi_W^{(1)}(t) = \frac{2}{(2 - t)^2}$$

$$\psi_W^{(2)}(t) = \frac{4}{(2 - t)^3}$$

$$\psi_W^{(k)}(t) = \frac{2 \cdot k!}{(2 - t)^{k+1}}, \qquad k = 1, 2, 3, \ldots.$$

Thus

$$\frac{1}{k!} \, \psi_W^{(k)}(0) = \frac{1}{2^k} = p_W(k), \qquad k = 1, 2, 3, \ldots$$

and

$$\psi_W^{(1)}(1) = 2 = \mu_W$$

$$\psi_W^{(2)}(1) = 4 = E[W^2] - \mu_W,$$

etc.

Let us close this section with a theorem which is one of the main reasons that moment generating functions are of interest in the study of random variables. Although the mathematics necessary to prove this result is beyond the level assumed in this text, we shall freely make use of it when required.

Theorem 3.4.1. Assume that X and Y are random variables with moment generating functions $m_X(t)$ and $m_Y(t)$, respectively. If $m_X(t) = m_Y(t)$ for all t in some neighborhood of 0, then $F_X(t) = F_Y(t)$ for all real t; that is X and Y have the same distribution.

EXERCISE 3.4

1. A single die is rolled 1 time. Let X be the number of 6's that occur. Compute $m_X(t)$ and use it to evaluate the first 3 moments of X.

2. Assume that Y, the number of minutes it takes you to eat lunch on an average day, is equally likely to lie in the interval from 30 to 40 minutes. Compute $m_Y(t)$ and $m_{Y-\mu_Y}(t)$.

3. The number of hours of satisfactory operation that a certain brand of TV set will give (without repair) is a random variable Z whose probability density function is

$$f_Z(z) = .0001e^{-.0001z}, \quad z > 0$$
$$= 0, \quad z \leq 0.$$

Derive $m_Z(t)$ and use it to compute μ_Z and σ_Z^2.

4. A fair coin is flipped 3 times. Let U be the total number of heads that occur. Derive $\psi_U(t)$ and use it to compute μ_U and σ_U^2.

5. One integer is selected at random from the set $\{1, 2, 3, \ldots, n\}$. Let V be the selected integer and derive $\psi_V(t)$.

6. Express ξ_1, ξ_2 and ξ_3 in terms of m_1, m_2 and m_3.

7. Express m_1, m_2 and m_3 in terms of ξ_1, ξ_2 and ξ_3.

8. Rachel is going to throw darts at a target until she hits the bull's-eye. Assume that she has constant probability .9 of hitting the bull's-eye each toss and that the tosses are independent. Let W be the number of tosses required. Derive $\psi_W(t)$ and use it to compute μ_W and σ_W^2.

9. Suppose that X is a random variable whose range is $\{-1, -2, \ldots, -n\}$. Show that $E[t^{-X}]$ would be a probability generating function for X.

10. Show that $m_X(0) = m_{X-\mu_X}(0) = 1$, for any random variable X.

11. Assume X is a (discrete) random variable with factorial moment generating function

$$\psi_X(t) = (1 - t^{n+1})/(n + 1)(1 - t).$$

Find $p_X(x)$ and $m_X(t)$.

12. Assume X is a (discrete) random variable with factorial moment generating function

$$\psi_X(t) = [1 - p(1 - t)]^n, \quad \text{where } 0 < p < 1.$$

Find $p_X(x)$ and $m_X(t)$. (X is a *binomial* random variable, as we shall see in Chapter 4.)

13. Assume X is a (discrete) random variable with probability generating function

$$\psi_X(t) = pt/[1 - (1 - p)t], \quad \text{where } 0 < p < 1.$$

Find $p_X(x)$ and $m_X(t)$. (As we shall see in Chapter 4, X is a *geometric* random variable.)

14. Another generating function which is sometimes useful (especially with *normal* random variables, see Section 4.5), is the *cumulant* generating function, $c_X(t)$. It is defined to be

$$c_X(t) = \log_e m_X(t).$$

Show that

$$c_X^{(1)}(0) = \mu_X$$

$$c_X^{(2)}(0) = \sigma_X^2,$$

i.e., that its first two derivatives at $t = 0$ give the mean and the variance of X.

15. Find the cumulant generating function of a binomial random variable (see Exercise 12) and use it to evaluate the binomial mean and variance.

16. Find the cumulant generating function for a geometric random variable (see Exercise 13) and use it to evaluate the geometric mean and variance.

17. Define $\delta_X(t) = \log_e \psi_X(t)$ and show that

$$\delta_X^{(1)}(1) = \mu_X,$$

$$\delta_X^{(2)}(1) = \sigma_X^2 - \mu_X.$$

3.5. Functions of a Random Variable

In many problems we may be interested in functions of a random variable and in the distributions of those functions. For example, a chemist may want to make statements about the temperature at which a certain reaction will occur, and to be able to translate such statements from, say, the Fahrenheit scale to the absolute scale. If a random variable is measuring time, we should be able to translate its values from minutes to hours or days if required; if a random variable has units of distance attached to it, we should be able to translate from inches to feet to yards to miles, if needed. Each of these are examples of linear transformations, but in many scientific fields nonlinear transformations are increasingly used, especially to simplify computations and interpretations. Physicists measure sound intensity on a logarithmic scale and social scientists frequently make use of factor analysis, which involves nonlinear transformations of several random variables simultaneously. We shall study the distributions of functions of random variables in this section.

Let us briefly discuss strictly monotonic functions before examining the theorem. We say that $y = g(x)$ is a strictly monotonic *increasing* function of x if and only if $g(x_1) > g(x_2)$ whenever $x_1 > x_2$; $y = g(x)$ is a strictly monotonic *decreasing* function of x if and only if $g(x_1) < g(x_2)$, whenever $x_1 > x_2$. Notice that a strictly monotonic function is always increasing or decreasing as a function of x; it does not reverse its direction and it does not remain constant for all x in any interval. For example, $y = a + bx$ is a strictly monotonic increasing function as long as $b > 0$; if $b < 0$ it is a strictly monotonic decreasing function. The functions $y = \log x$ and $y = \sqrt{x}$ for $x > 0$, are strictly monotonic increasing functions of x. The functions $y = 1/x$ and $y = e^{-x}$, for $x > 0$, are strictly monotonic decreasing functions of x. The function $y = x^2$, for $x > 0$, is strictly monotonic increasing and, for $x < 0$, is strictly monotonic decreasing.

As long as $y = g(x)$ is a strictly monotonic function of x, it is possible to find the (unique) value of x corresponding to any fixed value of y; this value will be denoted $g^{-1}(y)$ (g^{-1} is the inverse function and will always exist when g is strictly monotonic). For example, if $y = g(x) = a + bx$, then $x = g^{-1}(y) = (y - a)/b$, again a linear function. If $y = g(x) = \sqrt{x}$, where $x > 0$, then $x = g^{-1}(y) = y^2$, where $y > 0$. If $y = g(x) = 1/x$, where $x > 0$, then $x = g^{-1}(y) = 1/y$, where $y > 0$.

If X is a random variable defined on some sample space, then $Y = g(X)$ can be thought of as a random variable defined on the same sample space. For any outcome such that $X = x$, Y will equal $y = g(x)$. If g is a strictly monotonic function, the following theorem shows how to derive the distribution for Y from that of X.

Theorem 3.5.1.   Assume that X is a random variable with distribution function $F_X(t)$. Let $Y = g(X)$, where g is a strictly monotonic function. Then

$$F_Y(t) = F_X(g^{-1}(t)), \text{ if } g \text{ is increasing,}$$
$$F_Y(t) = 1 - F_X(g^{-1}(t)) + p_X(g^{-1}(t)), \text{ if } g \text{ is decreasing.}$$

Proof: Assume first that $Y = g(X)$ and that g is a strictly monotonic increasing function. Then

$$F_Y(t) = P(Y \le t)$$
$$= P(g(X) \le t)$$
$$= P(X \le g^{-1}(t))$$
$$= F_X(g^{-1}(t)),$$

since $g(X) \le t$ if and only if $X \le g^{-1}(t)$. Now assume that $Y = g(X)$ and that g is a strictly decreasing function. Then

$$
\begin{aligned}
F_Y(t) &= P(Y \le t) \\
&= P(g(X) \le t) \\
&= P(X \ge g^{-1}(t)) \\
&= 1 - P(X < g^{-1}(t)) \\
&= 1 - F_X(g^{-1}(t)) + p_X(g^{-1}(t))
\end{aligned}
$$

since $g(X) \le t$ if and only if $X \ge g^{-1}(t)$ and the result is established (see Figure 3.5.1). If X is continuous, note that $p_X(g^{-1}(t)) = 0$ for all t.   ◄

Let us immediately examine some examples using this result.

EXAMPLE 3.5.1.   Suppose that the temperature at which a particular reaction will occur is a random variable U with distribution function

$$
\begin{aligned}
F_U(t) &= 0, & t &< 170 \\
&= \tfrac{1}{2}\sqrt{t - 170}, & 170 &\le t \le 174 \\
&= 1, & t &> 174,
\end{aligned}
$$

where U is measured in degrees Fahrenheit. Let V be the temperature at which the reaction will occur, if V is now measured on the Kelvin (absolute) scale. Then $V = \tfrac{5}{9}U + 256\tfrac{7}{9}$ and

$$
\begin{aligned}
F_V(t) &= F_U\left(\frac{t - 256\tfrac{7}{9}}{\tfrac{5}{9}}\right) \\
&= F_U\left(\frac{9t - 2311}{5}\right) \\
&= 0, \quad \text{if } t < 351\tfrac{2}{9} \\
&= \frac{1}{2}\left[\frac{9t - 2311}{5} - 170\right]^{1/2} = \frac{1}{2}\left[\frac{9t - 3161}{5}\right]^{1/2} \quad \text{if } 351\tfrac{2}{9} \le t \le 353\tfrac{4}{9}, \\
&= 1, \quad \text{if } t > 353\tfrac{4}{9}.
\end{aligned}
$$

EXAMPLE 3.5.2.   Assume that the intensity of sound, measured directly, is a random variable X with distribution function

$$
\begin{aligned}
F_X(t) &= 0, & t &< 1, \\
&= \frac{t - 1}{9}, & 1 &\le t \le 10, \\
&= 1, & t &> 10.
\end{aligned}
$$

The log to the base 10 of the sound intensity gives measurements in decibels. Letting $Y = \log X = g(X)$, we have $X = g^{-1}(Y) = 10^Y$ and the distribution

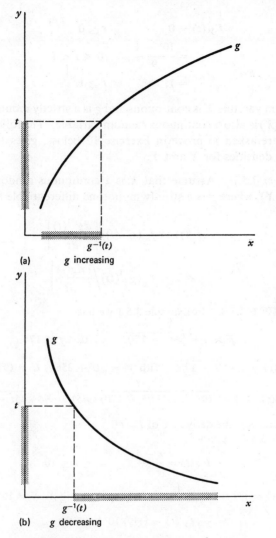

FIGURE 3.5.1   Strictly monotonic functions.

function for Y is

$$F_Y(t) = F_X(10^t)$$
$$= 0, \qquad 10^t \le 1$$
$$= \frac{10^t - 1}{9}, \qquad 1 < 10^t < 10$$
$$= 1, \qquad 10^t \ge 10,$$

that is

$$F_Y(t) = 0 \qquad t \leq 0$$
$$= \frac{10^t - 1}{9}, \qquad 0 < t < 1$$
$$= 1, \qquad t \geq 1.$$

If a random variable X is continuous and g is a strictly monotonic function, then $Y = g(X)$ is also a continuous random variable. The following theorem, which you are asked to prove in Exercise 12 below, gives the relationship between the densities for X and Y.

Theorem 3.5.2. Assume that X is a continuous random variable and that $Y = g(X)$, where g is a strictly monotone differentiable function. Then

$$f_Y(t) = f_X(g^{-1}(t)) \left| \frac{dg^{-1}(t)}{dt} \right|$$
$$= f_X(g^{-1}(t)) \Big/ \left| \frac{dg(t)}{dt} \right|$$

EXAMPLE 3.5.3. In Example 3.5.1 we had

$$F_U(t) = \tfrac{1}{2}\sqrt{t - 170}, \qquad 170 \leq t \leq 174$$

and thus $f_U(t) = 1/4\sqrt{t - 170}$. With $V = \tfrac{5}{9}U + 256\tfrac{7}{9}$, $U = (9V - 2311)/5$ and we find

$$f_V(t) = (1/4\sqrt{(9t - 2311)/5 - 170})\,(9/5) = 9/4\sqrt{45t - 15805}$$

which is the same as the derivative of $F_V(t)$.

In Example 3.5.2 we had

$$F_X(t) = \frac{t - 1}{9}, \qquad 1 \leq t \leq 10$$

and thus $f_X(t) = 1/9$ for $1 < t < 9$. Letting $Y = \log X$, $X = 10^Y$ and we find

$$f_Y(t) = (1/9)\,10^t \log_e 10,$$

which is the same as the derivative of $F_Y(t)$, derived earlier.

If X is a discrete random variable with range R_X and, again, g is a strictly monotonic function, then $Y = g(X)$ is a discrete random variable with range $R_Y = \{y: y = g(x)$ for $x \in R_X\}$. It is quite straightforward to derive the probability function for Y directly from that of X.

Each $y \in R_Y$ corresponds to exactly one x in R_X; thus the probability that $Y = y$ is given by $p_X(x)$, the probability that X equals the appropriate value in its range. The following example illustrates this fact.

EXAMPLE 3.5.4.   It is occasionally of use to standardize a random variable so that its range is a subset of the interval from 0 to 1. Suppose we are given a discrete random variable X with probability function

$$p_X(x) = \frac{2x}{n(n+1)}, \qquad \text{for } x = 1, 2, \ldots, n.$$

We could define $Y = X/n$ in which case we would have $R_Y = \{1/n, 2/n, \ldots, 1\}$, which is a subset of the interval from 0 to 1, for any n. Then

$$
\begin{aligned}
p_Y(y) &= P(Y = y) = P(X = ny) \\
&= p_X(ny) \\
&= \frac{2y}{n+1}, \qquad \text{for } y = 1/n, 2/n, \ldots, 1.
\end{aligned}
$$

The following theorem, which you are asked to prove in Exercise 13 below, shows that two options are available if we want to compute the expected value of a function of a random variable; we may either evaluate the probability law of the function and use it to compute the expected value, or we may compute it directly using the distribution of the original random variable. The theorem remains true also in the case in which g is not strictly monotonic, although we shall state it only for the more restrictive case.

Theorem 3.5.3.   Assume g is a strictly monotonic function and let

$$Y = g(X).$$

Then

$$E_Y[Y] = E_X[g(X)].$$

The subscripts are appended to the expectation symbols to stress that in the one case the expected value is taken with respect to the induced distribution for Y, in the other with respect to the original distribution for X. Thus, if all we want to evaluate is the expected value of $g(X)$, it is not necessary to derive the distribution for Y; the computation may be made from the distribution for X.

EXAMPLE 3.5.5.   Suppose X is a random variable with mean μ_X and standard deviation σ_X. The *standard* form for X is the random variable

$$Z = \frac{X - \mu_X}{\sigma_X}$$

Note that

$$
\begin{aligned}
E[Z] &= E\left[\frac{X - \mu_X}{\sigma_X}\right] \\
&= \frac{1}{\sigma_X} E[X - \mu_X] = 0
\end{aligned}
$$

and that

$$E[(Z - \mu_Z)^2] = E[Z^2]$$

$$= E\left[\left(\frac{X - \mu_X}{\sigma_X}\right)^2\right]$$

$$= \frac{1}{\sigma_X^2} E[(X - \mu_X)^2] = 1.$$

Thus, we can always transform an arbitrary random variable X with finite mean and variance into a new random variable with mean 0 and variance 1.

We have discussed the derivation of the distribution of $g(X)$, where g is assumed to be strictly monotonic. If we define $Y = g(X)$, where g is not strictly monotonic, then the probability that Y will lie between a and b, say, is the sum of the probabilities that X will assume the appropriate values to give a value for Y between a and b. Suppose X is a random variable whose range is the interval from -1 to 1 and we let $Y = g(X)$, where g is as pictured in Figure 3.5.2. Then the probability that Y will lie between a and

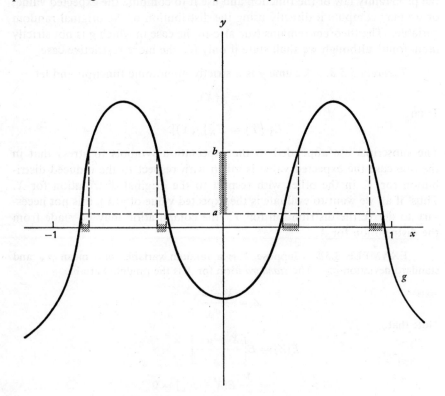

FIGURE 3.5.2   Nonmonotonic function.

b is given by the sum of the probabilities that X lies in any of the four intervals shaded on the x-axis. When g is not monotonic we need merely be careful that we sum together *all* the appropriate probabilities for X in evaluating probability statements about Y.

EXAMPLE 3.5.6.   Perhaps the most commonly occurring nonmonotonic function is the square of a random variable. Assume X has distribution function $F_X(t)$ and let $Y = X^2$. Then, clearly R_Y does not include any negative numbers and we have

$$F_Y(t) = 0 \quad \text{for } t < 0$$

For $t \geq 0$,

$$\begin{aligned}
F_Y(t) &= P(Y \leq t) \\
&= P(X^2 \leq t) \\
&= P(-\sqrt{t} \leq X \leq \sqrt{t}) \\
&= F_X(\sqrt{t}) - F_X(-\sqrt{t}) + p_X(-\sqrt{t}).
\end{aligned}$$

Notice then that

$$\begin{aligned}
P(1 < Y \leq 2) &= F_Y(2) - F_Y(1) \\
&= [F_X(\sqrt{2}) - F_X(-\sqrt{2}) + p_X(-\sqrt{2})] \\
&\quad - [F_X(1) - F_X(-1) + p_X(-1)] \\
&= [F_X(\sqrt{2}) - F_X(1)] \\
&\quad + [F_X(-1) - F_X(-\sqrt{2}) + p_X(-\sqrt{2}) - p_X(-1)] \\
&= P(1 < X \leq \sqrt{2}) + P(-\sqrt{2} \leq X < -1);
\end{aligned}$$

that is, the probability that Y lies between 1 and 2 is the sum of the two probabilities that X lies in the appropriate intervals. If X is continuous, so is $Y = X^2$ and we have

$$F_Y(t) = F_X(\sqrt{t}) - F_X(-\sqrt{t}),$$

so

$$\begin{aligned}
f_Y(t) &= \frac{d}{dt} F_Y(t) \\
&= \frac{1}{2\sqrt{t}} [f_X(\sqrt{t}) + f_X(-\sqrt{t})],
\end{aligned}$$

again illustrating the summing principle for nonmonotonic functions. The density for $Y = X^2$ at any point is the sum of the values of the density of X at two points.

EXERCISE 3.5

1. Let X be a random variable with distribution function $F_X(t)$ and let $Y = a + bX$ where $b < 0$. Derive the distribution function for Y.

2. Suppose that $b = 0$ in problem 1 above. Derive the distribution function for Y, defined as in that problem.

3. Given

$$F_X(t) = 0, \qquad\qquad t < -1$$
$$= \frac{t+1}{2}, \qquad -1 \le t \le 1$$
$$= 1, \qquad\qquad t > 1,$$

find the distribution function for $Y = 15 + 2X$ and the density function for Y.

4. Suppose that
$$F_W(t) = 0, \qquad t < 0$$
$$= t^3, \qquad 0 \le t \le 1$$
$$= 1, \qquad t > 1$$

and let $Z = W - 1$. Find $F_Z(t)$ and $f_Z(t)$.

5. If
$$F_X(t) = 0, \qquad\qquad t < -10$$
$$= \tfrac{1}{4}, \qquad -10 \le t < 0$$
$$= \tfrac{3}{4}, \qquad\quad 0 \le t < 10$$
$$= 1, \qquad\qquad t \ge 10,$$

find the distribution function for

$$U = 7X - 50 \quad \text{and} \quad p_U(u).$$

6. If
$$F_Y(t) = 1 - e^{-t}, \qquad t \ge 0$$
$$= 0, \qquad\qquad\quad t < 0,$$

find $F_X(t)$ and $f_X(t)$ where $X = 2Y - 7$.

7. If
$$F_X(t) = 0, \qquad t < 0$$
$$= t, \qquad 0 \le t \le 1$$
$$= 1, \qquad t > 1,$$

find the distribution function and the density function for the standard form of X.

8. If
$$F_Z(t) = 1 - e^{-t}, \qquad t > 0$$
$$= 0, \qquad\qquad\quad t \le 0,$$

find the distribution function and the density function for the standard form of Z.

9. Given

$$F_U(t) = 0, \qquad t < -1$$
$$= \tfrac{1}{3}, \qquad -1 \le t < 0$$
$$= \tfrac{2}{3}, \qquad 0 \le t < 1$$
$$= 1, \qquad t \ge 1,$$

find the distribution function for the standard form of U.

10. Let

$$F_X(t) = 0, \qquad t < 0$$
$$= t, \qquad 0 \le t \le 1$$
$$= 1, \qquad t > 1,$$

and find $F_Z(t)$ and $f_Z(t)$ where $Z = X^2$.

11. If

$$F_X(t) = 0, \qquad\qquad t < -1$$
$$= \frac{t + 1}{2}, \qquad -1 \le t \le 1$$
$$= 1, \qquad\qquad t > 1,$$

find the distribution function and the density function for $Z = X^2$.

12. Prove Theorem 3.5.2.

13. Prove Theorem 3.5.3.

14. Let Y have the distribution given in question 6 above and let $Z = e^{-Y}$. Find the distribution and density function for Z.

15. Let X have the distribution given in Exercise 7 and define $Y = - \log_e X$. Find the distribution and density function for Y.

16. (a) Assume Y is a continuous random variable with strictly monotonic distribution function F_Y. Show that $X = F_Y(Y)$ has the distribution function given in Exercise 7.

    (b) Let F_Y be a strictly monotonic distribution function and assume X has the distribution function given in Exercise 7. Show that $Y = F_Y^{-1}(X)$ has distribution function F_Y.

    The transformation given in part (b) is a special case of the *probability integral transform*.

17. A workman lays out a circular flower bed by cutting a length of rope, tying it to a stake, and then outlining the area for the bed. Assume that the length of rope used is equally likely to lie in the interval from $r - .1$ to $r + .1$. What is the distribution of X, the surface area of the flower bed? What is the probability that the surface area exceeds πr^2?

18. Assume in Exercise 17 that the length of rope used is equally likely to lie in the interval from $.9r$ to $1.1r$ and answer the questions posed there.

19. Assume X is a *positive random variable*. (This means $F_X(0) = 0$, that R_X has no negative elements.) Let $Y = \sqrt{X}$ and find $F_Y(t)$ in terms of $F_X(t)$.

20. Assume that X, in Exercise 19, is continuous and find $f_Y(t)$ in terms of $f_X(t)$.

21. Assume X is a positive random variable (see Exercise 19) and let $Y = \sqrt{X}$. Express the percentiles for Y in terms of the percentiles of X.

22. Assume X is a positive random variable (see Exercise 19) and let $V = X^2$. Express the percentiles for V in terms of the percentiles of X.

23. Assume a random variable Y is a function of a random variable X, i.e., $Y = f(X)$. If the function f can be expanded in a Taylor series about the constant point a, then we can get a simple approximation to the distribution of Y as follows: We assume that f is such that

$$f(x) = \sum_{i=0}^{\infty} f^{(i)}(a) \frac{(x-a)^i}{i!}, \quad \text{for each } x \in R_X,$$

where $f^{(0)}(a) = f(a), f^{(i)}(a) = \dfrac{d^i}{dx^i} f(x)\big|_{x=a}$. This is the Taylor series expansion of f. In general, after a certain number of terms, each succeeding term in the series gets smaller than its predecessor; thus we could expect to approximate f satisfactorily by keeping only a finite (sufficiently large) number of terms in the sum. In fact, if R_X is sufficiently small, and f sufficiently well behaved, we might keep only the first two terms in the series, to give a linear approximation to f:

$$y \doteq f(a) + (x-a)f^{(1)}(a).$$

Thus, to approximate the distribution of $Y = f(X)$ we might choose $a = \mu_X$ (since it is in the middle of R_X) and then use

$$Y = f(X) \doteq f(\mu_X) + f^{(1)}(\mu_X)(X - \mu_X)$$

to approximate the distribution for Y. Show that this procedure leads to

$$F_Y(t) \doteq F_X\left(\frac{t - f(\mu_X)}{f^{(1)}(\mu_X)} + \mu_X\right), \quad \text{if } f^{(1)}(\mu_X) > 0,$$

and

$$\mu_Y \doteq f(\mu_X)$$
$$\sigma_Y^2 \doteq [f^{(1)}(\mu_X)]^2 \sigma_X^2.$$

24. Assume

$$F_X(t) = 0, \qquad\qquad t < 100$$
$$= t - 100, \qquad 100 \le t \le 101$$
$$= 1, \qquad\qquad t > 101.$$

(a) Find μ_X and σ_X^2.
(b) Let $Y = X^2$ and find μ_Y, σ_Y^2.
(c) Apply the linear approximation given in Exercise 23, and compare these approximate results with the exact result in (b).

4

Some Standard Distributions

Some random variables occur frequently in many different contexts in applied problems. Because they do occur often in practice, these random variables are given special names; we shall study several of these random variables and their properties in this chapter.

4.1. Binomial and Bernoulli Random Variables

Before looking at the definitions of these two random variables we shall first define a Bernoulli trial.

DEFINITION 4.1.1. A *Bernoulli trial* is an experiment which has two possible outcomes, generally called success and failure.

The sample space for a Bernoulli trial will in general be written $S = \{s, f\}$. Many different examples of Bernoulli trials can be quoted: a flip of a single coin (head or tail), the flight of a missile (if we call it simply a success or not), performance of a student in a particular course (pass or fail), or performance of an athletic team (win or not win). Any chance mechanism whose outcomes are grouped into two classes can be looked at as being a Bernoulli trial.

A frequently used notation is

$$P(\{s\}) = p$$
$$P(\{f\}) = q = 1 - p,$$

125

which we shall also use; the quantity p is of course free to take on any value in the interval from 0 to 1, inclusive, for various types of Bernoulli trials. In Section 2.7 we briefly discussed the idea of independent trials (see Definition 2.7.4). We shall apply some of that discussion now. Suppose that we have an experiment which consists of n (a positive integer) repeated independent Bernoulli trials. The sample space for this experiment then is the Cartesian product of the sample spaces of the individual trials. Thus $S = S_1 \times S_2 \times \cdots \times S_n$ where $S_i = \{s, f\}$, $i = 1, 2, \ldots, n$, and $P_i(\{s\}) = p$ for all i. (See Definition 2.7.4.) The binomial random variable is defined as follows for this sample space.

DEFINITION 4.1.2.   Let X be the total number of successes in n repeated independent Bernoulli trials with probability p of success on a given trial. X is called the *binomial* random variable with parameters n and p.

The range of the random variable X is the integers $0, 1, 2, \ldots, n$; thus X is a discrete random variable and as such must have a probability function. The statement above that X has parameters n and p means that the probability function for X is completely specified if the values of n and p are known. This probability function is derived as follows.

Theorem 4.1.1.   If X is binomial with parameters n and p, then

$$p_X(x) = \binom{n}{x} p^x q^{n-x}, \qquad x = 0, 1, 2, \ldots, n$$

$$= 0, \qquad\qquad\qquad \text{otherwise.}$$

Proof: As mentioned earlier, the sample space for the experiment (of n trials) is $S = S_1 \times S_2 \times \cdots \times S_n$ where $S_i = \{s, f\}$ and $P_i(\{s\}) = p$, $i = 1, 2, \ldots, n$. Then, since the trials are independent, we assign probabilities to the single-element events of S by multiplying the values of the probabilities for the individual trials. Clearly, the random variable X assigns, to each element ω of S, the number of s's contained in ω. Then

$$p_X(k) = P[\{(x_1, x_2, \ldots, x_n): \text{exactly } k \ x_i\text{'s equal } s\}].$$

The probability on the right is evaluated by summing together the probabilities of the single-element events making it up. But every one of these single-element events must consist of a single n-tuple which contains exactly k s's. The probability assigned to every such single-element event is $p^k q^{n-k}$, using the product rule mentioned earlier. Thus, if we can merely count the number of n-tuples in S having exactly k s's, the product of this number with

$p^k q^{n-k}$ will give us $p_X(k)$. The number of n-tuples having exactly k s's is $\binom{n}{k}$. Thus,

$$p_X(k) = \binom{n}{k} p^k q^{n-k}, \qquad k = 0, 1, 2, \ldots, n$$

$$= 0, \qquad\qquad \text{otherwise.} \qquad \blacktriangleleft$$

We should first of all check to see that $p_X(k)$ is a probability function (see Equations 3.1 and 3.2). Obviously,

$$p_X(k) \geq 0, \qquad \text{for all } k.$$

The sum of $p_X(k)$ over the range of X is

$$\sum_{k=0}^{n} p_X(k) = \sum_{k=0}^{n} \binom{n}{k} p^k q^{n-k}$$

$$= (p + q)^n$$

$$= (p + 1 - p)^n = 1,$$

and thus $p_X(k)$ is a probability function.

Figure 4.1.1 presents graphs of four different binomial distributions (that is, four different pairs of values for n and p). These graphs are histograms or bar charts of the probability distributions; centered above each value in the range of the random variable is a bar whose base is one unit and whose height is proportional to the probability that the random variable equals that value. Thus the areas of the bars are proportional to the probabilities of occurrence of the various values in the range of the random variable. Note that if $p = .5$, then when n equals either 10 or 20 the bar chart is symmetric about a line passing through the mean of the random variable. Increasing p from .5 to .9 causes the majority of the area (probability) to be shifted to the right; but when $n = 20$, notice that the histogram is again close to being symmetric about a line through the mean. As we shall see in Chapter 5, the larger that n is, the closer the histogram is to this symmetric situation, no matter what the value of p.

EXAMPLE 4.1.1.   Suppose that a basketball player has probability $\frac{3}{4}$ of sinking a free throw and that his shots are independent. If he gets 5 free throws in a particular game, define X to be the number that he makes. Then X is binomial with parameters $n = 5$, $p = \frac{3}{4}$, and

$$p_X(k) = \binom{5}{k} \left(\frac{3}{4}\right)^k \left(\frac{1}{4}\right)^{5-k}, \qquad k = 0, 1, 2, 3, 4, 5$$

$$= 0, \qquad\qquad \text{otherwise.}$$

FIGURE 4.1.1   Binomial probability functions.

Thus, the probability that he makes them all is

$$p_X(5) = (\tfrac{3}{4})^5 = .237,$$

that he misses them all is

$$p_X(0) = (\tfrac{1}{4})^5 = .001,$$

and that he makes at least half of them is

$$P(X \geq 3) = \sum_{k=3}^{5} \binom{n}{k} \left(\frac{3}{4}\right)^k \left(\frac{1}{4}\right)^{5-k} = .8965.$$

EXAMPLE 4.1.2.   A student takes a multiple-choice exam which contains 8 questions, each with 3 alternative answers. Assume that he is guessing when answering each question. Then the probability that he answers a question correctly is $\frac{1}{3}$ for each question. Let Y be the total number he gets correct, and we have

$$p_Y(k) = \binom{8}{k}\left(\frac{1}{3}\right)^k\left(\frac{2}{3}\right)^{8-k}, \qquad k = 0, 1, 2, \ldots, 8$$
$$= 0, \qquad\qquad\qquad \text{otherwise.}$$

The probability that he misses them all is

$$p_Y(0) = \left(\tfrac{2}{3}\right)^8 = .039.$$

If he must get 6 or more correct to pass, then the probability that he passes is

$$\sum_{k=6}^{8} \binom{8}{k}\left(\frac{1}{3}\right)^k\left(\frac{2}{3}\right)^{8-k} = .020.$$

Since the binomial random variable is discrete, its distribution function is discontinuous; it is equal to

$$F_X(t) = \sum_{k \le t} \binom{n}{k} p^k q^{n-k}$$

where the summation is over all nonnegative integers k less than or equal to t. Table 1, in the Appendix, gives values of the distribution function for n increasing in steps of 1 from 2 to 20 and selected values of p between 0 and $\frac{1}{2}$. By evaluating the size of the jump in the distribution function, we can of course compute the individual binomial probabilities. Although tabular entries are given only for values of p between 0 and $\frac{1}{2}$, it should also be noted that by interchanging the roles of p and q the distribution function for a binomial random variable with p between $\frac{1}{2}$ and 1 may also be evaluated. Thus, if we denote the binomial distribution function by $F_X(k; n, p)$, then

$$F_X(k; n, p) = \sum_{x=0}^{k} \binom{n}{x} p^x q^{n-x},$$

where k is any integer between 0 and n inclusive. Now

$$\sum_{x=0}^{k} \binom{n}{x} p^x q^{n-x} = 1 - \sum_{x=k+1}^{n} \binom{n}{x} p^x q^{n-x}$$
$$= 1 - \sum_{n-x=0}^{n-k-1} \binom{n}{n-x} q^{n-x} p^{n-(n-x)}$$
$$= 1 - F_X(n - k - 1; n, q).$$

Thus, for example,

$$F_X(15; 20, .8) = 1 - F_X(4; 20, .2)$$
$$F_X(7; 14, .9) = 1 - F_X(6; 14, .1).$$

Let us derive the factorial moment generating function for the binomial random variable and use it to evaluate the mean and variance of X. By definition

$$\psi_X(t) = E(t^X)$$

$$= \sum_{k=0}^{n} t^k \binom{n}{k} p^k q^{n-k}$$

$$= \sum_{k=0}^{n} \binom{n}{k} (pt)^k q^{n-k}$$

$$= (q + pt)^n.$$

Then

$$\psi_X^{(1)}(t) = np(q + pt)^{n-1}$$

$$\psi_X^{(2)}(t) = n(n-1)p^2(q + pt)^{n-2}$$

and

$$E[X] = \psi_X^{(1)}(1) = np(q + p)^{n-1} = np$$

$$E[X(X-1)] = \psi_X^{(2)}(1) = n(n-1)p^2(q + p)^{n-2} = n(n-1)p^2$$

so that

$$\mu_X = np$$

$$\sigma_X^2 = E[X(X-1)] + np - (np)^2$$

$$= npq.$$

EXAMPLE 4.1.3.   The binomial random variable X discussed in Example 4.1.1 has mean value $\mu_X = 5 \cdot \frac{3}{4} = 3\frac{3}{4}$ and variance $\sigma_X^2 = 5 \cdot \frac{3}{4} \cdot \frac{1}{4} = \frac{15}{16}$. Thus, on the average, this player would make $3\frac{3}{4}$ shots out of 5, with a standard deviation of $\sqrt{15}/4$. The binomial random variable Y discussed in Example 4.1.2 has mean value $\mu_Y = 8 \cdot \frac{1}{3} = 2\frac{2}{3}$ and variance $8 \cdot \frac{1}{3} \cdot \frac{2}{3} = \frac{16}{9}$. Thus this student would, by guessing, answer $2\frac{2}{3}$ questions correctly, on the average, with a standard deviation of $\frac{4}{3}$ about that mean value.

The Bernoulli random variable is actually a special case of the binomial, as the following definition shows.

DEFINITION 4.1.3.   Suppose that X is a binomial random variable with parameters $n = 1$ and p. Then X is called the *Bernoulli* random variable with parameter p.

The Bernoulli random variable then is simply the number of successes we observe in a single Bernoulli trial and has probability function

$$p_X(x) = p, \quad \text{for } x = 1$$

$$= q, \quad \text{for } x = 0$$

$$= 0, \quad \text{otherwise.}$$

Its factorial moment generating function is

$$\psi_X(t) = (q + pt)^1 = q + pt$$

and it has $\mu_X = p$, $\sigma_X^2 = pq$.

EXERCISE 4.1

1. Five fair dice are rolled 1 time. Let X be the number of 1's that occur. Compute the mean of X, the variance of X, $P(1 \leq X < 4)$, and $P(X \geq 2)$.
2. An urn contains 8 red and 2 black balls. Twenty balls are drawn with replacement. Let Y be the number of red balls that occur; compute μ_Y, σ_Y, $P(Y = 16)$, $P(Y < 14)$, and $P(Y > 18)$.
3. Ten coins are tossed onto a table. Let Z be the number of coins that land head up. Compute $P(Z = 5)$ and μ_Z.
4. Suppose that it is known that 1% of the glasses made by a certain glass-blowing machine will be defective in some way. If we randomly select 10 glasses made by this machine, what is the probability that none of them is defective? How many would we expect to be defective?
5. Given that X is a binomial random variable with parameters n and p, derive directly the fact that $\mu_X = np$.
6. Derive the variance of the binomial random variable by evaluating $E[X(X - 1)]$ and then add $\mu_X - \mu_X^2$.
7. Y is known to be a binomial random variable with mean $\mu_Y = 6$ and variance $\sigma_Y^2 = 4$. Find the probability distribution for Y (that is, evaluate n and p).
8. The probability that an individual seed of a certain type will germinate is known to be .9. A nursery man wants to sell flats of this type of plant and will claim that each flat contains 100 plants. If he plants 110 seeds in a flat (which we assume will sprout independently), how many plants should we expect an "average" flat to contain? Is there any number of seeds he can plant in a flat in order to be certain that the flat will contain 100 plants?
9. Derive the moment generating function for a binomial random variable with parameters n and p.
10. At a county fair a ring toss game may be played for 25¢. You are given three rings and then attempt to toss them individually onto a peg. If you successfully get one ring on a peg you win a prize worth 50¢. If you get two on, you get a prize worth $1 and if you get all three on, you win a prize worth $5. Assuming the probability that you ring the peg is .1 each try, what is your expected gain if you play this game one time? Ten times?
11. Five fair dice are rolled one time. What is the probability of getting at least one 3? At least two 3's?

12. An investor buys five single-residence dwellings as an investment. He assumes that the probability he will make a profit on each is .9. Assuming independence,
    (a) What is the probability he makes a profit on every one?
    (b) What is the probability he takes a loss on each one?

13. A manufacturer of small parts ships his parts in lots of size 20 to his customers. Assume that each part is either defective or not, and that the probability an individual part is defective is .05.
    (a) What is the expected number of defectives per lot?
    (b) What is the probability that any particular lot contains no defectives?

14. Assume that you are a customer of the manufacturer mentioned in Exercise 13 and that you receive 10 lots from him.
    (a) What is the expected number of lots that have no defectives in them?
    (b) What is the probability that you have received no defectives in the 10 lots?

15. A rat maze consists of a straight corridor, at the end of which is a branch; at the branching point the rat must either turn right or left. Assume 10 rats are placed in the maze, one at a time. If each is choosing one of the two branches at random, what is the distribution of the number that turn right? That turn left? What is the probability that at least 9 will turn the same way?

16. A radio-television dealer extends credit to people buying his sets. Assume that in the past 10% of all those he extended credit to did not pay and he took a loss on each sale. The other 90% paid in full and he made a profit on those sales. Assume this dealer has 10 identical television sets and that he will sell them individually and independently on credit to 10 people. If the buyer does not pay he takes a loss of $200; if the buyer does pay in full he makes a profit of $100.
    (a) What is the distribution of the amount of profit he will make on these 10 sales?
    (b) What is his expected profit on these 10 sales?

4.2. Geometric and Hypergeometric Random Variables

We shall study two additional discrete random variables in this section, both of which are met with some frequency in applications. In studying the geometric random variable, some facility with geometric progressions is useful. It is easily shown that for any real number r

$$1 - r^{n+1} = (1 - r)(1 + r + r^2 + \cdots + r^n);$$

if $r \neq 1$, then

$$\frac{1 - r^{n+1}}{1 - r} = 1 + r + r^2 + \cdots + r^n.$$

The series $1 + r + r^2 + \cdots + r^n$ is called a *geometric progression;* each succeeding term is r (called the common ratio) times the term preceding.

If $|r| < 1$, then $\lim_{n \to \infty} r^{n+1} = 0$; thus we have

$$\sum_{i=0}^{\infty} r^i = \lim_{n \to \infty} \sum_{i=0}^{n} r^i = \lim_{n \to \infty} \frac{1 - r^{n+1}}{1 - r} = \frac{1}{1 - r},$$

so long as $|r| < 1$. Note as well that

$$\frac{d}{dr} \sum_{i=0}^{\infty} r^i = \sum_{i=0}^{\infty} i r^{i-1}$$

and

$$\frac{d}{dr}\left(\frac{1}{1 - r}\right) = \frac{1}{(1 - r)^2};$$

thus

$$\sum_{i=0}^{\infty} i r^{i-1} = \frac{1}{(1 - r)^2}, \qquad \text{if } |r| < 1.$$

By equating second derivatives we also find

$$\sum_{i=0}^{\infty} i(i - 1)r^{i-2} = \frac{2}{(1 - r)^3}, \qquad \text{if } |r| < 1.$$

These formulas for geometric progressions prove useful in studying the distribution of the *geometric* random variable. It is defined below.

DEFINITION 4.2.1. Independent Bernoulli trials are performed until we get a success. The probability of success on each trial is p where $0 < p \leq 1$. Let Y be the number of trials necessary (to get the first success). Y is called the *geometric* random variable with parameter p.

We see immediately that the range of Y is the set of positive integers, 1, 2, 3, ... , since the number of trials necessary may equal any of these values. (The reader might like to refer to Example 2.8.1 which is an example of this sort of experiment.) Thus Y is a discrete random variable. The probability function for the geometric random variable is given in the following theorem.

Theorem 4.2.1. If Y is a geometric random variable with parameter p, then

$$p_Y(k) = q^{k-1}p, \qquad k = 1, 2, 3, \ldots$$
$$= 0, \qquad \text{otherwise.}$$

Proof: The probability that $Y = 1$ is clearly p, because that is specified as the probability of a success on the first trial (as well as all others). Next, $Y = 2$ if and only if we get a failure on the first trial followed by a success on the second trial; since the trials are independent the probability of this occurring is qp. Thus, $p_Y(2) = qp$. $Y = 3$ if and only if we get failures on each of the first two trials followed by a success on the third; the probability of this occurring is $q \cdot q \cdot p$, so $p_Y(3) = q^2 p$.

In general, $Y = k$ if and only if we get failures on each of the first $k - 1$ trials followed by a success on the k-th trial; the probability of this happening is $q^{k-1}p$ and thus

$$p_Y(k) = q^{k-1}p, \qquad k = 1, 2, 3, \ldots$$
$$= 0, \qquad \text{otherwise.} \qquad \blacktriangleleft$$

The geometric random variable derives its name from the fact that its probability function forms a geometric progression. To check, first of all, that $p_Y(k)$ is a probability function, note that

1. $p_Y(k) \geq 0, \qquad$ for all k.

2. $\displaystyle\sum_{k=1}^{\infty} p_Y(k) = \sum_{k=1}^{\infty} q^{k-1}p$

$$= p \sum_{k=1}^{\infty} q^{k-1} = p \sum_{k-1=0}^{\infty} q^{k-1}$$

$$= p \frac{1}{1-q} = 1$$

since $q = 1 - p$.

EXAMPLE 4.2.1.   Suppose that Roger is going to position himself at the free-throw line on a basketball court and shoot until he makes a basket. If we assume that his shots are independent, and that he has constant probability .8 of making each shot, and if we let Y be the number of shots he takes until he makes a basket, then Y is a geometric random variable with parameter .8; its probability function is

$$p_Y(k) = (.2)^{k-1}(.8), \qquad k = 1, 2, 3, \ldots$$
$$= 0, \qquad \text{otherwise.}$$

Then, the probability that it takes him less than 5 shots is

$$P(Y < 5) = \sum_{k=1}^{4} (.2)^{k-1}(.8) = .9984.$$

The probability that it takes him an even number of shots is

$$\sum_{k=1}^{\infty} (.2)^{2k-1}(.8) = (.2)(.8) \sum_{k=1}^{\infty} (.2)^{2k-2}$$

$$= (.2)(.8) \sum_{k-1=0}^{\infty} [(.2)^2]^{k-1}$$

$$= (.2)(.8) \frac{1}{1 - (.2)^2} = \frac{1}{6}.$$

Let us derive the factorial moment generating function for the geometric random variable Y and use it to evaluate the mean and the variance of Y. By definition,

$$\psi_Y(t) = E(t^Y)$$

$$= \sum_{k=1}^{\infty} t^k q^{k-1} p$$

$$= pt \sum_{k=1}^{\infty} (qt)^{k-1}$$

$$= \frac{pt}{1 - qt}, \qquad \text{for } |qt| < 1, \qquad \text{i.e., } |t| < \frac{1}{q}.$$

Then

$$\psi_Y^{(1)}(t) = \frac{p}{1 - qt} + \frac{qpt}{(1 - qt)^2} = \frac{p}{(1 - qt)^2}$$

$$\psi_Y^{(2)}(t) = \frac{2qp}{(1 - qt)^3}$$

and thus

$$\mu_Y = \psi_Y^{(1)}(1) = \frac{p}{(1 - q)^2} = \frac{p}{p^2} = \frac{1}{p}$$

$$E[X(X - 1)] = E[X^2] - E[X] = \psi_Y^{(2)}(1) = \frac{2qp}{(1 - q)^3} = \frac{2q}{p^2}$$

so that

$$\sigma_Y^2 = \frac{2q}{p^2} + \frac{1}{p} - \frac{1}{p^2} = \frac{q}{p^2}.$$

Thus the average number of shots that Roger would have to make (in Example 4.2.1) is $\mu_Y = \dfrac{1}{.8} = 1\frac{1}{4}$ and the variance of his number of shots is $\sigma_Y^2 = \dfrac{.2}{(.8)^2} = \frac{5}{16}$. This relatively small variance implies that the probability of his number of shots being close to the average number of shots ($\mu_Y = 1\frac{1}{4}$) is quite high, as is verified by realizing that $p_Y(1) = .8$, $p_Y(2) = .16$.

EXAMPLE 4.2.2. Suppose that 1 copy in 10 of the *San Francisco Chronicle-Examiner* daily newspaper bears a special prize-winning number. Let Z be the number of papers you must buy to get 1 prize. Then Z is a geometric random variable with probability function

$$p_Z(z) = (.9)^{k-1}(.1), \qquad k = 1, 2, 3, \ldots$$

$$= 0, \qquad\qquad \text{otherwise,}$$

$p = .1$ $p = .9$

(a) (b)

FIGURE 4.2.1 Probability functions of geometric random variable.

and $\mu_Z = \dfrac{1}{.1} = 10$, $\sigma_Z^2 = \dfrac{.9}{(.1)^2} = 90$. Note that this geometric random variable has a much larger variance than does the one discussed in Example 4.2.1 (since it has a small value of p); thus we would expect much more variability in Z than we would in Y from one repetition of the experiment to another.

Figure 4.2.1 gives the histograms of the probability functions of two geometric random variables, one with $p = .1$ and the other with $p = .9$. Note that in both cases (as in all others) the probabilities of occurrence of succeedingly larger integers are decreasing geometrically. The difference between the binomial and the geometric random variables should be mentioned once more: in the binomial, the total number of trials is fixed and the random variable is the number of successes to occur; in the geometric, the number of successes to be observed (one) is fixed and the random variable is the number of trials required.

We shall now examine the hypergeometric distribution. This random variable derives its name from the fact that its probability function takes on values in a hypergeometric series. As we shall see, the hypergeometric random variable is similar to a binomial random variable.

DEFINITION 4.2.2. An urn contains M balls of which W are white. Define Z to be the number of white balls that occur in a sample of n balls drawn at random from the urn without replacement. Z is called the *hypergeometric* random variable.

Since the number of white balls drawn could take on only integer values, Z is another example of a discrete random variable. If the sampling from the urn had been done with replacement rather than without, Z would be a binomial random variable with parameters n and $p = W/M$; this is the similarity between the binomial and hypergeometric random variables mentioned above. The probability function for Z is derived in Theorem 4.2.2.

Theorem 4.2.2. If Z is the random variable given in Definition 4.2.2, then

$$p_Z(k) = \frac{\binom{W}{k}\binom{M-W}{n-k}}{\binom{M}{n}}, \qquad k = 0, 1, 2, \ldots, n,$$

where we are using the convention $\binom{b}{a} = 0$ for $a > b$.

Proof: The elements of the sample space for the experiment of selecting n balls without replacement from the urn will be the subsets of n balls that could occur; since the sampling is done at random, these subsets are equally likely to occur. Define:

A_k: exactly k white balls are drawn.

Then $P(A_k)$ is the ratio of the number of elements belonging to A_k to the number of elements belonging to S. The number of subsets of size n, each having exactly k white balls, is the product $\binom{W}{k}\binom{M-W}{n-k}$, since any k of the white balls could occur in conjunction with any $n-k$ nonwhite balls; the number of elements belonging to S is $\binom{M}{n}$. Thus,

$$P(A_k) = \frac{\binom{W}{k}\binom{M-W}{n-k}}{\binom{M}{n}}.$$

Since $Z = k$ if and only if A_k occurs, $P(Z = k) = P(A_k)$; i.e.,

$$p_Z(k) = \frac{\binom{W}{k}\binom{M-W}{n-k}}{\binom{M}{n}}, \qquad k = 0, 1, 2, \ldots, n$$

$$p_Z(k) = 0, \qquad\qquad \text{otherwise.} \qquad\blacktriangleleft$$

Obviously, $p_Z(k) \geq 0$ for all k. We can see that there are $\binom{M}{n}$ subsets of size n that could be selected from a set with M elements. Of these,

$\binom{W}{k} \binom{M - W}{n - k}$ contain exactly k from the W available of type I and $n - k$ from the $M - W$ available of type II. Thus

$$\sum_{k=0}^{n} \binom{W}{k} \binom{M - W}{n - k}$$

must give the total number of subsets; that is

$$\sum_{k=0}^{n} \binom{W}{k} \binom{M - W}{n - k} = \binom{M}{n}.$$

Dividing both sides by $\binom{M}{n}$ shows that

$$\sum_{k=0}^{n} p_Z(k) = 1$$

and thus $p_Z(k)$ satisfies the rules for a probability function.

EXAMPLE 4.2.3.   A milk case contains 5 quarts of milk, only 4 of which are fresh. If we randomly select 2 of these (without replacement) and let Z be the number of nonfresh quarts that we select, then Z is a hypergeometric random variable with

$$p_Z(k) = \frac{\binom{1}{k} \binom{4}{2 - k}}{\binom{5}{2}}, \qquad k = 0, 1, 2$$

$$= 0, \qquad\qquad\qquad \text{otherwise.}$$

(Note we have made an analogy between balls in an urn and milk quarts in the case; $M = 5$, $W = 1$, $n = 2$.)

The probability that we get 2 fresh quarts is

$$p_Z(0) = \frac{\binom{1}{0} \binom{4}{2}}{\binom{5}{2}} = \frac{3}{5};$$

the probability that the unfresh quart is one of those we selected is

$$p_Z(1) = \frac{\binom{1}{1} \binom{4}{1}}{\binom{5}{2}} = \frac{2}{5}.$$

EXAMPLE 4.2.4.   Assume that M people are eligible to vote in a certain election and that W of them favor proposition A while the remainder do not. If we select n of these voters at random without replacement and let X be the number of people in the sample that favor proposition A, then X is a hypergeometric random variable. The probability that a majority of people in the sample favor proposition A then is gotten by summing $p_X(k)$ from the smallest integer larger than $n/2$ on up. In such sample-survey problems, M and n usually are known, as is the observed value of X after the sample is taken, but W is generally unknown. A problem in statistical estimation, which we shall study in Chapter 7, is to use the observed value of X from the sample to guess the unknown value of W.

The mean value of a hypergeometric random variable Z is

$$\mu_Z = \sum_{k=0}^{n} k \, \frac{\binom{W}{k}\binom{M-W}{n-k}}{\binom{M}{n}} \; ;$$

this sum can be shown to equal $\mu_Z = n \, W/M$.

Similarly, the variance of Z can be shown to be

$$\sigma_Z^2 = n \, \frac{W}{M} \cdot \frac{(M-W)}{M} \cdot \frac{(M-n)}{M-1} .$$

Thus, for the random variable Z defined in Example 4.2.3, we find

$$\mu_Z = 2 \cdot \tfrac{1}{5} = \tfrac{2}{5}, \qquad \sigma_Z^2 = \tfrac{2}{5} \cdot \tfrac{4}{5} \cdot \tfrac{3}{4} = \tfrac{6}{25}.$$

Notice that if we set $n = M$, that is we sample all of the balls in the urn, and let Z be the number of white balls in the sample, then

$$\mu_Z = M \, \frac{W}{M} = W, \qquad \sigma_Z^2 = \frac{W}{M} \, \frac{(M-W)}{M} \, \frac{(M-M)}{M} = 0.$$

Thus the average number of white balls we get is W, as it should be, and the variance of Z about its mean (W) is zero (it never varies).

EXAMPLE 4.2.5.   A manufacturing firm receives a lot containing 100 items, of which five are defective. The firm regularly inspects incoming lots, to assess the quality of the material it receives. If the quality of an incoming lot is poor, it will send the full lot back to the supplier. Assume that this firm will randomly select a sample of three items from an incoming lot of 100. If all three are nondefective, the lot will be accepted; if any defects are found in the sample of three, the lot will be returned to the supplier. With a rule like this the probability that a lot of 100 items containing five defective items will be accepted is

$$\frac{\binom{95}{3}}{\binom{100}{3}} = .856.$$

Thus, if the supplier repeatedly sends lots of size 100, each containing five defectives, 85.6% of such lots will be accepted by the buyer and 14.4% will not be, using this plan. Notice that the average number of defective items that would be found in the sample of size three is $3 \cdot \dfrac{5}{100} = .15$ and the variance of the number of defectives in the sample of size three is $3 \cdot (.05)(.95)\left(\dfrac{97}{99}\right) = .140$.

It was mentioned earlier that if the sampling were done with replacement in Definition 4.2.2, then Z would be a binomial random variable with parameters n and $p = W/M$. As such, Z would then have mean $np = n\, W/M$ and variance $npq = n(W/M)(1 - W/M) = n(W/M)(M - W)/M$. We note then that the expected number of white balls in the sample is the same whether the sampling is done with or without replacement. The variance, though, is reduced by a factor of $(M - n)/(M - 1)$ if the sampling is done without replacement rather than with replacement. This is intuitively reasonable since, if the same ball cannot occur in the sample more than once, the number of balls to be sampled from is constantly decreasing. Thus the variability of the balls to be sampled from is decreased, as is reflected in the factor $(M - n)/(M - 1)$ by which the variance is reduced. It can be seen that as M increases (with n held constant) the variance of the hypergeometric gets closer and closer to the variance of the corresponding binomial random variable because $(M - n)/(M - 1)$ gets closer and closer to 1. In fact, it can be shown that if $M \to \infty$ and $W \to \infty$ in such a way that W/M remains constant at p (n is held constant), the distribution function of the hypergeometric random variable converges to the distribution function of the binomial random variable with parameters n and p. This is also intuitively reasonable, since the larger that M (the number of balls in the urn) becomes,

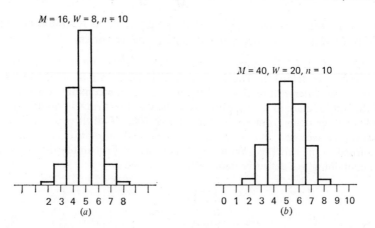

FIGURE 4.2.2   Probability functions of hypergeometric random variables.

the less difference it makes whether the balls are replaced or not before subsequent draws.

Figure 4.2.2 shows the histograms for two hypergeometric random variables; in part a we have $M = 16$, $W = 8$, $n = 10$; in part b $M = 40$, $W = 20$, $n = 10$. Note how similar the histogram in Figure 4.2.2(b) is to the histogram in Figure 4.1.1(a) which gives the probability distribution for a binomial random variable with $n = 10$, $p = .5$.

EXERCISE 4.2

1. A fair coin is flipped until a head occurs. What is the probability that less than 3 flips are required? That less than 4 flips are required?

2. An American roulette wheel commonly has 38 spots on it of which 18 are black, 18 are red and 2 are green. Let X be the number of spins necessary to get the first red number. Give the probability function for X and the mean for X.

3. Let Y be the number of spins necessary to observe the first green number for the roulette wheel mentioned in Exercise 2. What is the probability function, mean, and variance for Y?

4. Suppose an urn contains 10 balls of which 1 is black. Let Z be the number of draws, with replacement, necessary to observe the black ball. What is the probability function for Z? The mean of Z?

5. Suppose that the drawing in Exercise 4 is done without replacement. Find the probability function for Z and the mean of Z.

6. A box contains 5 marbles of which 3 are chipped. Two marbles are chosen randomly without replacement from the box. What is the probability function for the number of chipped marbles in the sample?

7. Thirteen cards are drawn randomly without replacement from a regular 52-card deck. What is the probability function for the number of red cards in the sample? What are the mean and the variance of the number of red cards?

8. A bag contains 10 flashbulbs, 8 of which are good. If 5 flashbulbs are chosen from the bag at random, what is the probability function for the number of good flashbulbs? For the number of bad flashbulbs?

9. An ice cream company makes chocolate-covered ice cream bars on sticks which sell for 10 cents. Suppose that they put a star on every 50-th stick; anyone who buys a bar with a starred stick gets a free ice cream bar. If you decide to buy ice cream bars until you get a free one, how much would you expect to spend before getting a free bar?

10. A panel of 7 judges is to decide which of 2 final contestants in a beauty contest will be declared the winner; a simple majority of the judges will determine the winner. Assume that 4 of the judges will vote for Marie and that the other 3 will vote for Sue. If we randomly select 3 of the judges and ask them who they are going to vote for, what is the probability that a majority of the judges in the sample will favor Marie?

11.   Repeated independent Bernoulli trials, each with probability p of success, are performed until $r \geq 1$ successes are obtained. Let X be the number of trials required and derive the probability function for X. (X is called the *negative binomial* or *Pascal* random variable.)

12.   Assume that every time you drive your car the probability is .001 that you will get a ticket for speeding. Also assume that you will lose your license once you have received three such tickets. Let X be the number of times you will drive your car until you get the third such ticket and derive the probability function for X (assume that you have the same probability, .001, of getting a ticket each time you drive and that the occurrences of such tickets are independent).

13.   Derive the mean and variance of a geometric random variable directly from its probability function. (*Hint:* Use the first two derivatives of the geometric progression.)

14.   A coin is flipped until the first head occurs. Assume the flips are independent and that the probability of a head occurring each time is p.
      (a) Show that the probability an odd number of flips is required is $p/(1 - q^2)$.
      (b) Find the value of p so that the probability is .6 an odd number of flips is required.
      (c) Can you find a value of p so that the probability is .5 that an odd number of flips is required?

15.   Show that the mean and variance of a negative binomial random variable (defined in Exercise 11) are r/p and rq/p^2, respectively. Note that if $r = 1$ these reduce to the mean and variance of a geometric random variable, as of course they must.

16.   If, in sampling balls without replacement from an urn, the sample size n is much smaller than the number M of balls in the urn, the exact hypergeometric distribution is well approximated by the corresponding binomial distribution with parameters n and $p = W/M$. Show that in Example 4.2.5 the binomial approximation gives .857 as the probability of there being no defectives in the sample of size 3.

17.   If, in Example 4.2.5, the firm receiving the lot will accept it if and only if there are no defectives in a sample of size n, what is the probability that a lot containing 5 defectives will be accepted? Evaluate this probability for $n = 4$ and 5.

18.   Suppose that, in Example 4.2.5, the firm receiving the lot will accept it if and only if there is 1 or fewer defectives in a sample of size 6. What is the probability a lot of 100 items, containing 5 defective items, will be accepted using this rule?

19.   Assume there is a constant probability p that a manufactured item is defective and that the occurrences of defectives are independent.
      (a) If these items are sold in lots of size M what is the distribution of the number of defectives per lot?

(b) A purchaser of a lot of size M will select a sample of n items from the lot without replacement. What is the expected number of defectives in the sample?

4.3. The Poisson Random Variable

As we shall see, the Poisson random variable has many important practical applications. Before defining the random variable, let us first discuss what is known as a Poisson process. Many problems consist of observing what are called discrete events in a continuous interval. For example, we might observe the arrival of cars at a supermarket between the hours of 10 a.m. and 11 a.m. on a specified day. The arrival of a car at the parking lot is a discrete event because its arrival time is a single point in our continuous 1-hour period. Or, we might observe the number of calls coming into the switchboard of a large corporation from 1 p.m. to 3 p.m. The particular calls that come in are the discrete events, since again the time of arrival of any single call is a single point in the 2-hour period. Or, if we were the manufacturers of insulated wire, we could count the number of defects in insulation that occurred in a particular 100-foot segment. In this case the continuous interval is the 100 feet of wire and the discrete events are the defects in insulation that occur, assuming that each defect could be thought of as occurring at a single point along the length of wire.

Many similar examples could be given. Each of these could be looked at as being a Poisson process if we are willing to assume that the occurrences of the discrete events are generated in such a manner that Definition 4.3.1 is satisfied.

DEFINITION 4.3.1. In a *Poisson process* with parameter λ, discrete events are generated in a continuous interval (time, length, etc.) in the following manner. (1) We can take a sufficiently short interval, of length h, such that: (i) the probability of exactly 1 occurrence in the interval is approximately λh, and (ii) the probability of 2 or more occurrences in the interval is approximately 0. (2) The occurrence of an event in 1 interval of length h has no effect on the occurrence or nonoccurrence in another nonoverlapping interval of length h. (The occurrences are statistically independent of one another.)

It would seem reasonable when observing the arrival of cars at a supermarket parking lot to assume that no more than 1 car would arrive in a millionth of a second, for example, and that the arrival or nonarrival of a car in the first millionth of a second would have no effect on whether or not a car will arrive in the next or any succeeding millionth of a second. Granted these assumptions, the arrivals would constitute a Poisson process. Similarly, it seems reasonable that no more than 1 call would arrive at the industrial switchboard during a very short period of time and that what happens in 1 short period of time would have no effect on succeeding short periods of time; this would also be an example of a Poisson process. Similarly, the occurrences of defects along the 100 foot length of insulated wire might well satisfy the requirements for a Poisson process.

If we observe a Poisson process for a unit length of time, the number of events that occur is a random variable. Assuming that they occur at a constant rate λ does not imply that exactly λ events occur in a unit interval but rather that the average number that occurs per unit is λ. The Poisson random variable assigns to any interval of length s the number of events to occur in that interval, as given in the following definition.

> DEFINITION 4.3.2.   A Poisson process with parameter λ is observed for s units of time. Let X be the number of events to occur. Then X is called the *Poisson* random variable with parameter λs.

Notice that the parameter of the Poisson process (λ) is multiplied by the length of time the process is observed (s) to give the parameter of X. If we expect λ events to occur on the average in an interval of unit length, then we would expect λs events to occur on the average in an interval of length s. Thus we might expect that the Poisson random variable X will have mean λs. The probability law for X is given as Theorem 4.3.1.

> *Theorem 4.3.1.*   If X is a Poisson random variable with parameter λs, then

$$p_X(k) = \frac{(\lambda s)^k}{k!}\, e^{-\lambda s}, \qquad k = 0, 1, 2, \ldots$$

$$= 0, \qquad\qquad \text{otherwise.}$$

Proof: A Poisson process with parameter λ is observed for s units of time and X is the number of events that occur. To derive the probability function for X, we assume that the interval of time is divided into $n = s/h$ nonoverlapping intervals, each of length h (h is very small). Then from Definition 4.3.1, it is reasonable to assume that either 0 or 1 event will occur in each of these short intervals and that the probability that exactly 1 event will occur is λh for each. Furthermore, the occurrence or nonoccurrence of an event in one of these short intervals has no effect on the occurrence or nonoccurrence of an event in another interval (thus they might be called independent). Thus the interval of time of length s has been subdivided into n repeated independent Bernoulli trials, each an interval of length h, with probability of success equal to λh. Then, if we let X be the total number of events occurring in the interval, X is approximately a binomial random variable with parameters

$n = s/h$ and $p = \lambda h = \lambda s/n$; then

$$p_X(k) \doteq \binom{n}{k}\left(\frac{\lambda s}{n}\right)^k\left(1 - \frac{\lambda s}{n}\right)^{n-k}, \qquad k = 0, 1, 2, \ldots, n$$

$$= 0, \qquad\qquad\qquad\qquad \text{otherwise.}$$

If we take the limit of this expression as $n \to \infty$ (thus the individual pieces are each shrinking to zero length), we get the exact probability function for the Poisson random variable. Now

$$\binom{n}{k}\left(\frac{\lambda s}{n}\right)^k\left(1 - \frac{\lambda s}{n}\right)^{n-k}$$

$$= \frac{(\lambda s)^k}{k!}\left(1 - \frac{\lambda s}{n}\right)^{-k}\left(1 - \frac{\lambda s}{n}\right)^n \frac{n(n-1)\cdots(n-k+1)}{n^k}$$

$$= \frac{(\lambda s)^k}{k!}\left(1 - \frac{\lambda s}{n}\right)^{-k}\left(1 - \frac{\lambda s}{n}\right)^n \cdot 1 \cdot \left(1 - \frac{1}{n}\right)\left(1 - \frac{2}{n}\right)\cdots\left(1 - \frac{k-1}{n}\right).$$

Since

$$\lim_{n\to\infty}\frac{(\lambda s)^k}{k!} = \frac{(\lambda s)^k}{k!}$$

$$\lim_{n\to\infty}\left(1 - \frac{\lambda s}{n}\right)^{-k} = 1$$

$$\lim_{n\to\infty}\left(1 - \frac{\lambda s}{n}\right)^n = e^{-\lambda s}$$

$$\lim_{n\to\infty} 1\left(1 - \frac{1}{n}\right)\left(1 - \frac{2}{n}\right)\cdots\left(1 - \frac{k-1}{n}\right) = 1,$$

we get

$$\lim_{n\to\infty}\binom{n}{k}\left(\frac{\lambda s}{n}\right)^k\left(1 - \frac{\lambda s}{n}\right)^{n-k} = \frac{(\lambda s)^k}{k!} e^{-\lambda s}.$$

Thus, the probability function for X is

$$p_X(k) = \frac{(\lambda s)^k}{k!} e^{-\lambda s}, \qquad k = 0, 1, 2, \ldots$$

$$= 0, \qquad\qquad \text{otherwise.} \qquad\qquad \blacktriangleleft$$

Before looking at some particular examples of the Poisson random variable, let us check to make sure that $p_X(k)$ is a probability function. First

$$\frac{(\lambda s)^k}{k!} e^{-\lambda s} > 0$$

for all values of k so $p_X(k) \geq 0$. Secondly,

$$\sum_{k=0}^{\infty} p_X(k) = \sum_{k=0}^{\infty} \frac{(\lambda s)^k}{k!} e^{-\lambda s}$$

$$= e^{-\lambda s} \sum_{k=0}^{\infty} \frac{(\lambda s)^k}{k!}$$

$$= e^{-\lambda s} e^{\lambda s} = 1.$$

(It will be recalled from calculus that $\sum_{k=0}^{\infty} \frac{x^k}{k!}$ converges to e^x, for any value of x.) Thus $p_X(k)$ does satisfy the requirements for a probability function.

EXAMPLE 4.3.1. Assume that the number of deaths by suicide in Manhattan is a Poisson process with parameter $\lambda = 2$ per day. Let X be the number of deaths by suicide that will occur in a week. Then X is a Poisson random variable with parameter

$$\lambda s = 2 \cdot 7 = 14.$$

and

$$p_X(k) = \frac{(14)^k}{k!} e^{-14}, \qquad k = 0, 1, 2, \ldots.$$

The probability that there will be at least 10 deaths by suicide in this period then is

$$P(X \geq 10) = \sum_{k=10}^{\infty} \frac{(14)^k}{k!} e^{-14} = .891.$$

(See Table 2 in the Appendix and the discussion below.) The probability of exactly 14 deaths is

$$P(X = 14) = F_X(14; \ 14) - F_X(13; \ 14) = .106.$$

The distribution function for the Poisson random variable is discontinuous since the Poisson is a discrete random variable. In fact, denoting the distribution function for a Poisson random variable X with parameter λs by $F_X(t; \lambda s)$, we have

$$F_X(t; \lambda s) = 0, \qquad t < 0$$

$$= \sum_{k \leq t} p_X(k), \qquad t \geq 0$$

where the sum is over all nonnegative integers less than or equal to t. Table 2 gives a table of F_X for selected values of λs. The number tabulated is the value of $F_X(t; \lambda s)$; to derive the value of the probability function at some integer k, we must compute the size of the jump in F_X at $t = k$. Thus,

$$p_X(k) = F_X(k; \lambda s) - F_X(k - 1; \lambda s).$$

For example,

$$F_X(9; 9) = .587$$

$$F_X(8; 9) = .456$$

and thus

$$P(X = 9) = .587 - .456 = .131 \qquad \text{(with } \lambda s = 9\text{)}.$$

Furthermore

$$F_X(2; 1) = .920$$

$$F_X(1; 1) = .736$$

so that

$$P(X = 2) = .920 - .736 = .184 \qquad \text{(with } \lambda s = 1\text{)}$$

and

$$P(X \geq 2) = 1 - F_X(1; 1) = .264$$

$$P(X \geq 3) = 1 - F_X(2; 1) = .080$$

Figure 4.3.1 shows the graphs of the probability functions for two Poisson random variables, one with $\lambda s = 1$, the other with $\lambda s = 10$. Note how much

FIGURE 4.3.1   Poisson probability functions.

more symmetric the histogram is with $\lambda s = 10$ than with $\lambda s = 1$. As we shall see later, this increased symmetry with increasing λs continues and gives a basis for approximations of the Poisson probability function.

EXAMPLE 4.3.2.   Suppose that the number of calls arriving at an industrial switchboard is a Poisson process with parameter $\lambda = 120$ calls per hour. Let X be the number of calls that arrive in a 1-minute period. Then X is a Poisson random variable with parameter $\lambda s = 2$ and

$$p_X(k) = \frac{2^k}{k!} e^{-2}, \qquad k = 0, 1, 2, \ldots.$$

The probability of no calls in this 1-minute interval is

$$p_X(0) = e^{-2} = .135;$$

the probability of between 1 and 5 calls, inclusive, arriving is

$$P(1 \le X \le 5) = F_X(5; \ 2) - F_X(0; \ 2)$$
$$= .983 - .135 = .848.$$

If we look at 3 successive, nonoverlapping 1-minute periods, the probability of between 1 and 5 calls, inclusive, in every 1-minute interval is $(.848)^3 = .610$, since the numbers of calls in these 1-minute intervals are independent. The probability of between 1 and 5 calls, inclusive, in exactly 2 of the 3 intervals is

$$\binom{3}{2} (.848)^2(.152) = .328$$

Let us now derive the factorial moment generating function for a Poisson random variable X with parameter λs and use it to evaluate μ_X and σ_X^2. We have, by definition,

$$\psi_X(t) = E[t^X]$$
$$= \sum_{k=0}^{\infty} t^k \frac{(\lambda s)^k}{k!} e^{-\lambda s}$$
$$= e^{-\lambda s} \sum_{k=0}^{\infty} \frac{(\lambda s t)^k}{k!}$$
$$= e^{-\lambda s} e^{\lambda s t}.$$

Then

$$\psi_X^{(1)}(t) = \lambda s e^{-\lambda s} e^{\lambda s t}$$
$$\psi_X^{(2)}(t) = (\lambda s)^2 e^{-\lambda s} e^{\lambda s t},$$

so that

$$\psi_X^{(1)}(1) = \lambda s$$
$$\psi_X^{(2)}(1) = (\lambda s)^2$$

and thus

$$\mu_X = \psi_X^{(1)}(1) = \lambda s$$
$$\sigma_X^2 = \psi_X^{(2)}(1) + \mu_X - \mu_X^2 = \lambda s.$$

Thus, in Example 4.3.1, the expected number of suicides in the 1-week period is 14, as is the variance of the number of suicides. In Example 4.3.2, the expected number of calls and the variance of the number of calls in a 1-minute period are 2.

EXAMPLE 4.3.3.   It is not necessary that the events observed be occurrences in time for the Poisson random variable to be useful. Assume that molecules of a rare gas occur at an average rate of 3 per cubic foot of air. Then, if it is reasonable to assume that the molecules of this particular gas are distributed independently and randomly in air, the number of molecules we would find in a cubic foot sample of air is a Poisson random variable with parameter 3, and the probability function for X is

$$p_X(k) = \frac{3^k}{k!} e^{-3}, \qquad k = 0, 1, 2, \dots .$$

Thus, the probability we would find no molecules of this gas in our sample of air is $p_X(0) = .050$, the probability we would find exactly 1 molecule is $p_X(1) = .149$, etc. The mean number we would find is $\mu_X = 3$ and the variance of the number we would find is $\sigma^2_X = 3$.

Suppose we wanted to take a sufficiently large amount of air (say s cubic feet) such that the probability of our finding at least one molecule of this gas in the sample is at least .99. How large must s be? If we let Y be the number of molecules of this gas in a sample of s cubic feet, then Y is a Poisson random variable with parameter $3s$, and

$$P(Y \geq 1) = 1 - P(Y = 0)$$
$$= 1 - e^{-3s};$$

then if we are to have at least probability .99 of getting 1 or more molecules, we must have $e^{-3s} \leq .01$; that is

$$-3s \leq \log_e(.01) = -4.605$$

or $s \geq 1.535$. Thus, if we select a sample of 1.535 cubic feet of air (or more) the probability is at least .99 that we will find one or more molecules of the gas in the sample.

In certain circumstances the Poisson probability function gives a good approximation to the binomial probability function. As we saw in the proof of Theorem 4.3.1, the limit of the binomial probability function, where $p = \lambda s/n$, is actually given by the Poisson probability function (for the same number of successes, k). Since p is inversely related to n, then, as $n \to \infty$, the parameter $p \to 0$. Thus, we might expect that if n is large and p is small, the Poisson probability function with $\lambda s = np$ should give a good approximation to the binomial probability function. This is in fact the case. The following example illustrates this idea.

EXAMPLE 4.3.4.   A college professor, based on his past experience, feels that there is a probability of .001 that he will be late to any given class and that his being late or not for any class has no effect on whether he is late or not for any other class. Then the number of times X that he will be late to his next 100 classes is a binomial random variable with parameters $n = 100$ and $p = .001$, and the exact probabilities of his being late exactly 0 times and exactly 1 time are, respectively,

$$P(X = 0) = (.999)^{100} = .9048$$

$$P(X = 1) = \binom{100}{1}(.999)^{99}(.001) = .0906.$$

Note that in this case n is fairly large and p is rather small so that we would expect the above probabilities to be well approximated by those for a Poisson random variable with $\lambda s = np = 100(.001) = .1$. The Poisson values are

$$\frac{(.1)^0}{0!} e^{-.1} = .9048$$

$$\frac{(.1)^1}{1!} e^{-.1} = .0905$$

and the approximation is quite good.

A number of rules-of-thumb have been given which describe when this Poisson approximation is accurate; many texts say that if $n \geq 20$ and $p \leq .05$, it is quite accurate. It is very good if $n \geq 100$ and $np \leq 10$.

EXERCISE 4.3

1. It has been observed that cars pass a certain point on a rural road at the average rate of 3 per hour. Assume that the instants at which the cars pass are independent and let X be the number that pass this point in a 20-minute interval. Compute $P(X = 0)$, $P(X \geq 2)$.

2. It has been observed empirically that deaths per hour, due to traffic accidents, occur at a rate of 8 per hour on long holiday weekends in the United States. Assuming that these deaths occur independently, compute the probability that a 1-hour period would pass with no deaths; that a 15-minute period would pass with no deaths; that 4 consecutive, nonoverlapping 15-minute periods would pass with no deaths.

3. It has been observed that packages of Hamm's beer are removed from the shelf of a particular supermarket at a rate of 10 per hour during rush periods. What is the probability that at least 1 package is removed during the first 6 minutes of a rush period? What is the probability that at least 1 is removed from the shelf during each of 3 consecutive, nonoverlapping 6-minute intervals?

4. At a certain manufacturing plant, accidents have been occurring at the rate of 1 every 2 months. Assuming that the accidents occur independently, what is the expected number of accidents per year? What is the standard deviation of the number of accidents per year? What is the probability of there being no accidents in a given month?

5. If X is a Poisson random variable with parameter λs (a positive integer), show that $p_X(\lambda s) = p_X(\lambda s - 1)$.

6. If X is a Poisson random variable with parameter λs (a positive integer), show that the largest value which $p_X(k)$ takes on is $p_X(\lambda s)$.
    (Hint: Look at $p_X(k)/p_X(k-1)$, $k = 1, 2, \ldots$.)

7. Derive the moment generating function for the Poisson random variable.

8. Assume that 1 baby in 10,000 is born blind. If a large city hospital has 5000 births in 1970, approximate the probability that none of the babies born that year was blind at birth. Also approximate the probabilities that exactly 1 is born blind and that at least 2 are born blind. (Use the Poisson approximation.)

9. Assume that 1 new tire in 1000 has a weak spot in its side wall. If you buy 4 new cars, approximate the probability that none of your cars has a tire with a weak spot and compare this with the exact binomial value.

10. A bakery makes chocolate chip cookies; a batch consists of 1000 cookies. A total of 3000 chocolate chips is added to the batter for each batch and the batter is well mixed. If we select 1 cookie at random from a batch, what is the probability that it contains no chocolate chips? That it contains exactly 3 chips? How many cookies with exactly 1 chocolate chip each would you expect in a batch?

11. Derive the cumulant generating function for a Poisson random variable and use it to evaluate μ_X and σ_X^2.

12. Verify that the derivatives of the Poisson probability generating function, at $t = 0$, give the values of the Poisson probability function.

13. Assume that the sales made by a used car salesman occur like events in a Poisson process with parameter $\lambda = 1$ per week.
    (a) What is the probability he makes (exactly) 3 sales in a two-week period? At least 3? At most 3?
    (b) What is the probability of three 2-week periods in a row with no sales?

14. Assume that a printed page in a book contains 40 lines, and each line contains 75 positions (each of which may be left blank or filled with some symbol). Thus, each page has 3000 positions to be set. Assume that a particular typesetter makes one error per 6000 positions on the average.
    (a) What is the distribution for X, the number of errors per page?
    (b) Compute the probability that a page contains no errors.
    (c) What is the probability a 16-page chapter contains no errors?

15. In a given semester, a large university will process 100,000 grades. In the past, .1 % of all grades have been erroneously reported. Assume you are taking five courses at this university in one semester. What is the probability all of your grades are correctly reported?

4.4. Uniform and Exponential Random Variables

In this section we shall discuss two commonly used continuous random variables. The simplest possible continuous random variable is called the uniform random variable; it is defined as follows.

Definition 4.4.1. X is a *uniform* random variable on the interval (a, b) if

1. The range of X is the interval (a, b).
2. $f_X(x)$ is constant for $x \in R_X$.

(X is *uniformly distributed* on (a, b).)

We have, in fact, already seen some examples of a uniform random variable (see Examples 2.8.3 and 2.8.4 and also problems 5 through 8 in Exericse 2.8). In these cases we did not formally define a random variable but could easily have done so by defining $X(\omega) = \omega$. Then the fact that the sample outcome is equally likely to take on any value in a certain interval implies that X is equally likely to take on values in that same interval. (See also Examples 3.1.3, 3.2.5, and 3.2.7.) If X takes on only values in the interval (a, b), then its density function must be zero outside this interval (in order that the probability of X lying outside the interval be zero). The density function of a uniform random variable is given as Theorem 4.4.1.

Theorem 4.4.1.   If X is uniformly distributed on (a, b), then

$$f_X(x) = \frac{1}{b-a}, \qquad a < x < b$$
$$= 0, \qquad\qquad \text{otherwise.}$$

Proof:

$$f_X(x) = c, \qquad a < x < b$$
$$= 0, \qquad \text{otherwise.}$$

The condition

$$\int_{-\infty}^{\infty} f_X(x)\, dx = 1$$

then implies

$$\int_a^b c\, dx = 1;$$

that is

$$c\int_a^b dx = c(b-a) = 1.$$

Thus $c = 1/(b-a)$ and the theorem is proved.   ◀

The uniform random variable gets its name from the fact that its density function is uniform (constant) on the interval (a, b). The distribution

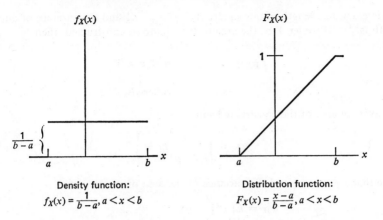

Density function:

$$f_X(x) = \frac{1}{b-a}, a < x < b$$

Distribution function:

$$F_X(x) = \frac{x-a}{b-a}, a < x < b$$

FIGURE 4.4.1   Uniform density and distribution functions.

function for a uniform random variable is

$$F_X(t) = \int_{-\infty}^{t} f_X(x)\, dx$$

$$= 0, \qquad t < a$$

$$= \frac{t-a}{b-a}, \qquad a \le t \le b$$

$$= 1, \qquad t > b;$$

thus F_X is increasing linearly on the interval (a, b). Figure 4.4.1 shows graphs of the density function and the distribution function of a uniform random variable on the interval (a, b).

Suppose that X is uniformly distributed on (a, b). Then

$$\mu_X = E[X] = \int_a^b \frac{x}{b-a}\, dx = \frac{b+a}{2}$$

$$E[X^2] = \int_a^b \frac{x^2}{b-a}\, dx = \frac{b^2 + ab + a^2}{3},$$

so that

$$\sigma_X^2 = E[X^2] - \mu_X^2 = \frac{(b-a)^2}{12}.$$

EXAMPLE 4.4.1.   Suppose that X is uniform distributed on the interval $(0, 1)$ and we construct a square having side length X. Then

$$f_X(x) = 1, \qquad 0 < x < 1$$

$$= 0, \qquad \text{otherwise.}$$

The average side length of our square then is $\mu_X = \frac{1}{2}$ and the variance of our side length is $\frac{1}{12}$. If we let Y be the area of the square so constructed, then

$$f_Y(y) = \frac{1}{2\sqrt{y}} \qquad 0 < y < 1$$
$$= 0, \qquad \text{otherwise.}$$

The average area of the constructed square is

$$\mu_Y = \int_0^1 \frac{y}{2\sqrt{y}} \, dy = \frac{1}{3}$$

(note that $\mu_Y \neq \mu_X^2$) and the variance of the area of the square is

$$\sigma_Y^2 = E[Y^2] - \mu_Y^2$$
$$= \int_0^1 \frac{y^2}{2\sqrt{y}} \, dy - \frac{1}{9} = \frac{4}{45}.$$

Another frequently occurring continuous random variable is the exponential random variable. Although it may occur in many different contexts, we shall introduce it through a Poisson process (see Definition 4.3.1) as follows.

DEFINITION 4.4.2. Given a Poisson process with parameter λ, we designate by zero the time at which we begin observing the process. Let T be the time that passes until the first event occurs. T is called the *exponential* random variable with parameter λ.

Since time is measured continuously (and positively), it is immediately apparent that T is a continuous random variable with its range being positive numbers. The form of the distribution function and density function for T is given as Theorem 4.4.2.

Theorem 4.4.2. Suppose that T is an exponential random variable with parameter λ. Then

$$F_T(s) = 0, \qquad\qquad s < 0$$
$$= 1 - e^{-\lambda s}, \qquad s \geq 0$$

and

$$f_T(s) = \lambda e^{-\lambda s}, \qquad s > 0$$
$$= 0, \qquad\qquad \text{otherwise.}$$

Proof: Since time is measured positively, it is immediately apparent that $P(T \leq s) = 0$ if $s < 0$; thus $F_T(s) = 0$, $s < 0$. Consider some time $s \geq 0$. The time to the first event will *exceed* s if and only if there are no events in the interval $(0, s)$. But the probability of no events in $(0, s)$ we have seen is $e^{-\lambda s}$ (see Theorem 4.3.1, set $k = 0$); thus $P(T > s) = e^{-\lambda s}$. But $P(T \leq s) = 1 - P(T > s)$ and therefore $F_T(s) = 1 - e^{-\lambda s}$ for $s \geq 0$. The density function for T is given by the derivative of F_T; since

$$\frac{d}{ds} F_T(s) = \lambda e^{-\lambda s}, \qquad s > 0$$

$$= 0, \qquad s < 0,$$

we have

$$f_T(s) = \lambda e^{-\lambda s}, \qquad s > 0$$

$$= 0, \qquad s \leq 0,$$

which completes the proof. ◀

It is easily shown that F_T is, in fact, a distribution function and that f_T is a density function. The moment generating function for T is

$$m_T(t) = E[e^{tT}]$$

$$= \int_0^\infty e^{ts} \lambda e^{-\lambda s} \, ds$$

$$= \int_0^\infty \lambda e^{-s(\lambda - t)} \, ds$$

$$= \frac{\lambda}{\lambda - t}, \qquad \text{for } t < \lambda.$$

Then

$$m_T^{(1)}(t) = \frac{\lambda}{(\lambda - t)^2}$$

$$m_T^{(2)}(t) = \frac{2\lambda}{(\lambda - t)^3}$$

so that

$$\mu_T = m_T^{(1)}(0) = \frac{1}{\lambda}$$

$$\sigma_T^2 = m_T^{(2)}(0) - \mu_T^2 = \frac{1}{\lambda^2}.$$

Figure 4.4.2 shows graphs of the density function and the distribution function for an exponential random variable X with parameter λ.

EXAMPLE 4.4.2. Breakdowns in equipment at a large industrial plant have been observed to be approximately a Poisson process with parameter $\lambda = \frac{1}{2}$ per hour (1 every 2 hours). If we arrive at this plant at 9 a.m. on a Monday morning and let T be the time (from our arrival) until the first breakdown, then T has density

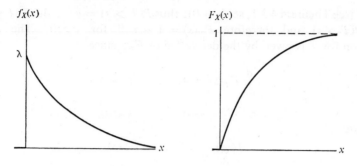

Density function: Distribution function:
$f_X(x) = \lambda e^{-\lambda x}, x > 0$ $F_X(x) = 1 - e^{-\lambda x}, x > 0$

FIGURE 4.4.2   Exponential density and distribution functions.

function

$$f_T(s) = \tfrac{1}{2}e^{-s/2}, \qquad s > 0$$
$$= 0, \qquad s \leq 0.$$

The probability that it is at least 1 hour until the first breakdown is

$$P(T > 1) = \int_1^\infty \tfrac{1}{2}e^{-s/2}\,ds = e^{-1/2} = .6065;$$

the probability that it is no more than 4 hours to the first breakdown is

$$P(T \leq 4) = \int_0^4 \tfrac{1}{2}e^{-s/2}\,ds = 1 - e^{-2} = .8647.$$

The average time to the first breakdown is $\mu_T = 1/\lambda = 2$ hours; the probability that the time to the next breakdown is greater than the average (2 hours) is

$$P(T > 2) = \int_2^\infty \tfrac{1}{2}e^{-s/2}\,ds = e^{-1} = .3679.$$

EXAMPLE 4.4.3.   It is commonly assumed that failures of electron tubes of various sorts occur like events in a Poisson process; specifically, if a tube is put into operation, the time T until it fails is an exponential random variable with parameter λ. Let us suppose that a particular device contains 5 electron tubes; the time (in hours) until failure for each of the 5 is assumed to be an exponential random variable with parameter $\lambda = \frac{1}{1000}$. Suppose further that the device works only so long as all 5 tubes are working and that the tubes operate (and fail) independently. We might then ask what is the probability that the device will operate at least a hours. Each tube has density function

$$f_T(s) = \tfrac{1}{1000}e^{-s/1000}, \qquad s > 0$$
$$= 0, \qquad s \leq 0.$$

The probability that a particular tube will last at least a hours is

$$P(T > a) = \int_a^\infty \tfrac{1}{1000} e^{-s/1000} \, ds = e^{-a/1000}.$$

Then, since the tubes are operating independently and the device continues working only so long as all the tubes do, we find that the probability the device will work at least a hours is $(e^{-a/1000})^5 = e^{-a/200}$. Thus, the probability that it will operate at least 100 hours is $e^{-1/2} = .6065$; the probability that it will operate at least 1000 hours is $e^{-5} = .0067$.

There are a number of similarities between the discrete geometric random variable and the continuous exponential random variable. It will be recalled that the geometric random variable is the number of trials X until the first success in repeated independent Bernoulli trials with parameter p. As we have seen, the exponential random variable is the length of time T until the first event (which could be called a success) in a Poisson process with parameter λ. The mean value of X was seen to be $1/p$; the mean value of T is $1/\lambda$. Theorem 4.4.3 gives a particular property of the exponential random variable which is also shared by the geometric (as you are asked to prove in problem 12 below).

Theorem 4.4.3. If T is an exponential random variable, then

$$P(T > a + b \mid T > a) = P(T > b).$$

Proof: Let us first get a clear understanding of the notation used. Let the event that $T > a$ be denoted by A, the event that $T > b$ be denoted by B, and the event that $T > a + b$ be denoted by C. Then we are to prove that $P(C \mid A) = P(B)$. By definition, $P(C \mid A) = P(C \cap A)/P(A)$. The event C, that $T > a + b$, is a subset of the event A, that $T > a$. Thus, $C \cap A = C$ and we have $P(C \mid A) = P(C)/P(A)$. Now

$$P(A) = P(T > a) = \int_a^\infty \lambda e^{-\lambda s} \, ds = e^{-\lambda a}$$

$$P(B) = P(T > b) = \int_b^\infty \lambda e^{-\lambda s} \, ds = e^{-\lambda b}$$

$$P(C) = P(T > a + b) = \int_{a+b}^\infty \lambda e^{-\lambda s} \, ds = e^{-\lambda(a+b)}$$

and we see that

$$P(C \mid A) = \frac{P(C)}{P(A)} = \frac{e^{-\lambda(a+b)}}{e^{-\lambda a}} = e^{-\lambda b} = P(B),$$

as was to be proved.   ◀

Suppose, for example, that the time to failure of an electron tube is an exponential random variable. Then Theorem 4.4.3 shows that if the tube has already been used for 100 hours without failure, the probability that it will last another 50 hours is the same as the probability of its lasting 50 hours when it was new. Since this property is also shared by the geometric random variable, we can say that if we have already had 10 trials without a success, the probability of 5 more trials without a success is identical with the original probability of having 5 trials with no success when we first started. Thus, both the exponential random variable and the geometric random variable can be said to have no memory; the fact that they have already achieved a certain value has no effect on the probability of achieving a larger value (relative to the original probability of equalling the incremental amount). This property is a direct consequence of the independence properties assumed: that the Bernoulli trials are independent and that the events in the Poisson process occur independently.

EXAMPLE 4.4.4. Studies have shown that the occurrence of accidents to individuals (broken legs, bumped heads, car accidents, etc.) appear to be well described by Poisson processes; that is, the occurrences of accidents to individuals seem to occur in a manner consistent with the Poisson process assumptions. People who have a relatively large value of λ (the rate of occurrence) are called accident prone. Suppose you are a person to whom accidents occur at a rate of 1 per year in a Poisson manner. Then Theorem 4.4.3 says that even if it has been 10 years since your last accident, your probability of having an accident within the next year is no higher than it was during any preceding year. Small comfort, perhaps, but comfort nonetheless.

EXERCISE 4.4

1. Suppose that X is uniformly distributed on the interval $(1, 2)$ and we construct a square having sides of length X. Derive the probability density function of $Y = X^2$, the area of the square and compute $P(Y > 2)$.

2. If X is uniformly distributed on the interval $(1, 4)$, derive the density function of $Z = X^{1/2}$.

3. Derive the moment generating function for a random variable X which is uniformly distributed on the interval (a, b) and use it to evaluate μ_X and σ^2_X.

4. Derive the factorial moment generating function for a random variable X uniformly distributed on (a, b).

5. Suppose that quarter-pound bars of butter are cut from larger slabs by a machine. We assume that the larger slabs are quite uniform in density; if the length of the bar is exactly $3\frac{3}{8}$ inches, then the bar will weigh $\frac{1}{4}$ pound. Suppose that the true length X of a bar cut by this machine is equally likely to lie in the

interval from 3.35 inches to 3.45 inches. Assuming that the lengths of bars cut by this machine are independent, what is the probability that all 4 bars in a particular pound package of butter will weigh at least $\frac{1}{4}$ pound? That exactly 3 will weigh at least $\frac{1}{4}$ pound?

6. X is uniformly distributed on $(0, 2)$ and Y is exponential with parameter λ. Find the value of λ such that $P(X < 1) = P(Y < 1)$.

7. Calls arrive at a switchboard according to a Poisson process with parameter $\lambda = 5$ per hour. If we are at the switchboard, what is the probability that it is at least 15 minutes until the next call? That it is no more than 10 minutes? That it is exactly 5 minutes?

8. A newsboy is selling papers on a busy street. The papers he sells are events in a Poisson process with parameter $\lambda = 50$ per hour. If we have just purchased a paper from him, what is the probability that it will be at least 2 minutes until he sells another? If it is already 5 minutes since his last sale, what is the probability it will be at least 2 more minutes until his next sale?

9. X is a uniform on $(-1, 3)$ and Y is exponential with parameter λ. Find λ such that $\sigma_X^2 = \sigma_Y^2$.

10. X is geometric with parameter p and Y is exponential with parameter λ. Find λ such that $P(X > 1) = P(Y > 1)$.

11. We are given a Poisson process with parameter λ. We begin observing the process at time zero; let S be the time until the second event occurs. Derive the probability density function for S.

12. Prove Theorem 4.4.3 for a geometric random variable with parameter p.

13. Generalize Exercise 11. Assume we are given a Poisson process with parameter λ and let X be the time until the r-th occurrence, where $r \geq 1$. This is an example of a *gamma* random variable. (**Hint:** The event, $X > t$, is equivalent to the event: We observe $r - 1$ or fewer occurrences in the fixed interval $(0, t)$.)

14. Find the cumulant generating function for an exponential random variable and use it to evaluate the mean and variance of the exponential distribution. (See Exercise 3.4.14.)

15. By using repeated integration by parts show that

$$\Gamma(n) = \int_0^\infty x^{n-1} e^{-x} \, dx = (n - 1)!$$

if n is a positive integer. This is called the *gamma function*.

16. Evaluate the r-th moment of an exponential random variable directly, without using a generating function. Use these moments to verify the mean and variance, derived in the text.

17. The exponential distribution has found many applications in *reliability* theory, the modelling of how reliable (or unreliable) components and systems may be. The reliability of a component in a specific environment for a period of length t is defined to be the probability its time to failure exceeds t (thus it

has worked satisfactorily for the period of length t); that is

$$R(t) = P(X > t) = 1 - F_X(t).$$

(a) If X, the time to failure, is an exponential random variable, evaluate $R(t)$
(b) Nonmechanical components, which do not suffer friction or some wearout mechanism, are frequently assumed to have lifetimes described by the exponential distribution. Considering $R(t)$, and Theorem 4.4.3, why might this seem reasonable?

18. Some items are repeatedly turned on and off when in use. Let Y be the number of on-off cycles a particular item goes through until it fails. Then its reliability for t cycles can be defined by $R(t) = P(Y > t)$, $t = 1, 2, 3, \ldots$. If Y is a geometric random variable with parameter p, evaluate $R(t)$. Why might this distribution be appropriate if the item does not suffer any wear in being cycled on and off?

4.5. Normal Random Variable

The normal probability density function is undoubtedly the most frequently used of all probability laws, both because the normal random variable does frequently occur in practical problems and, as we shall see in the next chapter, because it provides an accurate approximation to a large number of other probability laws. As can be seen from Definition 4.5.1, the density function for a normal random variable is a symmetric, bell-shaped curve, familiar to many people as the curve that is used when tests are "graded-on-the-curve." Due to the bell-shape of the density function, a normally distributed random variable has the biggest probability of taking on a value close to μ (in a sense) and correspondingly less of taking on values further from μ (on either side). The formal definition follows.

DEFINITION 4.5.1.   A random variable X is *normally* distributed if and only if its probability density function is

$$f_X(x) = \frac{1}{\sigma\sqrt{2\pi}} e^{-(x-\mu)^2/2\sigma^2}$$

for all real x. The parameter μ can equal any real number while the parameter σ must be positive.

Since the exponential function is nonnegative, we obviously have $f_X(x) \geq 0$ for all x. You are asked to show in problem 7 below that $\int_{-\infty}^{\infty} f_X(x)\,dx = 1$; granting that this is the case, $f_X(x)$ is a density function.

Figure 4.5.1 shows the graphs of some typical density functions for various values of the parameters μ and σ^2. Notice that in any event the density function is a symmetric, bell-shaped curve centered at μ (thus we would guess that μ is, in fact, the mean of the random variable; this is proved below). The parameter σ controls the relative flatness of the bell. Keeping μ

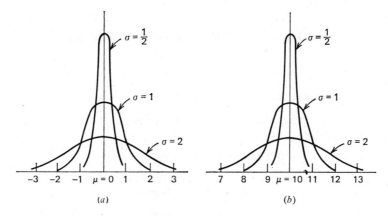

FIGURE 4.5.1   Normal density functions.

constant and decreasing σ causes the density function to become more sharply peaked, thus giving higher probabilities of X being close to μ. Increasing σ causes the density function to flatten, thus giving lower probabilities of X being close to μ. If σ is held constant and μ is varied, the density function's shape is held constant with its midpoint moving to the location of μ.

Let us derive the moment generating function for a normal random variable.

Theorem 4.5.1.   If X is a normal random variable with probability density function given in Definition 4.5.1, then the moment generating function for X is

$$m_X(t) = e^{t\mu + t^2\sigma^2/2}.$$

Proof: By definition,

$$m_X(t) = E[e^{tX}] = \int_{-\infty}^{\infty} e^{tx} \frac{1}{\sigma\sqrt{2\pi}} e^{-(x-\mu)^2/2\sigma^2} \, dx$$

$$= \frac{1}{\sigma\sqrt{2\pi}} \int_{-\infty}^{\infty} e^{tx-(x-\mu)^2/2\sigma^2} \, dx.$$

Completing the square in the exponent, we have

$$tx - \frac{(x-\mu)^2}{2\sigma^2} = -\frac{1}{2\sigma^2}[x^2 - 2\mu x + \mu^2 - 2\sigma^2 tx]$$

$$= -\frac{1}{2\sigma^2}[(x - \mu - \sigma^2 t)^2 - 2\sigma^2 t\mu - \sigma^4 t^2]$$

$$= -\frac{1}{2\sigma^2}[(x - \mu - \sigma^2 t)^2] + t\mu + \frac{\sigma^2 t^2}{2}.$$

Thus

$$m_X(t) = e^{t\mu + \sigma^2 t^2/2} \int_{-\infty}^{\infty} \frac{1}{\sigma\sqrt{2\pi}} e^{-[x-\mu-\sigma^2 t]^2/2\sigma^2} \, dx$$

$$= e^{t\mu + \sigma^2 t^2/2},$$

since the integral has value 1 (see problem 7 below).   ◀

The cumulant generating function (see Exercise 3.4.14) for a normal random variable is

$$c_X(t) = \log_e m_X(t) = t\mu + t^2\sigma^2/2;$$

thus

$$\mu_X = c_X^{(1)}(0) = \mu$$

$$\sigma_X^2 = c_X^{(2)}(0) = \sigma^2$$

that is, the parameter μ in the density function is the mean of the random variable X and the parameter σ^2 in the density is the variance of X. The distribution function for a normal random variable X is

$$F_X(t) = \int_{-\infty}^{t} \frac{1}{\sigma\sqrt{2\pi}} e^{-(x-\mu)^2/2\sigma^2} \, dx.$$

Unfortunately, this integration cannot be carried out in closed form. Numerical techniques could be used, of course, to evaluate the integral for specific values of μ and σ^2. Luckily, it is necessary only to use numerical integration to tabulate the distribution function of the standard normal random variable. Once such a table is available for the standard normal distribution function, it can be used to evaluate the distribution function for a normal random variable with any mean μ and any variance σ^2. Since the standard normal distribution function is of such importance, we shall use a special symbol for it, as given in Definition 4.5.2.

DEFINITION 4.5.2. If Z is a normal random variable with $\mu = 0$, $\sigma^2 = 1$, then Z is called the *standard normal* random variable. Its density function is

$$n_Z(z) = \frac{1}{\sqrt{2\pi}} e^{-z^2/2},$$

and its distribution function is

$$N_Z(t) = \int_{-\infty}^{t} \frac{1}{\sqrt{2\pi}} e^{-z^2/2} \, dz.$$

Table 3 in the Appendix gives a table of values of $N_Z(t)$. Let us consider some examples of its use.

EXAMPLE 4.5.1   If Z is a standard normal random variable, let us compute the following probabilities.

$$P(Z < 1) = N_Z(1) = .8413$$
$$P(1 < Z < 2) = N_Z(2) - N_Z(1) = .9773 - .8413 = .1360$$
$$P(|Z| < \tfrac{1}{2}) = P(-\tfrac{1}{2} < Z < \tfrac{1}{2}) = N_Z(\tfrac{1}{2}) - N_Z(-\tfrac{1}{2})$$
$$= .6915 - .3085 = .3830.$$

EXAMPLE 4.5.2.   We can also use Table 3 to answer questions such as these. Find the number b such that $P(Z > b) = .25$. Since $P(Z > b) = 1 - N_Z(b)$, we are asked to find b such that $N_Z(b) = .75$; by scanning the values of N_Z, we find that $b = .674$. We note that for a standard normal random variable, since its density function is symmetric about zero, we have $P(Z > a) = P(Z < -a)$ for any $a > 0$. That is $1 - N_Z(a) = N_Z(-a)$. Then suppose that we want to find the number c such that $P(|Z| < c) = .9$. That is, since

$$P(|Z| < c) = P(-c < Z < c) = N_Z(c) - N_Z(-c)$$
$$= 2N_Z(c) - 1,$$

we want to find the point c such that $N_Z(c) = .95$. By scanning the values of N_Z, we find that $c = 1.645$.

It was mentioned above that the standard normal distribution function could be used to evaluate the distribution function for a normal random variable with arbitrary μ and σ. The actual technique by which this is done is given as Theorem 4.5.2.

Theorem 4.5.2.   If X is a normal random variable with mean μ and variance σ^2, then

$$F_X(t) = N_Z\left(\frac{t - \mu}{\sigma}\right).$$

Proof: By definition

$$F_X(t) = \int_{-\infty}^{t} \frac{1}{\sigma\sqrt{2\pi}} e^{-(x-\mu)^2/2\sigma^2} \, dx;$$

in this integral we change to z as the variable of integration where

$$z = \frac{x - \mu}{\sigma}, \qquad dz = \frac{dx}{\sigma}.$$

The limits for x were $-\infty < x < t$ which implies

$$-\infty < \frac{x - \mu}{\sigma} = z < \frac{t - \mu}{\sigma}.$$

Then we have

$$F_X(t) = \int_{-\infty}^{(t-\mu)/\sigma} \frac{1}{\sqrt{2\pi}} e^{-z^2/2} \, dz = N_Z\left(\frac{t-\mu}{\sigma}\right),$$

as was to be proved   ◀

EXAMPLE 4.5.3.   It is frequently assumed that measurements of length, by a sufficiently finely calibrated instrument, are normally distributed about the "true" length of the object measured. Specifically, assume the true length of an object is, say, 9 feet, and that a measurement made of this object, a random variable X, is normal with mean 9 and standard deviation .02 (both in feet). Then the probability that the measured length will lie between 8.99 and 9.02 feet is

$$P(8.99 < X < 9.02) = F_X(9.02) - F_X(8.99)$$

$$= N_Z\left(\frac{9.02 - 9}{.02}\right) - N_Z\left(\frac{8.99 - 9}{.02}\right)$$

$$= N_Z(1) - N_Z(-\tfrac{1}{2})$$

$$= .8413 - .3085 = .5328.$$

If this same object is measured three times, independently, and each measurement has the normal distribution described above, then the probability all three measurements will lie between 8.99 and 9.02 is $(.5328)^3 = .1512$.

EXAMPLE 4.5.4.   Assume that the distance X that a particular athlete will be able to put a shot (on his first try) is a normal random variable with parameters $\mu = 50$ feet, $\sigma = 5$ feet. Compute the probability that he tosses it no less than 55 feet and the probability that his toss travels between 50 feet and 60 feet.

$$P(X > 55) = 1 - F_X(55)$$

$$= 1 - N_Z\left(\frac{55 - 50}{5}\right)$$

$$= 1 - N_Z(1)$$

$$= .1587.$$

$$P(50 < X < 60) = F_X(60) - F_X(50)$$

$$= N_Z(2) - N_Z(0)$$

$$= .4773.$$

When we get into statistical inference problems, we shall derive the distribution of what is called a χ^2 (chi-square) random variable (see Chapter 6). The χ^2 random variable has a single parameter called its degrees of freedom; let us note now that the square of a standard normal random variable Z is a χ^2 random variable with 1 degree of freedom. That is, if we let $W = Z^2$, then W is a χ^2 random variable with 1 degree of freedom. We can in fact evaluate the distribution function of W from the distribution function for Z. As was derived in Example 4.5.2, if Z is standard normal, then

$$N_Z(-a) = 1 - N_Z(a).$$

If $W = Z^2$, then

$$F_W(t) = N_Z(\sqrt{t}) - N_Z(-\sqrt{t}), \quad \text{for } t > 0;$$

thus, if W is a χ^2 random variable with one degree of freedom,

$$F_W(t) = N_Z(\sqrt{t}) - [1 - N_Z(\sqrt{t})]$$
$$= 2N_Z(\sqrt{t}) - 1.$$

Our table of N_Z then enables us to evaluate probabilities for W as well.

EXAMPLE 4.5.5.   If X is a normal random variable with parameters μ and σ, the distribution function of the χ^2 random variable just discussed can be used to evaluate probabilities that X would differ from μ by more than $k\sigma$. Now

$$P(|X - \mu| > k\sigma) = P\left(\frac{|X - \mu|}{\sigma} > k\right)$$

$$= P\left(\frac{|X - \mu|^2}{\sigma^2} > k^2\right)$$

$$= P(W > k^2),$$

where W is a χ^2 random variable with 1 degree of freedom since $(X - \mu)/\sigma$ is a standard normal random variable. Then, for example,

$$P(|X - \mu| > \sigma) = P(W > 1) = 1 - F_W(1)$$
$$= 2 - 2N_Z(1) = .3174$$
$$P(|X - \mu| > 2\sigma) = P(W > 4) = 1 - F_W(4)$$
$$= 2 - 2N_Z(2) = .0454$$
$$P(|X - \mu| < 3\sigma) = P(W < 9) = F_W(9)$$
$$= 2N_Z(3) - 1 = .9974$$

EXERCISE 4.5

1. Assume that the time X required for a distance runner to run a mile is a normal random variable with parameters $\mu = 4$ minutes, 1 second and $\sigma = 2$ seconds. What is the probability that this athlete will run the mile in less than 4 minutes? In more than 3 minutes, 55 seconds?

2. The length X of an adult rock cod caught in Monterey Bay is a normal random variable with parameters $\mu = 30$ inches and $\sigma = 2$ inches. If you catch one of these fish, what is the probability that it will be at least 31 inches long? That it will be no more than 32 inches long? That its length will be between 24 inches and 28 inches?

3. If Z is a standard normal random variable and we define $U = |Z|$, then U is called the *folded* standard normal variable. Express $F_U(t)$ in terms of $N_Z(t)$.

4. Suppose that we are given a target with a vertical straight line drawn through its center. Let us assume that if we throw a dart at this target and measure the distance Z between the point we hit and the center line, then Z is a standard normal random variable (if the dart lands right of the center line the measurement is positive, if it lands to the left of the center line the measurement is negative). Then, the distance from the point we hit to the center line is $|Z| = U$, the folded normal random variable defined in problem 3. Compute $P(U > 1)$ and $P(U < \frac{1}{2})$.

5. Find the median and the interquartile range for a standard normal random variable Z.

6. Find the median and the interquartile range for a normal variable X with parameters μ and σ.

7. Show that

$$\int_{-\infty}^{\infty} \frac{1}{\sigma\sqrt{2\pi}} e^{-(x-\mu)^2/2\sigma^2} \, dx = 1$$

for any μ and $\sigma > 0$. (*Hint:* If

$$A = \int_{-\infty}^{\infty} \frac{1}{\sigma\sqrt{2\pi}} e^{-(x-\mu)^2/2\sigma^2} \, dx,$$

then

$$A^2 = \int_{-\infty}^{\infty} \frac{1}{2\pi\sigma^2} e^{-(x-\mu)^2/2\sigma^2-(y-\mu)^2/2\sigma^2} \, dx \, dy;$$

let $u = (x - \mu)/\sigma$, $v = (y - \mu)/\sigma$ and transform to polar coordinates to show $A^2 = 1$ which implies $A = 1$.)

8. If W is a χ^2 random variable with 1 degree of freedom, find the median and the interquartile range for W.

9. Evaluate $f_W(t)$ where W is a χ^2 random variable with 1 degree of freedom.

10. What is the moment generating function for a standard normal random variable?

11. Derive the density function for a folded standard normal random variable.

12. The height a university high-jumper will clear, each time he jumps, is a normal random variable with mean six feet and standard deviation 2.4 inches.
(a) What is the greatest height he will jump with probability .95?
(b) What is the height he will clear only 10% of the time?

13. Scores on a national educational achievement test are assumed to be normally distributed with $\mu = 500$, $\sigma = 100$.
(a) What is the 90-th percentile for this distribution?
(b) What are the quartiles of this distribution?
(c) Assume you take this exam and make a score of 585. What proportion of all scores are smaller than yours?

14. Show that the points of inflection of the normal density function occur at $\pm\sigma$.

15. The 90-th percentile of a normal distribution is known to equal 50 while its 15-th percentile is 25.
    (a) Find μ and σ.
    (b) What is the value of the 40-th percentile?

16. Express the percentiles of the χ^2, 1 degree of freedom, distribution in terms of the percentiles of a standard normal distribution.

17. Express the percentiles of a folded standard normal distribution in terms of the percentiles of the standard normal distribution.

18. What is the relationship between the χ^2, 1 degree of freedom, random variable and the folded standard normal random variable?

5

Jointly Distributed Random Variables

Many important problems require the simultaneous treatment of several random variables. In all practical statistical problems, the observed values of several jointly distributed random variables are available for making inferences. Thus, to be able to examine statistical methods, we must have some acquaintance with the behavior of jointly distributed random variables. As we shall see, even if several random variables have been defined on the same experiment, each random variable can be considered by itself and any techniques already discussed are then appropriate in describing its individual behavior. What is new, then, in this chapter will be the definitions and concepts that are useful in describing the simultaneous, joint behavior of several random variables.

5.1. Two-Dimensional Random Variables

The simplest possible example of jointly distributed random variables is given by the two-dimensional case. The extension from a single random variable to two is the biggest step however; the extension from two to an arbitrary n is relatively straightforward.

DEFINITION 5.1.1. Given an experiment, the pair (X, Y) is called a *two-dimensional* random variable (or vector) if each of X and Y associates a real number with every element of S.

169

Thus, (X, Y) is a pair of real-valued functions defined on S; as such each (of X and Y) is then a random variable in its own right.

EXAMPLE 5.1.1.   Suppose a fair die is rolled 2 times. Let X be the number that occurs on the first roll and let Y be the number that occurs on the second roll. Then (X, Y) is a two-dimensional random variable.

EXAMPLE 5.1.2.   One student is selected at random from the student body at San Francisco State. Let U be the weight of the selected student and let V be the height of the selected student. Then (U, V) is a two-dimensional random variable.

As we might guess on the basis of our previous probability theory, if we have a two-dimensional random variable, it is possible that both are discrete, both are continuous, or one is discrete and the other continuous. We shall, in the main, be concerned only with the first two cases; it will be quite easy to formulate results for the third case based on a good understanding of the first two.

In the case of a single random variable X, we could think of the values X takes on as being points on a line. Thus we defined the probability function for X or the density function for X, both of which are functions of a real variable, depending on whether X was discrete or continuous. We derived these one-dimensional functions (at least in theory) for computing probabilities about X from the probability measure defined on the subsets of the sample space S. In the case of a two-dimensional random variable (X, Y), it is convenient to think of the values the (two-dimensional) random variable takes on as being points in a Cartesian plane. We formally think of deriving the (two-dimensional) probability function or density function for computing probabilities about (X, Y) again from the probability measure defined on the elements of S. This derivation is illustrated in the next two examples.

EXAMPLE 5.1.3.   An urn contains 3 balls numbered 1, 2, 3, respectively. We draw 2 balls at random, with replacement, from the urn. Let X be the number on the first ball drawn and let Y be the number on the second ball drawn. The collection of 2-tuples (pairs of numbers) that could be drawn would be a reasonable

TABLE 5.1

y	x		
	1	2	3
1	$\frac{1}{9}$	$\frac{1}{9}$	$\frac{1}{9}$
2	$\frac{1}{9}$	$\frac{1}{9}$	$\frac{1}{9}$
3	$\frac{1}{9}$	$\frac{1}{9}$	$\frac{1}{9}$

sample space for the experiment. Moreover, since the balls are drawn at random, the single-element events (2-tuples) are equally likely to occur. The range of X is $\{1, 2, 3\}$, as is the range of Y. Table 5.1 has as entries the probabilities of occurrence of the possible pairs of values for (X, Y). The probability of occurrence of any other pair of values (x, y) is zero.

EXAMPLE 5.1.4.   Suppose that we draw 2 balls without replacement from the same urn discussed in Example 5.1.3 and again let X be the number on the first ball drawn and let Y be the number on the second ball. Again the ranges of the two random variables are each $\{1, 2, 3\}$, except now it is not possible for both X and Y to equal the same number simultaneously. The entries in Table 5.2 are the probabilities

TABLE 5.2

		x	
y	1	2	3
1	0	$\frac{1}{6}$	$\frac{1}{6}$
2	$\frac{1}{6}$	0	$\frac{1}{6}$
3	$\frac{1}{6}$	$\frac{1}{6}$	0

of occurrence of the possible pairs of values for (X, Y), and again the probability of occurrence of any other pair (x, y) is 0.

The function which we have tabled in each of the two preceding examples is actually the two-dimensional probability function for (X, Y); the definition of this function follows.

DEFINITION 5.1.2. (X, Y) is a two-dimensional *discrete* random variable if the range of (X, Y) is a discrete set $R_{X,Y}$. The probability function for a discrete two-dimensional random variable is defined to be

$$p_{X,Y}(x, y) = P(X = x, Y = y)$$

for all real pairs (x, y).

Two-dimensional probability functions follow the same rules as do their one-dimensional analogs, namely

$$p_{X,Y}(x, y) \geq 0, \qquad \text{for all } (x, y)$$

$$\sum_{\substack{\text{range} \\ \text{of } (X,Y)}} p_{X,Y}(x, y) = 1.$$

Furthermore, any function of a pair of real variables that satisfies these two equations can be thought of as being the probability function for some two-dimensional random variable.

Given $p_{X,Y}(x, y)$, we can easily evaluate $p_X(x)$ and $p_Y(y)$, the probability functions for X and Y individually. It will be recalled

$$p_X(x) = P(X = x)$$

and that the event $\{X = x\}$ is the collection of all elements belonging to S such that X takes on the value x. If it happens that Y may take on several values, say $y_1, y_2, \ldots, y_k$, when $X = x$ then

$$\{X = x\} = \{X = x, Y = y_1\} \cup \{X = x, Y = y_2\} \cup \cdots \cup \{X = x, Y = y_k\}$$

so that

$$p_X(x) = P(X = x) = \sum_{y \in R_{Y|X=x}} P(X = x, Y = y)$$

$$= \sum_{y \in R_{Y|X=x}} p_{X,Y}(x, y)$$

where $R_{Y|X=x}$ is the range of values for Y, given that $X = x$. Note that this range may be different for different values of x. (See Example 5.1.4, where $R_{Y|X=1} = \{2, 3\}$, $R_{Y|X=2} = \{1, 3\}$.) Similar reasoning, interchanging the roles of X and Y, gives

$$p_Y(y) = \sum_{x \in R_{X|Y=y}} p_{X,Y}(x, y)$$

The two one-dimensional functions, $p_X(x)$ and $p_Y(y)$, are called the *marginal* probability functions for X and for Y, respectively. The word marginal is used because the values of these functions are given by the totals at the margins of the rows and columns of a tabular presentation of $p_{X,Y}(x, y)$, as in Tables 5.1 and 5.2. Notice that in each of these tables if we sum across a row (y fixed) the total is $p_Y(y)$; if we sum down a column (x fixed) the column total is simply $p_X(x)$. It is important to realize that knowledge of the marginals, $p_X(x)$ and $p_Y(y)$, generally is not equivalent to knowledge of $p_{X,Y}(x, y)$. This is immediately apparent by referring again to Tables 5.1 and 5.2; the marginals are the same in both cases but the two-dimensional probability functions are not.

To compute probabilities about two-dimensional continuous random variables, we must integrate two-dimensional density functions. These are defined as follows.

DEFINITION 5.1.3.   If each of X and Y are defined on the elements of the same sample space and each (ignoring the other) is a continuous random variable, then (X, Y) is a two-dimensional *continuous* random variable. The density function, $f_{X,Y}(x, y)$, for (X, Y) is defined over the real plane and

$$P((X, Y) \in R) = \iint_R f_{X,Y}(x, y)\, dx\, dy,$$

where R is any region in the (x, y) plane.

As in the one-dimensional case, it can be shown that a density function must satisfy two requirements:

$$1. \quad f_{X,Y}(x, y) \geq 0 \qquad \text{for all } (x, y)$$

$$2. \quad \int_{-\infty}^{\infty} \int_{-\infty}^{\infty} f_{X,Y}(x, y)\, dx\, dy = 1.$$

Furthermore, any function of two variables that satisfies these two requirements can be thought of as being the density function for some two-dimensional random variable.

EXAMPLE 5.1.5. Suppose we select one point at random from the unit square (the set $\{(x, y): 0 < x < 1, 0 < y < 1)\}$).

Let X be the x-coordinate of the selected point and let Y be the y-coordinate of the selected point. Then (X, Y) is a two-dimensional continuous random variable. Since we are selecting the single point from this square at random we must have

$$f_{X,Y}(x, y) = c, \qquad 0 < x < 1, \qquad 0 < y < 1$$
$$= 0, \qquad \text{otherwise}$$

because, presumably, all points have the same chance of occurring. Furthermore,

$$\int_0^1 \int_0^1 c\, dx\, dy = 1$$

which implies that $c = 1$; thus the density function for (X, Y) is

$$f_{X,Y}(x, y) = 1, \qquad 0 < x < 1, \qquad 0 < y < 1$$
$$= 0, \qquad \text{otherwise.}$$

For this density, notice then that the probability (X, Y) takes on a pair of values in any region inside the unit square is simply the area of the region.

We have seen, for (X, Y) discrete, that the joint probability function may be used to derive the marginal, one-dimensional probability functions for X and for Y. In a similar manner, the marginal density functions for X and for Y may be derived from their joint density, if (X, Y) is continuous. Indeed

$$P(a < X < b) = P(a < X < b, -\infty < Y < \infty)$$

$$= \int_{x=a}^{b} \left\{ \int_{y=-\infty}^{\infty} f_{X,Y}(x, y)\, dy \right\} dx.$$

The quantity in braces is what must be integrated from a to b to compute $P(a < X < b)$, thus it must be $f_X(x)$. That is, for any fixed value x,

$$f_X(x) = \int_{-\infty}^{\infty} f_{X,Y}(x, y)\, dy.$$

Similarly, for any fixed y

$$f_Y(y) = \int_{-\infty}^{\infty} f_{X,Y}(x, y)\, dx.$$

As with the discrete case, we can derive $f_X(x)$ and $f_Y(y)$ from $f_{X,Y}(x, y)$, but knowledge of $f_X(x)$ and $f_Y(y)$ alone does not specify $f_{X,Y}(x, y)$.

EXAMPLE 5.1.6.   Suppose that X and Y are jointly continuous with density function

$$f_{X,Y}(x, y) = 2(x + y - 3xy^2), \qquad 0 < x < 1, \qquad 0 < y < 1$$
$$= 0, \qquad\qquad\qquad\qquad \text{otherwise.}$$

The marginal for X is

$$f_X(x) = \int_0^1 2(x + y - 3xy^2)\, dy = 1, \qquad 0 < x < 1$$
$$= 0, \qquad\qquad\qquad\qquad \text{otherwise;}$$

the marginal for Y is

$$f_Y(y) = \int_0^1 2(x + y - 3xy^2)\, dx = 1 + 2y - 3y^2, \qquad 0 < y < 1$$
$$= 0, \qquad\qquad\qquad\qquad\qquad \text{otherwise.}$$

The distribution function for a two-dimensional random variable is defined quite analogously to its one-dimensional counterpart.

DEFINITION 5.1.4.   The *distribution function* (written $F_{X,Y}(t_1, t_2)$) for a two-dimensional random variable (X, Y) is defined to be $F_{X,Y}(t_1, t_2) = P(X \le t_1, Y \le t_2)$, for all real pairs (t_1, t_2).

Thus, if (X, Y) is discrete, we can derive the distribution function for (X, Y) by summing values of $p_{X,Y}(x, y)$; if (X, Y) is continuous, we can derive its distribution function by integrating $f_{X,Y}(x, y)$. If (X, Y) is a discrete random variable, then $F_{X,Y}(t_1, t_2)$ is a discontinuous function of t_1 and t_2; if (X, Y) is a continuous random variable, then $F_{X,Y}(t_1, t_2)$ is a continuous function of t_1 and t_2.

EXERCISE 5.1

1. A fair coin has a 1 painted on one side and a 2 painted on the other. The coin is flipped twice. Let X be the sum of the two numbers that occur and let Y be

the difference of the two (the first minus the second). Compute $p_{X,Y}(x, y)$, $p_X(x)$ and $p_Y(y)$.

2. The fair coin mentioned in problem 1 is flipped 3 times. Let X be the sum of the first two numbers and let Y be the sum of the last two numbers and compute $p_{X,Y}(x, y)$.

3. Two cards are drawn at random (without replacement) from a standard 52-card bridge deck. Let X be the number of aces that occur and let Y be the number of spades that occur. Derive $p_{X,Y}(x, y)$ and compute $P(X > Y)$.

4. A bag contains 3 black and 7 red pieces of candy. We select 2 pieces without replacement. Let X be the number of black pieces we draw and let Y be the number of red pieces we draw. Derive $p_{X,Y}(x, y)$ and compute $P(X \leq Y)$.

5. Suppose that we select a point at random from the interior of the circle centered at the origin, with radius 1. Let X be the x-coordinate and Y be the y-coordinate of the selected point and derive $f_{X,Y}(x, y)$, $f_X(x)$ and $f_Y(y)$.

6. What must A equal if

$$f_{X,Y}(x, y) = A \frac{x}{y}, \qquad 0 < x < 1, \qquad 1 < y < 2$$

$$= 0, \qquad \text{otherwise}$$

is to be a density function?

7. A family has 2 young boys. Let X be the adult height of the older boy and let Y be the adult height of the younger boy. Suppose (X, Y) is equally likely to fall in the rectangle with corners (66, 68), (66, 72), (71, 68), (71, 72). Compute the probability that the older boy will be taller than the younger as adults.

8. Show that

$$P(a < X \leq b, Y \leq d) = F_{X,Y}(b, d) - F_{X,Y}(a, d).$$

(*Hint:* Draw a picture.)

9. Show that

$$P(a < X \leq b, c < Y \leq d) = F_{X,Y}(b, d) - F_{X,Y}(a, d) - F_{X,Y}(b, c) + F_{X,Y}(a, c).$$

(*Hint:* Use problem 8 and a new picture.)

10. If (X, Y) is a continuous two-dimensional random variable, show that

$$f_{X,Y}(t_1, t_2) = \frac{\partial^2}{\partial t_1 \, \partial t_2} F_{X,Y}(t_1, t_2).$$

11. Assume that (X, Y) is a two-dimensional continuous random variable with density

$$f_{X,Y}(x, y) = \frac{1}{x}, \qquad 0 < y < x, \qquad 0 < x < 1$$

$$= 0, \qquad \text{otherwise}$$

(a) Find the marginal densities for X and Y.

(b) Compare $f_X(x)$ with the marginal density for X in Example 5.1.5.

12. Assume that (X, Y) has density

$$f_{X,Y}(x, y) = \frac{1}{2}$$

for (x, y) inside the square with corners (a, a), $(a, -a)$, $(-a, a)$, $(-a, -a)$ and that $f_{X,Y}(x, y)$ is zero, otherwise.
(a) Find a.
(b) Find the marginal densities for X and Y.

13. Assume that (X, Y) has density

$$f_{X,Y}(x, y) = \frac{1}{2}$$

over the square with corners $(a, 0)$, $(-a, 0)$, $(0, a)$, $(0, -a)$ and that $f_{X,Y}(x, y)$ is zero, otherwise.
(a) Find a.
(b) Find the marginal densities for X and Y and compare them with the marginals found in 12 above.

14. Assume that (X, Y) is a two-dimensional continuous random variable with distribution function

$$F_{X,Y}(t_1, t_2) = 0 \quad \text{if } t_1 < 0 \text{ or } t_2 < 0$$
$$= 1 - e^{-\lambda t_1} - e^{-\lambda t_2} + e^{-\lambda(t_1+t_2)} \quad \text{if } t_1 \geq 0, t_2 \geq 0.$$

Find $f_{X,Y}(t_1, t_2)$. (See Exercise 10 above.)

5.2. Expected Values and Moments

Given an experiment, and a two-dimensional random variable defined on that experiment, it will be useful to refer to the expected value of a function of the two random variables. For example, if two fair dice are rolled (one red and one green) and we let X be the number on the red die, while Y is the number on the green die, we might want to compute the expected value of $H(X, Y) = X + Y$. Again, to get an expected value of a function we simply weight the various values it may equal by their probabilities of occurrence, and sum them all together. The formal definition is given below.

DEFINITION 5.2.1. Suppose that (X, Y) is a two-dimensional random variable and $H(X, Y)$ is a function of X and Y.
(1) If X and Y are jointly discrete, the *expected value* of $H(X, Y)$ is defined to be

$$E[H(X, Y)] = \sum_{\substack{\text{range} \\ \text{of } (X,Y)}} H(x, y)p_{X,Y}(x, y),$$

so long as the sum is absolutely convergent.

(2) If X and Y are jointly continuous, the *expected value* of $H(X, Y)$ is defined to be

$$E[H(X, Y)] = \int_{-\infty}^{\infty} \int_{-\infty}^{\infty} H(x, y) f_{X,Y}(x, y) \, dx \, dy,$$

so long as the integral is absolutely convergent.

Then, if we know the probability law of (X, Y), we can use it to compute the expected value of such things as $X + Y$, $X - Y$, and XY.

EXAMPLE 5.2.1.   A pair of fair dice (1 red, 1 green) is rolled 1 time. Let X be the number on the red die and let Y be the number on the green die. Then

$$p_{X,Y}(x, y) = \tfrac{1}{36}, \qquad x = 1, 2, \ldots, 6, \qquad y = 1, 2, \ldots, 6$$
$$= 0, \qquad \text{otherwise.}$$

The expected value of the sum of the 2 numbers is

$$E[X + Y] = \sum_{x=1}^{6} \sum_{y=1}^{6} (x + y)\tfrac{1}{36} = 7;$$

the expected value of the product of the 2 numbers is

$$E[XY] = \sum_{x=1}^{6} \sum_{y=1}^{6} xy\tfrac{1}{36} = 12\tfrac{1}{4}.$$

EXAMPLE 5.2.2.   People who buy and sell items in a free market must expect fluctuations both in the amount they must pay for an article and in the amount they will be able to sell it for. Suppose, in suitably normalized units, a dealer pays an amount X for an item and then is able to sell the same item for an amount Y; we arbitrarily assume (X, Y) to be continuous (to simplify computations) and that their joint density function is

$$f_{X,Y}(x, y) = 2, \qquad x < y < 1, \qquad 0 < x < 1$$
$$= 0, \qquad \text{otherwise.}$$

The region R where this density is positive is shown in Figure 5.2.1. Notice, then, that the amount he receives will always exceed the amount he paid for the same item. The profit he will make on a single item, then, is $Y - X$, again a random variable (assuming he has no fixed costs to be charged against this type of item). The expected profit per item then is

$$E[Y - X] = \int_{x=0}^{1} \int_{y=x}^{1} (y - x)2 \, dy \, dx$$
$$= \tfrac{2}{3} - \tfrac{1}{3} = \tfrac{1}{3}.$$

The following theorem gives a result which frequently makes the computations of some expected values easier.

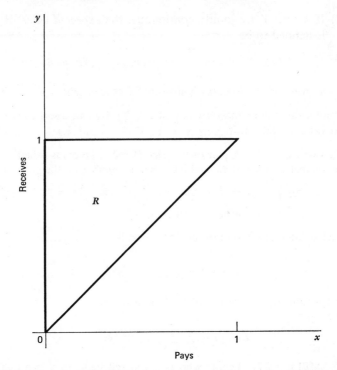

FIGURE 5.2.1   Region R for which $f_{X,Y}(x, y)$ is positive. (See Example 5.2.2.)

Theorem 5.2.1.   If (X, Y) is a two-dimensional random variable, then $E[G(X) + H(Y)] = E[G(X)] + E[H(Y)]$.

Proof: We shall give the proof only for the continuous case. The discrete case follows in exactly the same way, with summation replacing integration. By definition, if X and Y are jointly continuous,

$$E[G(X) + H(Y)] = \int_{-\infty}^{\infty} \int_{-\infty}^{\infty} [G(x) + H(y)]f_{X,Y}(x, y)\, dx\, dy$$

$$= \int_{-\infty}^{\infty} G(x)\left\{\int_{-\infty}^{\infty} f_{X,Y}(x, y)\, dy\right\} dx$$

$$+ \int_{-\infty}^{\infty} H(y)\left\{\int_{-\infty}^{\infty} f_{X,Y}(x, y)\, dx\right\} dy$$

$$= \int_{-\infty}^{\infty} G(x)f_X(x)\, dx + \int_{-\infty}^{\infty} H(y)f_Y(y)\, dy$$

$$= E[G(X)] + E[H(Y)]. \qquad \blacktriangleleft$$

We note immediately then that we can in general compute $E[G(X) + H(Y)]$ from knowledge only of the two marginals $f_X(x)$ and $f_Y(y)$; we do not need to know the joint density function of (X, Y). This theorem pertains only to additive functions, not to multiplicative ones. In general, we find

$$E[G(X)H(Y)] \neq E[G(X)]E[H(Y)].$$

(An exception to this occurs in the case of independent random variables; see Section 5.3.)

In Chapter 3 we discussed the moments of a single random variable X and found them useful in describing the distribution for X. We shall also find that the moments of two-dimensional (and higher) random variables give useful descriptions of their distributions. The joint moments of (X, Y) are defined as follows.

DEFINITION 5.2.2. The ij-th *joint moment* of (X, Y) is defined to be

$$m_{ij} = E[X^i Y^j], \qquad i = 0, 1, 2, \ldots, \qquad j = 0, 1, 2, \ldots$$

(i and j are not simultaneously zero).
Then, for example,

$$m_{11} = E[XY], \qquad m_{02} = E[X^0 Y^2] = E[Y^2]$$

$$m_{12} = E[XY^2], \qquad \text{etc.}$$

Note in particular that m_{i0}, $i = 1, 2, 3, \ldots$, are simply the moments of X and m_{0j}, $j = 1, 2, 3, \ldots$, are the moments of Y. Thus the joint moments of X and Y include as special cases the moments of X and the moments of Y; these special moments then describe the marginals of X and of Y, respectively. The mixed moments m_{ij}, for which neither i nor j is zero, measure various aspects of the joint distribution of X and Y. Just as in the one-dimensional case, we can define a function that generates the moments of (X, Y) through differentiation; this is called the moment generating function for (X, Y) and is defined as follows.

DEFINITION 5.2.3. The function

$$m_{X,Y}(t_1, t_2) = E[e^{t_1 X + t_2 Y}]$$

is called the *moment generating* function for (X, Y) (so long as the expectation exists).

That this function does generate moments may be seen by taking partial derivatives of $m_{X,Y}$ with respect to t_1 and t_2 and then evaluating these derivatives at the origin. Thus,

$$\frac{\partial}{\partial t_1} m_{X,Y}(t_1, t_2) = \frac{\partial}{\partial t_1} E[e^{t_1 X + t_2 Y}]$$

$$= E[X e^{t_1 X + t_2 Y}]$$

$$\frac{\partial}{\partial t_2} m_{X,Y}(t_1, t_2) = E[Y e^{t_1 X + t_2 Y}]$$

and, in general,

$$\frac{\partial^{i+j}}{\partial t_1^i \, \partial t_2^j} m_{X,Y}(t_1, t_2) = E[X^i Y^j e^{t_1 X + t_2 Y}].$$

And by setting $(t_1, t_2) = (0, 0)$, we see that we do generate the moments of (X, Y).

EXAMPLE 5.2.3. Suppose that (X, Y) is continuous with density function

$$f_{X,Y}(x, y) = e^{-x-y}, \qquad x > 0, \qquad y > 0$$

$$= 0, \qquad \text{otherwise.}$$

The moment generating function for (X, Y) is

$$m_{X,Y}(t_1, t_2) = E[e^{t_1 X + t_2 Y}]$$

$$= \int_0^\infty \int_0^\infty e^{t_1 x + t_2 y} e^{-x-y} \, dx \, dy$$

$$= \int_0^\infty e^{-x(1-t_2)} \, dx \int_0^\infty e^{-y(1-t_2)} \, dy$$

$$= \frac{1}{(1 - t_1)(1 - t_2)}.$$

Then for example,

$$\frac{\partial}{\partial t_1} m_{X,Y}(t_1, t_2) = \frac{1}{(1 - t_1)^2(1 - t_2)}$$

$$\frac{\partial}{\partial t_2} m_{X,Y}(t_1, t_2) = \frac{1}{(1 - t_1)(1 - t_2)^2}$$

$$\frac{\partial^2}{\partial t_1 \, \partial t_2} m_{X,Y}(t_1, t_2) = \frac{1}{(1 - t_1)^2(1 - t_2)^2},$$

so that $m_{10} = 1$, $m_{01} = 1$, $m_{11} = 1$.

EXAMPLE 5.2.4. Suppose that (X, Y) is continuous with density

$$f_{X,Y}(x, y) = x + y, \qquad 0 < x < 1, \qquad 0 < y < 1$$

$$= 0, \qquad \text{otherwise.}$$

Let us find the moments of (X, Y) directly, without using the moment generating function. By definition

$$m_{ij} = E[X^i Y^j]$$

$$= \int_0^1 \int_0^1 x^i y^j (x + y) \, dx \, dy$$

$$= \frac{1}{(i + 2)(j + 1)} + \frac{1}{(i + 1)(j + 2)}, \quad i = 0, 1, 2, \ldots, \quad j = 0, 1, 2, \ldots$$

Thus,

$$E[X] = \frac{1}{3 \cdot 1} + \frac{1}{2 \cdot 2} = \frac{7}{12}$$

$$E[XY] = \frac{1}{3 \cdot 2} + \frac{1}{2 \cdot 3} = \frac{1}{3}$$

$$E[X^2] = \frac{1}{4 \cdot 1} + \frac{1}{3 \cdot 2} = \frac{5}{12},$$

so that $\sigma_X^2 = E[X^2] - (E[X])^2 = \frac{11}{144}$.

EXAMPLE 5.2.5. Suppose that (X, Y) is discrete with probability function

$$p_{X,Y}(x, y) = \tfrac{1}{3}, \quad \text{for } (x, y) = (0, 0), (-1, 0), (0, 1)$$

$$= 0, \quad \text{otherwise.}$$

Then the ij-th moment of (X, Y) is

$$m_{i0} = E(X^i) = \sum_{\substack{\text{range} \\ \text{of } (X,Y)}} x^i p_{X,Y}(x, y)$$

$$= (-1)^i (\tfrac{1}{3}), \quad \text{for } i = 1, 2, 3, \ldots.$$

$$m_{0j} = E(Y^j) = \sum_{\substack{\text{range} \\ \text{of } (X,Y)}} y^j p_{X,Y}(x, y)$$

$$= \tfrac{1}{3}, \quad \text{for } j = 1, 2, 3, \ldots.$$

$$m_{ij} = E(X^i Y^j) = \sum_{\substack{\text{range} \\ \text{of } (X,Y)}} x^i y^j p_{X,Y}(x, y)$$

$$= 0, \quad \text{for all } i = 1, 2, 3, \ldots, \quad j = 1, 2, 3, \ldots.$$

Thus $\mu_X = -\tfrac{1}{3}$, $\mu_Y = \tfrac{1}{3}$, $\sigma_X^2 = \sigma_Y^2 = \tfrac{2}{9}$.

The most frequently used mixed moment is m_{11} which occurs in the definition of both the covariance and the correlation of (X, Y). The covariance is defined as follows.

DEFINITION 5.2.4. The *covariance* of X and Y is defined to be

$$\sigma_{XY} = E[(X - \mu_X)(Y - \mu_Y)].$$

The covariance gives a measure of how X and Y tend to vary together. Since it is the expected value of the product of the deviation of X from its mean times the deviation of Y from its mean, we can see that σ_{XY} will be positive if, when X is larger than its mean, so is Y; if $X - \mu_X$ and $Y - \mu_Y$ have opposite signs with high probability, then σ_{XY} will be negative. If $X - \mu_X$ is negative as frequently as it is positive when $Y - \mu_Y$ is positive or negative, we would expect $\sigma_{XY} = 0$. It is in this general sense that σ_{XY} measures the covariability of X and Y. The actual magnitude of σ_{XY} does not have much meaning without knowledge of σ_X^2 and σ_Y^2 as well; for one probability law we might have $\sigma_{XY} = 1$ and for another $\sigma_{XY} = 100$. We cannot say that the tendency for X and Y to vary together is greater in the latter case than in the former because X and Y themselves might be extremely variable in the latter case, which tends to increase the numerical value of σ_{XY}. The correlation coefficient ρ_{XY}, defined below, gives a measure of covariability which also accounts for the variances of X and of Y.

DEFINITION 5.2.5. The *correlation* between X and Y is defined to be

$$\rho_{XY} = \frac{\sigma_{XY}}{\sigma_X \sigma_Y}$$

(ρ_{XY} is called the correlation coefficient of X and Y.)

A simpler computational formula for σ_{XY} (and thus also for ρ_{XY}) than is given by the definition is derived as follows.

$$\begin{aligned}
\sigma_{XY} &= E[(X - \mu_X)(Y - \mu_Y)] \\
&= E[XY - \mu_X Y - X\mu_Y + \mu_X \mu_Y] \\
&= E[XY] - \mu_X E[Y] - \mu_Y E[X] + \mu_X \mu_Y \\
&= E[XY] - \mu_X \mu_Y.
\end{aligned}$$

Note the similarity between this equation and the computational formula for a variance derived in Chapter 3.

EXAMPLE 5.2.6. For the random variables X and Y discussed in Example 5.2.3, we found $E[XY] = E[X] = E[Y] = 1$ and thus, for that case, $\sigma_{XY} = 1 - 1 \cdot 1 = 0$ so that $\rho_{XY} = 0$. For the random variables discussed in Example 5.2.4, we find that $E[XY] = \frac{1}{3}$, $E[X] = E[Y] = \frac{7}{12}$, $E[X^2] = E[Y^2] = \frac{5}{12}$ so that

$$\sigma_{XY} = \tfrac{1}{3} - \tfrac{49}{144} = -\tfrac{1}{144}$$

$$\sigma_X^2 = \tfrac{5}{12} - \tfrac{49}{144} = \tfrac{11}{144} = \sigma_Y^2$$

$$\rho_{XY} = \frac{-\tfrac{1}{144}}{\sqrt{\tfrac{11}{144} \cdot \tfrac{11}{144}}} = -\tfrac{1}{11}.$$

For random variables X and Y discussed in Example 5.2.5, we have $\sigma_{XY} = 0 - (-\frac{1}{3})(\frac{1}{3}) = \frac{1}{9}$, $\sigma_X^2 = \sigma_Y^2 = \frac{2}{9}$, $\rho_{XY} = \frac{1}{2}$.

The reader will note that, of the three correlation coefficients derived in Example 5.2.6, one was negative, one was positive, and one was zero. That $|\rho_{XY}| \leq 1$ for all random variables is proved as Theorem 5.2.2.

Theorem 5.2.2. For any random variables X and Y, $-1 \leq \rho_{XY} \leq 1$.

Proof: The proof is a special case of the Schwarz inequality which finds application in many branches of mathematics. Suppose that W and Z are random variables. Then $(aW - Z)^2$ is a random variable which takes on only nonnegative values, no matter what the (real) value of a. Then its mean is nonnegative:

$$E[(aW - Z)^2] \geq 0 \cdot$$

that is,

$$E[a^2 W^2 - 2aWZ + Z^2] \geq 0$$

so that

$$a^2 E[W^2] - 2aE[WZ] + E[Z^2] \geq 0, \qquad \text{for all } a.$$

If in particular

$$a = \frac{E[WZ]}{E[W^2]},$$

this becomes

$$\frac{-(E[WZ])^2}{E[W^2]} + E[Z^2] \geq 0;$$

then

$$1 \geq \frac{(E[WZ])^2}{E[W^2]E[Z^2]}.$$

If we now let $W = X - \mu_X$, $Z = Y - \mu_Y$, we have

$$1 \geq \frac{(E[(X - \mu_X)(Y - \mu_Y)])^2}{E[(X - \mu_X)^2]E[(Y - \mu_Y)^2]};$$

that is

$$1 \geq \frac{\sigma_{XY}^2}{\sigma_X^2 \sigma_Y^2} = \rho_{XY}^2,$$

which implies $|\rho_{XY}| \leq 1$.   ◀

If in fact $|\rho_{XY}| = 1$, then there is a perfect linear relation between X and Y; i.e., Y is a linear function of X. If the equality holds in the proof of Theorem 5.2.2, then $E[(aW - Z)^2] = 0$. But the mean value of a nonnegative random variable can only be zero if the random variable is equal to zero with probability 1. That is, we must have $P[(aW - Z)^2 = 0] = 1$

which implies that $P[aW - Z = 0] = 1$, or $P[Z = aW] = 1$, and thus

$$P[Y - \mu_Y = a(X - \mu_X)] = 1$$
$$P[Y = \mu_Y - a\mu_X + aX] = 1,$$

as was claimed. The fact that $Y = \mu_Y - a\mu_Y + aX$ with probability 1 means that in the (x, y) plane of possible values for (X, Y), only those points lying on $y = \mu_Y - a\mu_X + ax$ can have positive probability; all points off this line have probability zero. Since the sign of ρ_{XY} is determined by the sign of

$$\sigma_{XY} = E[(X - \mu_X)(Y - \mu_Y)] = aE[(X - \mu_X)^2],$$

we find that $\rho_{XY} = 1$ if a is positive and $\rho_{XY} = -1$ if a is negative (see Figure 5.2.2). The "closer" that Y is to a linear function of X, that is, the

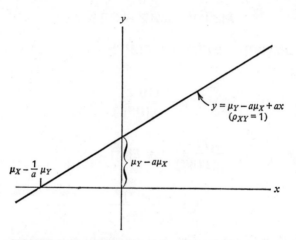

FIGURE 5.2.2 Y is a linear function of X, $a > 0$.

"closer" that all of the probability is to lying on a straight line in the (x, y) plane, the closer that $|\rho_{XY}|$ is to 1. (See problems 7 and 8 below.) On the other hand, $\rho_{XY} = 0$ only if $\sigma_{XY} = 0$, which would imply the the co-variability of X and Y is zero. (That is, X and Y do not tend to vary together.) Note, however, that σ_{XY} is a measure of only linear covariability and it is in fact possible that Y can be completely determined by X (nonlinearly) and still the correlation is zero. Example 5.2.7 presents a case of this sort.

EXAMPLE 5.2.7.   Suppose that X is a discrete random variable with probability function

$$p_X(x) = \tfrac{1}{4}, \quad \text{at } x = -2, -1, 1, 2$$
$$= 0, \quad \text{otherwise.}$$

Define $Y = X^2$ and then Y certainly is a function of X and all the probability will lie on a line in the (x, y) plane but not on a straight line; in fact, we have

$$p_{Y,Y}(x, y) = \tfrac{1}{4}, \qquad \text{at } (x, y) = (-2, 4), (-1, 1), (1, 1), (2, 4)$$

$$= 0, \qquad \text{otherwise.}$$

We find $\mu_X = 0$, $\mu_Y = 2\tfrac{1}{2}$, $E[XY] = 0$ and thus $\sigma_{XY} = 0$ and $\rho_{XY} = 0$, despite the fact that Y is a quadratic function of X.

EXERCISE 5.2

1. Suppose that $+1$ is painted on one side of a fair coin and -1 is painted on the other. The coin is flipped 2 times; let X be the number that occurs on the first flip and let Y be the number that occurs on the second. Derive $p_{X,Y}(x, y)$, μ_X, μ_Y and $E\left[\dfrac{X}{Y}\right]$.

2. Students at a large university are required to take an entrance exam when entering the school (scored on a basis of 100 points possible). Let X be the score made by a particular incoming student (who does go on to graduate) and let Y be his graduating quality point ratio (4 points $= A$). The joint density function of X and Y is observed to be

$$f_{X,Y}(x, y) = \tfrac{1}{50}, \qquad \text{for } 2 < y < 4, \quad 25(y - 1) < x < 25y$$

$$= 0, \qquad \text{otherwise.}$$

   Find μ_X, μ_Y and $m_{11} = E(XY)$.

3. In problem 2, compute σ_X^2, σ_Y^2, σ_{XY} and ρ_{XY}.

4. Given

$$p_{W,Z}(w, z) = \frac{1}{n}, \qquad \text{for } (w, z) = (1, 1), (2, 2), \ldots, (n, n)$$

$$= 0, \qquad \text{otherwise,}$$

   evaluate $E\left[\dfrac{W}{Z}\right], E\left[\dfrac{Z}{W}\right], E[W^2 + Z^2]$, and $E\left[\dfrac{W^3}{Z^2}\right]$.

5. Given

$$f_{U,V}(u, v) = 6(1 - u - v), \qquad 0 < u < 1, \quad 0 < v < 1 - u$$

$$= 0, \qquad \text{otherwise,}$$

   compute $E[U]$, $E[V]$, $E[UV]$ and σ_{UV}.

6. If X and Y are jointly distributed random variables, show that $m_{X,Y}(t, 0) = m_X(t)$ and $m_{X,Y}(0, t) = m_Y(t)$.

7. X and Y are jointly discrete random variables with probability function

$$p_{X,Y}(x, y) = \tfrac{1}{4}, \qquad \text{at } (x, y) = (-3, -5), (-1, -1), (1, 1), (3, 5)$$
$$= 0, \qquad \text{otherwise.}$$

Compute ρ_{XY}.

8. Suppose that X and Y are jointly discrete with probability function

$$p_{X,Y}(x, y) = \tfrac{1}{4}, \qquad \text{at } (x, y) = (-3, -4), (-1, -1), (1, 1), (3, 4)$$
$$= 0, \qquad \text{otherwise.}$$

Compute ρ_{XY}. (Compare the plot of points with positive probability for this Exercise versus Exercise 7. Note that the correlation is higher in Exercise 8 because the points with positive probability are closer to lying on a straight line.)

9. Define a function $\psi_{X,Y}(t_1, t_2)$ which would generate factorial moments for (X, Y). (Note that you should have $\psi_{X,Y}(t, 1) = \psi_X(t)$ and $\psi_{X,Y}(1, t) = \psi_Y(t)$.)

10. (a) Find ρ_{XY} if X and Y have the joint density given in Exercise 5.1.12.
    (b) Find ρ_{XY} if X and Y have the joint density given in Exercise 5.1.13.

11. Assume (X, Y) has the probability function

$$p_{X,Y}(x, y) = \tfrac{1}{3} \quad \text{at } (x, y) = (-1, -1), (1, 1), (a, b).$$

(a) Find $\mu_X, \mu_Y, \sigma_X^2, \sigma_Y^2, \sigma_{XY}$.
(b) Show that $\rho_{XY} = 1$ if $a = b$.

12. A fair coin is flipped 3 times. Let X be the number of heads that occur and let Y be the number of tails that occur. Find the joint probability function for (X, Y) and evaluate ρ_{XY}.

13. Assume that X and Y are jointly distributed random variables and define $U = 2X, V = 2Y$.
    (a) Show that $\mu_U = 2\mu_X, \mu_V = 2\mu_Y$.
    (b) Show that $\sigma_U^2 = 4\sigma_X^2, \sigma_V^2 = 4\sigma_Y^2, \sigma_{UV} = 4\sigma_{XY}$.
    (c) Show that $\rho_{UV} = \rho_{XY}$.

14. Assume (X, Y) has the density $f_{X,Y}(x, y) = c$ for (x, y) in the quadrilateral with corners $(0, 0), (1, 1), (a, 1 - a), (1 - a, a)$, where $0 \le a \le \tfrac{1}{2}$.
    (a) Find c.
    (b) Evaluate ρ_{XY} if $a = 0$ and if $a = \tfrac{1}{2}$.

15. As in Example 5.2.2, let X be the amount paid for an item and let Y be the amount received for the same item. Assume (X, Y) is discrete with probability function

$$p_{X,Y}(x, y) = \tfrac{1}{24}, \qquad (x, y) = (\tfrac{1}{4}, \tfrac{1}{2}), (\tfrac{1}{4}, \tfrac{3}{4}), (\tfrac{1}{4}, 1),$$
$$(1, 1\tfrac{1}{4}), (1, 1\tfrac{1}{2}), (1, 1\tfrac{3}{4})$$
$$= \tfrac{1}{8}, \qquad (x, y) = (\tfrac{1}{2}, \tfrac{3}{4}), (\tfrac{1}{2}, 1), (\tfrac{1}{2}, 1\tfrac{1}{4})$$
$$(\tfrac{3}{4}, 1), (\tfrac{3}{4}, 1\tfrac{1}{4}), (\tfrac{3}{4}, 1\tfrac{1}{2}).$$

Find the expected profit in buying and selling one of these items.

16. Assume X and Y are jointly distributed random variables and define $Z = X + Y$. Show that
    (a) $\mu_Z = \mu_X + \mu_Y$
    (b) $\sigma_Z^2 = \sigma_X^2 + \sigma_Y^2 + 2\sigma_{XY}$

17. Let X and Y be jointly distributed random variables and define $U = X + Y$, $V = X - Y$. Show that $\rho_{UV} = 0$ if $\sigma_X^2 = \sigma_Y^2$.

18. In 17, assume $\sigma_{XY} = 0$ and show that

$$\rho_{UV} = \frac{\sigma_X^2 - \sigma_Y^2}{\sigma_X^2 + \sigma_Y^2}.$$

5.3. Conditional Distributions and Independent Random Variables

Given a pair of events, A and B, recall that we discussed the conditional probability of one event occurring, given that the other had occurred. This idea of conditional probability statements can be extended to the conditional distribution of a random variable X, given that a second random variable Y has taken on a particular value y, say. Indeed, assume X and Y are jointly discrete random variables and let A be the event $X = x$, while B is the event $Y = y$. Then $P(A \mid B) = \dfrac{P(A \cap B)}{P(B)}$, that is

$$P(X = x \mid Y = y) = \frac{P(X = x, Y = y)}{P(Y = y)}$$

$$= \frac{p_{X,Y}(x, y)}{p_Y(y)}.$$

The ratio $p_{X,Y}(x, y)/p_Y(y)$ is called the conditional probability function for X, given that $Y = y$; it would be appropriate for evaluating any conditional probability statements about X, given that Y has been observed to equal y. The following definition defines both conditional probability functions (for the discrete case) and conditional density functions (for the continuous case).

DEFINITION 5.3.1. (1) If X and Y are jointly discrete random variables, the *conditional probability function* for Y, given $X = x$, is defined to be

$$p_{Y|X}(y \mid x) = \frac{p_{X,Y}(x, y)}{p_X(x)}, \qquad \text{for } p_X(x) > 0.$$

$p_{X|Y}(x \mid y)$ is defined analogously.

(2) If X and Y are jointly continuous random variables, the *conditional density* for Y, given $X = x$, is defined to be

$$f_{Y|X}(y \mid x) = \frac{f_{X,Y}(x, y)}{f_X(x)}, \qquad \text{for } f_X(x) > 0.$$

$f_{X|Y}(x \mid y)$ is defined analogously.

The first part of this definition, as discussed above, follows quite naturally from the conditional probability of one event happening, given that another has occurred. To see that the second part of the definition, for the continuous case, is also reasonable, let us assume (X, Y) is continuous and that its joint density function is positive over a region R. Then, if we are given that $X = 1$, say, we are certain that (X, Y) has taken on a value somewhere on the line $x = 1$ in the plane (the y-coordinate is still unspecified). To compute the conditional probability that Y lies between a and b, say, it would seem reasonable to refer to the joint density, $f_{X,Y}(x, y)$, over the line $x = 1$, that is to examine $f_{X,Y}(1, y)$. This will in general be a 1-dimensional function of y. The total area under $f_{X,Y}(1, y)$ is

$$\int_{-\infty}^{\infty} f_{X,Y}(1, y) \, dy = f_X(1),$$

the marginal density of X evaluated at 1. Then if we form the ratio

$$f_{X,Y}(1, y)/f_X(1)$$

we have a 1-dimensional density which parallels the joint density over the line $x = 1$. This ratio is the conditional density of Y given $X = 1$, from the above definition. It is used to evaluate conditional probability statements about Y, given $X = 1$. The same line of reasoning is appropriate whatever value we are given for X, so long as it is a number x such that $f_X(x) > 0$ (since these are the only values that could be observed for X). The conditional probability that Y lies between a and b, then, given $X = x$, is equal to the area under the joint density over the line $X = x$, normalized by the total area under this curve.

EXAMPLE 5.3.1.   Assume that during basketball practice, Joe tries three shots from the free throw line; assume further that he has probability .6 of making each shot and that the attempts are independent. Let X be the number of baskets he makes in the first two shots and let Y be the number of baskets he makes in the last two shots. Then, the marginal distributions for X and Y are each binomial with $n = 2$ and $p = .6$. The joint distribution for X and Y can be constructed from Table 5.3.1. The joint probability function for X and Y is given in Table 5.3.2,

TABLE 5.3.1
Probability Assignments for Example 5.3.1

Outcome	HHH	HHM	HMH	HMM	MHH	MHM	MMH	MMM
Probability	$(.6)^3$	$(.6)^2(.4)$	$(.6)^2(.4)$	$(.6)(.4)^2$	$(.6)^2(.4)$	$(.6)(.4)^2$	$(.6)(.4)^2$	$(.4)^3$
Value for X	2	2	1	1	1	1	0	0
Value for Y	2	1	1	0	2	1	1	0

TABLE 5.3.2
Joint Probability Function for X and Y

		x			
		0	1	2	
	0	.064	.096	0	.160
y	1	.096	.240	.144	.480
	2	0	.144	.216	.360
		.160	.480	.360	

and the conditional probability functions for Y, given the various possible values for X, are given in Table 5.3.3. Notice that the conditional probability function for Y is different for the different possible given values of x; in particular, the conditional probability that he makes exactly one of his last two shots is maximized (at .6) if we are given that he missed both of his first two shots. We can also see

TABLE 5.3.3
Conditional Probability Functions for
Y given $X = x$

		x		
		0	1	2
	0	.4	.2	0
y	1	.6	.5	.4
	2	0	.3	.6
		1.0	1.0	1.0

that the probability he makes both his last two shots, given he made exactly one of the first two, is only .3. Does this say, having observed Joe miss his first shot and make his second, that he has only probability .3 of making his third shot?

EXAMPLE 5.3.2.   Suppose U and V are jointly continuous random variables with density

$$f_{U,V}(u, v) = \frac{1}{\pi}, \qquad u^2 + v^2 \leq 1$$

$$= 0, \qquad \text{otherwise.}$$

Then we say that (U, V) is uniformly distributed on the circle with radius 1 centered at the origin. We find then that the marginal density for U is

$$f_U(u) = \frac{2\sqrt{1 - u^2}}{\pi} \quad \text{for } -1 < u \leq 1$$

$$= 0, \qquad \text{otherwise,}$$

and thus the conditional density of V, given $U = u$, is

$$f_{V\,|\,U}(v \mid u) = f_{U,V}(u, v)/f_U(u)$$

$$= \frac{1}{2\sqrt{1 - u^2}}, \quad -\sqrt{1 - u^2} < v < \sqrt{1 - u^2}, \; -1 < u < 1.$$

That is, the conditional density of V given $U = u$ is uniform, where the length of the interval that V is distributed over depends on the given value u. If $u = \frac{1}{2}$, the conditional density for V is uniform over the interval $\left(-\dfrac{\sqrt{3}}{2}, \dfrac{\sqrt{3}}{2}\right)$ whereas if $u = .8$, the conditional density for V is uniform over the interval $(-.6, .6)$.

The mean of the conditional distribution for Y, given $X = x$, is denoted by $\mu_{Y|X}$; this conditional mean is called the *regression* of Y on X. It may or may not be a function of x, the given value of X, even if the conditional distribution of Y given X depends on x. Note in Example 5.3.1, above, that $\mu_{Y|X} = .6 + x/2$ and thus the regression of Y on X is linear in that case. In Example 5.3.2, we find $\mu_{V|U} = 0$, no matter what the given value of u, even though the conditional density does depend on the given value for u. In general, the regression of one variable on another may or may not depend on the value of the given variable; if it does depend on it, the dependence may or may not be linear. In any case, if $\mu_{Y|X}$ is averaged with respect to the distribution of X (acting as though the given value for X is a random variable), the result is $E[Y]$, the unconditional mean value for Y, since

$$E_X[E[Y \mid X = x]] = \int_{-\infty}^{\infty} \mu_{Y|X} f_X(x) \, dx$$

$$= \int_{-\infty}^{\infty} \int_{-\infty}^{\infty} \frac{y f_{X,Y}(x, y) \, dy}{f_X(x)} f_X(x) \, dx$$

$$= \int_{-\infty}^{\infty} \int_{-\infty}^{\infty} y f_{X,Y}(x, y) \, dy \, dx$$

$$= E[Y].$$

The same result obtains in the discrete case. The following example illustrates the construction of the joint distribution for X and Y, given the conditional distribution of Y given X and the marginal for X, as well as the above fact that the average value of the regression of Y on X is the unconditional mean value for Y.

EXAMPLE 5.3.3. This example illustrates how conditional probability functions and marginal probability functions can be used to construct a joint probability function as well as the computation of the regression of one variable on the other. Suppose the number of telephone calls that a busy executive receives per hour is a Poisson random variable X with mean $\lambda = 10$. Also, this executive is busy with internal office affairs and, on the average, is away from his phone 15 minutes out of every hour. Thus he is willing to assume that if a call comes in for him the probability is $\frac{45}{60} = \frac{3}{4}$ that he will receive it. Furthermore, he is willing to assume that the fact he does or does not receive any particular call has no effect on whether or not he receives any other call (the receipts are independent). This last assumption would only be reasonable if he were popping in and out of the office for very short periods of time; if he were gone for relatively large blocks of time, he would certainly miss all arriving calls in that period and their receipts or nonreceipts would not be independent.

With this independence assumption, it is clear that Y, the number of calls he receives per hour, given that $X = x$ (x calls were made to him), is a binomial random variable with parameters x and $p = \frac{3}{4}$. Thus,

$$p_{Y|X}(y \mid x) = \binom{x}{y} \left(\frac{3}{4}\right)^x \left(\frac{1}{4}\right)^{x-y}, \qquad y = 0, 1, 2, \ldots, x.$$

The mean number of calls he would receive per hour then, given x were made, is the regression of Y on X and we obviously have

$$E[Y \mid X = x] = \tfrac{3}{4}x.$$

The joint probability function for X and Y is

$$
\begin{aligned}
p_{X,Y}(x, y) &= p_{Y|X}(y \mid x) p_X(x) \\
&= \binom{x}{y} \left(\frac{3}{4}\right)^y \left(\frac{1}{4}\right)^{x-y} \frac{10^x}{x!} e^{-10}, \quad x = 0, 1, 2, \ldots; \quad y = 0, 1, \ldots, x \\
&= \frac{1}{y!} \left(\frac{30}{4}\right)^y e^{-10} \frac{1}{(x-y)!} \left(\frac{10}{4}\right)^{x-y}.
\end{aligned}
$$

Then, summing to get the marginal for y,

$$
\begin{aligned}
p_Y(y) &= \sum_{x=y}^{\infty} \frac{1}{y!} \left(\frac{30}{4}\right)^y e^{-10} \frac{1}{(x-y)!} \left(\frac{10}{4}\right)^{x-y} \\
&= \frac{1}{y!} \left(\frac{30}{4}\right)^y e^{-10} \sum_{x-y=0}^{\infty} \frac{1}{(x-y)!} \left(\frac{10}{4}\right)^{x-y} \\
&= \frac{1}{y!} \left(\frac{30}{4}\right)^y e^{-10} e^{10/4} \\
&= \frac{1}{y!} \left(\frac{30}{4}\right)^y e^{-30/4}, \qquad y = 0, 1, 2, \ldots.
\end{aligned}
$$

Thus, Y itself is a Poisson random variable with parameter $\frac{30}{4} = 7\frac{1}{2}$ and we have

$$E[Y] = 7\frac{1}{2}.$$

As was shown to always be the case above,

$$\begin{aligned} E[E[Y \mid X]] &= E[\tfrac{3}{4}X] \\ &= \tfrac{3}{4}E[X] \\ &= 7\tfrac{1}{2}. \end{aligned}$$

It will be recalled from Chapter 2 that, if $P(A \mid B) = P(A)$, then the two events A and B are independent. In a similar way it may happen that the conditional density (or probability function) for Y is equal to the marginal density (or probability function) for Y, no matter what the given value for X; thus, in such a case the knowledge that X has taken on a particular value does not change or affect the probability distribution of Y. As with the case of two events, we shall say that the two random variables are independent. The definition of independent random variables is given below.

DEFINITION 5.3.2.   (1) X and Y will be called *independent* discrete random variables if and only if

$$p_{X,Y}(x, y) = p_X(x)p_Y(y), \qquad \text{for all } (x, y).$$

(2) X and Y will be called *independent* continuous random variables if and only if

$$f_{X,Y}(x, y) = f_X(x)f_Y(y), \qquad \text{for all } (x, y).$$

Notice first of all that the idea of independence of X and Y requires that the joint density or probability function of X and Y will factor into the product of the marginals. Thus, if X and Y are independent, knowledge of their marginals is equivalent to knowledge of their joint distribution, a statement which is not generally true for all random variables as we have seen. In practice, then, if we are able to ascertain that the two random variables are independent, we need only investigate their individual marginal distributions to be able to construct their joint distribution.

It follows then, that if X and Y are independent

$$\begin{aligned} f_{X \mid Y}(x \mid y) &= f_{X,Y}(x, y)/f_Y(y) = f_X(x)f_Y(y)/f_Y(y) \\ &= f_X(x); \end{aligned}$$

that is, the conditional density of X, given $Y = y$, is equal to its marginal density. The analogous result is true for the discrete case.

As we shall see, independent random variables play a central role in most statistical methods and thus this definition will be frequently used in the remainder of this book.

EXAMPLE 5.3.4. If X and Y are jointly discrete with probability function

$$p_{X,Y}(x, y) = \tfrac{1}{4}, \quad \text{for } (x, y) = (-1, -1), (-1, 1), (1, -1), (1, 1)$$
$$= 0, \quad \text{otherwise,}$$

then

$$p_X(x) = \tfrac{1}{2}, \quad \text{for } x = -1, 1$$
$$= 0, \quad \text{otherwise}$$

and

$$p_Y(y) = \tfrac{1}{2}, \quad \text{for } y = -1, 1$$
$$= 0, \quad \text{otherwise.}$$

Thus

$$p_{X,Y}(x, y) = p_X(x)p_Y(y), \quad \text{for all } (x, y)$$

and X and Y are independent. However, if

$$p_{X,Y}(x, y) = \tfrac{1}{4}, \quad \text{for } (x, y) = (-1, -1), (-1, 1), (1, -1), (1, 2),$$
$$= 0, \quad \text{otherwise,}$$

then

$$p_X(x) = \tfrac{1}{2}, \quad x = -1, 1$$
$$= 0, \quad \text{otherwise}$$

and

$$p_Y(y) = \tfrac{1}{2}, \quad y = -1$$
$$= \tfrac{1}{4}, \quad y = 1, 2$$
$$= 0, \quad \text{otherwise.}$$

In this case X and Y are not independent since $p_{X,Y}(1, 1) \neq p_X(1)p_Y(1)$. (We have found one particular point in the plane for which the joint probability function does not equal the product of the marginals; this is sufficient to show that they are not independent.)

EXAMPLE 5.3.5. We saw in Example 5.3.2 above that, if U and V are uniformly distributed on the circle with radius 1, centered at the origin, then

$$f_U(u) = \frac{2\sqrt{1 - u^2}}{\pi} \text{ for } -1 < u < 1$$
$$= 0, \quad \text{otherwise.}$$

Because of the symmetry (in u and v) of the joint density, $f_V(v) = f_U(v)$ and thus U and V are not independent random variables. If, on the other hand, U and V are uniformly distributed on the square with corners $(u, v) = (-1, -1)$, $(-1, 1)$, $(1, -1)$, $(1, 1)$, it is easily verified that U and V are independent.

One of the most frequent uses of the concept of independent random variables is the fact that the joint distribution is given by the product of the marginals. Thus, if it seems certain that the random variables must be independent, knowledge of the marginal distributions is sufficient to specify the values of any probabilities of interest. The following example illustrates this type of idea.

EXAMPLE 5.3.6.   One electronic device in a missile consists of 2 integrated circuits; in order that the missile perform successfully, it is necessary that this particular device operate correctly for 2 minutes. Assume that the 2 circuits in the device operate independently (that is, the failure of the first circuit has no effect on the second and vice versa). Let S be the time to failure of circuit 1 and T be the time to failure of circuit 2. It is known that S is an exponential random variable with parameter λ and T is an exponential random variable with parameter β. Then, since the 2 integrated circuits operate independently in the electronic device, the joint distribution of the 2 failure times is

$$
\begin{aligned}
f_{S,T}(s, t) &= f_S(s)f_T(t) \\
&= \lambda e^{-\lambda s}\beta e^{-\beta t}, \qquad s > 0, \qquad t > 0 \\
&= 0, \qquad\qquad\quad \text{otherwise.}
\end{aligned}
$$

If we are given that the device fails if either of the 2 circuits fail, then the probability that the device operates correctly for at least 2 minutes is

$$
\begin{aligned}
P(S > 2, T > 2) &= \int_2^\infty \int_2^\infty \lambda e^{-\lambda s}\beta e^{-\beta t}\, ds\, dt \\
&= e^{-2\lambda} \cdot e^{-2\beta}\,.
\end{aligned}
$$

If λ and β were known, we could then get a numerical answer for this probability; the way in which we might realistically gather information on λ and β is a statistical question taken up in succeeding chapters.

Theorem 5.3.1 gives a very important consequence of the independence of two random variables.

Theorem 5.3.1.   If X and Y are independent random variables, then $E[H(X)G(Y)] = E[H(X)]E[G(Y)]$ where H and G are any two functions of a real variable.

Proof: Let us indicate the proof for jointly continuous random variables. If X and Y are jointly discrete, the reasoning is identical, integrations being replaced by summations. By definition,

$$
\begin{aligned}
E[H(X)G(Y)] &= \int_{-\infty}^\infty \int_{-\infty}^\infty H(x)G(y)f_{X,Y}(x, y)\, dx\, dy \\
&= \int_{-\infty}^\infty \int_{-\infty}^\infty H(x)G(y)f_X(x)f_Y(y)\, dx\, dy \\
&= \left\{\int_{-\infty}^\infty H(x)f_X(x)\, dx\right\}\left\{\int_{-\infty}^\infty G(y)f_Y(y)\, dy\right\} \\
&= E[H(X)]E[G(Y)]. \qquad\qquad\qquad\quad \blacktriangleleft
\end{aligned}
$$

There are a number of important consequences to Theorem 5.3.1. Note, first of all, that any joint moment is the product of the appropriate marginal moments if X and Y are independent, since

$$m_{ij} = E[X^i Y^j] = E[X^i]E[Y^j] = m_{X_i} m_{Y_j};$$

indeed, the joint moment generating function for independent random variables is the product of the marginal moment generating functions, since

$$m_{X,Y}(t_1, t_2) = E[e^{t_1 X + t_2 Y}]$$
$$= E[e^{t_1 X} e^{t_2 Y}]$$
$$= E[e^{t_1 X}]E[e^{t_2 Y}]$$
$$= m_X(t_1)m_Y(t_2).$$

If X and Y are independent random variables then their covariance (and thus the correlation as well) is zero since

$$\sigma_{XY} = E[XY] - \mu_X \mu_Y = \mu_X \mu_Y - \mu_X \mu_Y = 0.$$

It is important for some applications to remember that the converse of this statement is not true in general; that is, a zero correlation (or covariance) between X and Y does not in general imply that X and Y are independent random variables. It will be recalled that Example 5.2.7 presented a situation in which $Y = X^2$, so that all the probability in the plane is assigned to points on the line $y = x^2$. Certainly X and Y are not independent (in fact, if we know $X = x$, then $Y = y = x^2$ with probability 1) yet $\rho_{XY} = 0$.

EXERCISE 5.3

1. Given X and Y are jointly discrete random variables with

$$p_{X,Y}(x, y) = \frac{1}{n^2}, \quad x = 1, 2, \ldots, n, \quad y = 1, 2, \ldots, n$$
$$= 0, \quad \text{otherwise,}$$

verify that X and Y are independent.

2. Given X and Y are jointly discrete random variables with

$$p_{X,Y}(x, y) = \frac{2}{n(n+1)}, \quad x = 1, 2, \ldots, n, \quad y = 1, 2, \ldots, x$$
$$= 0, \quad \text{otherwise,}$$

show that X and Y are not independent. Compute the regression of Y on X and the regression of X on Y.

3. X and Y are jointly continuous with density function

$$f_{X,Y}(x, y) = 4, \qquad 0 < x < 1, \qquad 0 < y < \tfrac{1}{4}$$
$$= 0, \qquad \text{otherwise.}$$

Verify that X and Y are independent.

4. Suppose that X and Y have joint density

$$f_{X,Y}(x, y) = \tfrac{3}{2}, \qquad 0 < x < 1, \qquad -(x - 1)^2 < y < (x - 1)^2$$
$$= 0, \qquad \text{otherwise.}$$

Show that X and Y are not independent and compute the regression of Y on X and the regression of X on Y.

5. Consider the population of people living in California. Suppose we select a person at random from California and let X be the weight and Y be the height of the selected person. Do you think X and Y are independent random variables?

6. Show that if X and Y are independent, then the regression of Y on X is a constant (not depending on x) and evaluate that constant value. If the regression of Y on X is a constant, are X and Y necessarily independent? If both the regression of Y on X and the regression of X on Y are constants, are X and Y necessarily independent?

7. Verify that the conditional densities and probability functions, as defined in Definition 5.3.1, satisfy the requirements for density and probability functions, respectively.

8. If the joint probability for (X, Y) is nonzero at exactly 3 points, what must be true for X and Y to be independent?

9. If X and Y are independent, jointly distributed, binomial random variables, each with parameters n and p, write down their joint probability function $p_{X,Y}(x, y)$.

10. Assume that the number of deaths from automobile accidents in Montana is a Poisson random variable with parameter $\lambda = 2$ per day. If we let X be the number of deaths that occur on Monday of a certain week and let Y be the number of deaths that occur on Tuesday of the same week, what is the joint probability function for X and Y?

11. Assuming that X and Y are independent normal random variables with parameters μ_X, σ_X, μ_Y, and σ_Y, respectively, what is the joint moment generating function for (X, Y)?

12. Let X be the time you get out of bed in the morning (measured in fractions of an hour past 7 a.m.) and let Y be the length of time it takes you to get to your place of business (in fractions of an hour) after getting up. Assume that the conditional density of Y given $X = x$ is

$$f_{Y|X}(y \mid x) = \frac{2y}{(1 - x)^2}, \qquad 0 < x < \tfrac{2}{3}, \qquad 0 < y < 1 - x$$
$$= 0, \qquad \text{otherwise}$$

while the marginal of X is

$$f_X(x) = \tfrac{81}{26}(1 - x)^2, \qquad 0 < x < \tfrac{2}{3}$$

$$= 0, \qquad\qquad \text{otherwise.}$$

Given that it took you 30 minutes to get to your place of business one morning (thus $Y = \tfrac{1}{2}$), compute the probability that you were out of bed by 7:15 a.m. that morning. Given that it took you 50 minutes to get there, what is the probability that you got up later than 7:20 a.m.?

13. Assume X and Y are jointly distributed random variables and show that

$$Z = X + Y$$

has variance $\sigma_X^2 + \sigma_Y^2 + 2\sigma_{XY}$.

14. Assume X and Y are jointly distributed random variables and show that

$$V = X - Y$$

has variance $\sigma_X^2 + \sigma_Y^2 - 2\sigma_{XY}$.

15. Is it possible that $Z = X + Y$ may have smaller variance than either X or Y alone?

16. Answer 15 if X and Y are independent.

17. Assume that X_1 and X_2 are independent random variables, each with mean zero and variance σ^2. Define

$$Y = a_1 X_1 + a_2 X_2$$

$$Z = b_1 X_1 + b_2 X_2$$

and compute μ_Y, μ_Z, σ_Y^2, σ_Z^2, σ_{YZ}, and the correlation between Y and Z.

18. It was shown in the text that

$$E_X[\mu_{Y|X}] = \mu_Y$$

Is it also true that

$$E_X[\sigma_{Y|X}^2] = \sigma_Y^2 ?$$

19. Assume X is an exponential random variable with parameter λ; let a be an arbitrary constant. What would you use as the conditional density for X, given $X > a$? What are the values of the mean and variance of this distribution?

5.4. *n*-Dimensional Random Variables and the Multinomial Random Variable

As mentioned earlier, once the extension from one random variable to two has been made, the additional extension to any number, n, of jointly distributed random variables is easily made.

DEFINITION 5.4.1.   Given an experiment, the n-tuple or vector

$$(X_1, X_2, \ldots, X_n)$$

is called an *n-dimensional random variable* if each X_i, $i = 1, 2, \ldots, n$, associates a real number with every experimental outcome $\omega \in S$.

Thus, an n-dimensional random variable is simply a rule associating an n-tuple of real numbers with every $\omega \in S$ (or a collection of n rules each associating a single number with every $\omega \in S$). The study of n-dimensional random variables is in itself a large body of material. We shall concentrate mainly on the idea of independent random variables because these are very useful in our study of statistics. As we might imagine, we can speak of n-dimensional discrete random variables or n-dimensional continuous random variables (mixtures of the two can easily be defined but we shall not be concerned with them). The n-dimensional discrete random variable will be described by an n-dimensional probability function, denoted by

$$p_{X_1, X_2, \ldots, X_n}(x_1, x_2, \ldots, x_n)$$

(defined quite analogously to Definition 5.1.2), while the n-dimensional continuous random variable is described by a density function

$$f_{X_1, X_2, \ldots, X_n}(x_1, x_2, \ldots, x_n)$$

(analogous to Definition 5.1.3).

We shall continue to subscript the f and p with symbols representing the appropriate random variables, but to conserve space this notation will be abbreviated as follows: $\mathbf{X}$ will be written instead of $X_1, X_2, \ldots, X_n$ as a subscript for the f or p. The number of variables $\mathbf{X}$ represents will be clear from the context. Similarly, instead of writing $(x_1, x_2, \ldots, x_n)$ as the n arguments of $f_{\mathbf{X}}$ or $p_{\mathbf{X}}$, we shall use $(\mathbf{x})$. Thus in the remainder of this book

$$f_{\mathbf{X}}(\mathbf{x}) = f_{X_1, X_2, \ldots, X_n}(x_1, x_2, \ldots, x_n)$$

$$p_{\mathbf{X}}(\mathbf{x}) = p_{X_1, X_2, \ldots, X_n}(x_1, x_2, \ldots, x_n)$$

where n will be clear from the context. As before, probability statements about discrete random variables are evaluated by summing the probability function appropriately, while probability statements about continuous random variables are evaluated by integrating the density function. The operation of computing the expected value of a function of an n-dimensional random variable extends readily from the two-dimensional case as well. Let us illustrate some of the foregoing discussion by considering the multinomial random variable.

DEFINITION 5.4.2. A *multinomial* trial, with parameters $p_1, p_2, \ldots, p_k$, is a trial which results in one of k possible outcomes (these outcomes are called classes). The probability of the i-th class occurring on a single trial is p_i, $i = 1, 2, \ldots, k$; thus $0 \leq p_i \leq 1$, $i = 1, 2, \ldots, k$, and $\sum_{i=1}^{k} p_i = 1$.

A single roll of a single die is a multinomial trial (with $k = 6$) since every roll results in 1 of the 6 faces being uppermost. The grade a student gets in a statistics course can be thought of as a multinomial trial with $k = 5$ (if the only grades he may receive are A, B, C, D, and F). Clearly, a multinomial trial is simply a generalization of a binomial trial, having an arbitrary k rather than just 2 possible outcomes. The multinomial random variable is defined as follows:

DEFINITION 5.4.3. Given an experiment which consists of n repeated, independent, multinomial trials with parameters p_i, $i = 1, 2, \ldots, k$, let X_i be the number of trials which result in outcomes in the i-th class, $i = 1, 2, \ldots, k$. $(X_1, X_2, \ldots, X_k)$ is called the *multinomial* random variable with parameters $n, p_1, p_2, \ldots, p_k$.

Notice that we have allowed some redundancy in the definition of the multinomial random variable. Since every trial must result in exactly 1 of the k possible classes occurring, $\sum_{i=1}^{k} X_i = n$, so that inclusion of all k is redundant. If we know the number of trials, n, and the values of $X_1, X_2, \ldots, X_{k-1}$, the value of X_k is specified. (The listing of both the number of successes and the number of failures for repeated Bernoulli trials would be redundant in the same manner.)

Theorem 5.4.1. If $(X_1, X_2, \ldots, X_k)$ is the multinomial random variable with parameters $n, p_1, p_2, \ldots, p_k$, then

$$p_X(x) = \frac{n!}{x_1! \, x_2! \cdots x_k!} \, p_1^{x_1} p_2^{x_2} \cdots p_k^{x_k},$$

for $x_i = 0, 1, 2, \ldots, n$, $i = 1, 2, \ldots, k$, such that $\sum_{i=1}^{k} x_i = n$ and the probability function is zero, otherwise. (Note that here $\mathbf{X}$ stands for X_1, $X_2, \ldots, X_k$ and $\mathbf{x}$ stands for $x_1, x_2, \ldots, x_k$.)

Proof: We are given an experiment consisting of n repeated, independent, multinomial trials. Let $s_1, s_2, \ldots, s_k$ represent the outcome of a single trial falling into classes 1 through k, respectively. Then, if we define $S_i = \{s_1, s_2, \ldots, s_k\}$, $i = 1, 2, \ldots, n$, S_i is clearly a sample space for a single trial and the sample space for the n trials is $S = S_1 \times S_2 \times \cdots \times S_n$; the probability assigned to a single n-tuple in S is the product of the probabilities of the appropriate outcomes on the individual trials. Let X_i be the number of outcomes in the i-th class, $i = 1, 2, \ldots, k$. Then

$$P(X_1 = x_1, X_2 = x_2, \ldots, X_k = x_k) = p_{\mathbf{X}}(\mathbf{x})$$

is gotten by counting the number of n-tuples in S containing exactly x_1 s_1's, x_2 s_2's, $\ldots$ and exactly x_k s_k's and multiplying by $p_1^{x_1} p_2^{x_2} \cdots p_k^{x_k}$, the common probability of any such n-tuple occurring (given by our product rule for assigning probabilities to single-element events in S). The total number of such n-tuples is given by counting the number of ways we could lay out n things in a row of which x_1 are of one kind, x_2 are of a second kind, $\ldots x_k$ are of a k-th kind. The number of ways we could choose x_1 positions for the s_1's is $\binom{n}{x_1}$; after having put the s_1's in their positions, the number of ways could we choose positions for the s_2's is $\binom{n - x_1}{x_2}$, etc. The total number of n-tuples with x_1 s_1's, x_2 s_2's, $\ldots$, x_k s_k's then is

$$\binom{n}{x_1}\binom{n - x_1}{x_2}\binom{n - x_1 - x_2}{x_3} \cdots \binom{n - x_1 - x_2 \cdots - x_{k-1}}{x_k}$$

$$= \frac{n!}{x_1! \, x_2! \cdots x_k!}$$

and we have

$$p_{\mathbf{X}}(\mathbf{x}) = \frac{n!}{x_1! \, x_2! \cdots x_k!} \, p_1^{x_1} p_2^{x_2} \cdots p_k^{x_k},$$

as was to be proved.   ◄

EXAMPLE 5.4.1.   Suppose 6 fair dice are rolled 1 time. Let X_i be the number of i's that occur, $i = 1, 2, \ldots, 6$. Then $(X_1, X_2, \ldots, X_6)$ is the multinomial random variable with parameters

$$n = 6, \qquad p_1 = p_2 = \cdots = p_6 = \tfrac{1}{6}.$$

The probability that each number occurs exactly once then is

$$p_{\mathbf{X}}(1, 1, \ldots, 1) = \frac{6!}{(1!)^6}\left(\frac{1}{6}\right)^6 = \frac{6!}{6^6} = \frac{5}{324}.$$

The probability that exactly four 1's and two 2's occur is

$$p_{\mathbf{X}}(4, 2, 0, 0, 0, 0) = \frac{6!}{4! \, 2! \, (0!)^4}\left(\frac{1}{6}\right)^4\left(\frac{1}{6}\right)^2 = \frac{5}{15{,}552};$$

the probability that exactly two 3's, two 4's and two 5's occur is

$$p_{\mathbf{X}}(0, 0, 2, 2, 2, 0) = \frac{6!}{(2!)^3(0!)^3}\left(\frac{1}{6}\right)^2\left(\frac{1}{6}\right)^2\left(\frac{1}{6}\right)^2 = \frac{5}{2592}.$$

EXAMPLE 5.4.2.   A college sophomore is taking 5 courses in a semester. He is willing to assume that he will get an A with probability .1, a B with probability .8, and a C with probability .1 in each of the courses. He defines X_1, X_2, X_3 to be the number of A's, B's, and C's, respectively, that he gets in this semester. Then X_1, X_2, X_3 is a multinomial random variable with parameters $n = 5$, $p_1 = .1$, $p_2 = .8$, $p_3 = .1$, and

$$p_{\mathbf{X}}(\mathbf{x}) = \frac{5!}{x_1!\, x_2!\, x_3!}\, (.1)^{x_1}(.8)^{x_2}(.1)^{x_3}.$$

The probability that he gets all B's then is

$$p_{\mathbf{X}}(0, 5, 0) = (.8)^5;$$

the probability that he gets 2 A's and 3 B's is

$$p_{\mathbf{X}}(2, 3, 0) = \frac{5!}{2!\, 3!}\, (.1)^2(.8)^3.$$

The probability he gets no C's (all A's and B's) is $(.9)^5$.

Theorem 5.4.2.   If $X_1, X_2, \ldots, X_k$ is a multinomial random variable with parameters $n, p_1, p_2, \ldots, p_k$, then X_1 is a binomial random variable with parameters n and p_1; similarly X_i is binomial with parameters n and p_i, $i = 2, 3, \ldots, k$.

Proof: Our experiment consists of n repeated, independent trials, each with probability p_1 of class 1 occurring (class 1 is a success). Thus X_1, the number of trials resulting in a success, is a binomial random variable with parameters n and p_1. The marginal distributions of $X_2, X_3, \ldots, X_k$ are derived in the same manner.   ◄

As mentioned earlier, the concept of independent, *n*-dimensional random variables is quite important. These are defined as follows.

DEFINITION 5.4.4.   (1) $X_1, X_2, \ldots, X_n$ are *independent* discrete *random variables* if and only if

$$p_{\mathbf{X}}(\mathbf{x}) = \prod_{i=1}^{n} p_{X_i}(x_i), \qquad \text{for all } (\mathbf{x}).$$

(2) $X_1, X_2, \ldots, X_n$ are *independent* continuous *random variables* if and only if

$$f_{\mathbf{X}}(\mathbf{x}) = \prod_{i=1}^{n} f_{X_i}(x_i), \qquad \text{for all } (\mathbf{x}).$$

This is a direct extension of the definition of independence of two random variables; their joint probability or density function must be equal to the product of their marginals. Thus, if $X_1, X_2, \ldots, X_n$ are independent, we need only know their marginal probability laws to be able to construct their joint probability law.

EXAMPLE 5.4.3.   The number of telephone calls a particular individual receives per day is a Poisson random variable with parameter $\lambda s = 8$. Let X_1 be the number of calls he receives on Monday, X_2 the number of calls he receives on Tuesday, and X_3 the number of calls he receives on Wednesday of a certain week. Since the numbers of occurrences in a Poisson process for nonoverlapping intervals are independent (see Definition 4.3.1), X_1, X_2, and X_3 are independent random variables; their joint probability function then is

$$p_\mathbf{X}(\mathbf{x}) = p_{X_1}(x_1)p_{X_2}(x_2)p_{X_3}(x_3)$$

$$= \frac{8^{x_1}}{x_1!} e^{-8} \frac{8^{x_2}}{x_2!} e^{-8} \frac{8^{x_3}}{x_3!} e^{-8}$$

$$= \frac{8^{x_1+x_2+x_3}}{x_1!\,x_2!\,x_3!} e^{-24}, \qquad \text{for } x_1 = 0, 1, 2, \ldots ;$$

$$x_2 = 0, 1, 2, \ldots ; \quad x_3 = 0, 1, 2, \ldots .$$

The probability that he receives exactly 1 call on each of these 3 days then is

$$p_\mathbf{X}(1, 1, 1) = \frac{8^3}{(1!)^3} e^{-24} = .193 \times 10^{-7};$$

the probability that he gets exactly 7 calls on Monday, 8 calls on Tuesday, and 9 calls on Wednesday is

$$p_\mathbf{X}(7, 8, 9) = \frac{8^{24}}{7!\,8!\,9!} e^{-24} = .00242.$$

EXAMPLE 5.4.4.   Assume that you buy four 60-watt bulbs; on the package it says "average life—1000 hours." If X is the lifetime of a single one of these bulbs, then, the manufacturer is claiming $E[X] = 1000$. Suppose we arbitrarily label the bulbs 1, 2, 3, 4 and let X_1, X_2, X_3, X_4 represent their individual lifetimes. If we assume each bulb has a lifetime selected from a symmetric distribution with mean equal to 1000, the probability that all 4 bulbs will have lifetimes exceeding 1000 hours is $(\frac{1}{2})^4$ (assuming independence of failure time from one bulb to another). If we make the stronger assumption that lifetimes of bulbs of this type are selected from a normal distribution with $\mu = 1000$, $\sigma = 100$ (and still their lifetimes are independent) the joint distribution for X_1, X_2, X_3, X_4 is

$$f_\mathbf{X}(\mathbf{x}) = \prod_{i=1}^{4} \frac{1}{\sigma\sqrt{2\pi}} e^{-(x_i-\mu)^2/2\sigma^2}$$

$$= \left(\frac{1}{\sigma\sqrt{2\pi}}\right)^4 \exp -\left[\sum_{i=1}^{4} (x_i - \mu)^2/2\sigma^2\right]$$

where $\mu = 1000$, $\sigma = 100$. Using this distribution we could then evaluate any probability statements of interest about X_1, X_2, X_3, X_4. For example, the probability that all four bulbs last at least 900 hours is

$$P(X_1 \geq 900, X_2 \geq 900, X_3 \geq 900, X_4 \geq 900) = \prod_{i=1}^{4} P(X_i \geq 900)$$

$$= \left[1 - N_Z\left(\frac{900 - 1000}{100}\right) \right]^4$$

$$= [.8413]^4 = .5010.$$

The probability they all fail in less than 1050 hours is

$$P(X_1 < 1050, X_2 < 1050, X_3 < 1050, X_4 < 1050) = \prod_{i=1}^{4} P(X_i < 1050)$$

$$= \left[N_Z\left(\frac{1050 - 1000}{100}\right) \right]^4$$

$$= [.6915]^4 = .2286.$$

EXAMPLE 5.4.5. We saw in Theorem 5.4.2 that if $X_1, X_2, \ldots, X_k$ is the multinomial random variable with parameters $n, p_1, p_2, \ldots, p_k$, then X_i is a binomial random variable with parameters n and p_i, $i = 1, 2, \ldots, k$. Then we know immediately that $X_1, X_2, \ldots, X_k$ are not independent random variables since

$$p_{\mathbf{X}}(\mathbf{x}) = \frac{n!}{x_1! \cdots x_k!} \, p_1^{x_1} \cdots p_1^{x_k}$$

$$\neq \binom{n}{x_1} p_1^{x_1}(1 - p_1)^{n-x_1} \cdots \binom{n}{x_k} p_k^{x_k}(1 - p_k)^{n-x_k}$$

$$= \prod_{i=1}^{k} p_{X_i}(x_i).$$

In Examples 5.4.3 and 5.4.4, the random variables (within the same example) had the same marginal probability laws. These are examples of identically distributed random variables; the definition of this concept follows.

DEFINITION 5.4.5. (1) $X_1, X_2, \ldots, X_n$ are *identically distributed* discrete *random variables* if and only if

$$p_{X_i}(x) = p_X(x), \qquad \text{for all } x, \qquad i = 1, 2, \ldots, n.$$

(2) $X_1, X_2, \ldots, X_n$ are *identically distributed* continuous *random variables* if and only if

$$f_{X_i}(x) = f_X(x), \qquad \text{for all } x, \qquad i = 1, 2, \ldots, n.$$

Thus, in brief, a group of random variables are identically distributed if and only if each of their marginals is the same (then all of their moment generating functions are also equal). When we consider the concept of sampling, identically distributed random variables will play a central role.

The following theorem gives the form of the moment generating function of the sum of independent random variables.

Theorem 5.4.3. Suppose that $X_1, X_2, \ldots, X_n$ are independent (discrete or continuous) random variables and define $Y = \sum_{i=1}^{n} X_i$. The moment generating function for Y then is

$$m_Y(t) = \prod_{i=1}^{n} m_{X_i}(t).$$

Proof: We have, by definition,

$$m_Y(t) = E[e^{tY}]$$

$$= E[e^{t \sum_{i=1}^{n} X_i}] = E[e^{tX_1} e^{tX_2} \cdots e^{tX_n}]$$

$$= E[e^{tX_1}] E[e^{tX_2}] \cdots E[e^{tX_n}]$$

$$= m_{X_1}(t) m_{X_2}(t) \cdots m_{X_n}(t). \quad \blacktriangleleft$$

Using this result, coupled with Theorem 3.4.1 (which says that the moment generating function uniquely characterizes the distribution function for the random variable), we can easily establish several distributional results for sums of independent random variables.

Theorem 5.4.4. $X_1, X_2, \ldots, X_n$ are independent Bernoulli random variables, each with parameter p. Then $Y = \sum_{i=1}^{n} X_i$ is a binomial random variable with parameters n and p.

Proof: The Bernoulli moment generating function is $m_X(t) = q + pe^t$ (see Section 4.1). Then, from Theorem 5.4.3,

$$m_Y(t) = [m_X(t)]^n = (q + pe^t)^n.$$

But this is the moment generating function of the binomial random variable with parameters n and p (see Section 4.1). Thus Y is a binomial random variable with parameters n and p. $\quad \blacktriangleleft$

Theorem 5.4.5. $X_1, X_2, \ldots, X_n$ are independent normal random variables; X_i has parameters μ_i, σ_i^2, $i = 1, 2, \ldots, n$. Then $Y = \sum_{i=1}^{n} X_i$ is a normal random variable with mean $\sum_{i=1}^{n} \mu_i$ and variance $\sum_{i=1}^{n} \sigma_i^2$.

Proof: The moment generating function for X_i is

$$m_{X_i}(t) = e^{t\mu_i + t^2 \sigma_i^2/2}, \qquad i = 1, 2, \ldots, n.$$

Then, by Theorem 5.4.3

$$m_Y(t) = \prod_{i=1}^{n} m_{X_i}(t)$$

$$= \prod_{i=1}^{n} e^{t\mu_i + t^2\sigma_i^2/2}$$

$$= e^{t\Sigma\mu_i + (t^2/2)(\Sigma\sigma_i^2)};$$

thus Y has the normal distribution as claimed.   ◀

The result given in Theorem 5.4.5 is a special case of the *reproductive* property of the normal distribution; sums of normal random variables reproduce the same form of probability distribution. This property is shared by several random variables, including the Poisson and, in a restricted sense, the binomial, as you are asked to show in the problems below.

EXAMPLE 5.4.6. Suppose that, as in Example 5.4.4, lifetimes of 60-watt bulbs are normally distributed with $\mu = 1000$, $\sigma = 100$. If you buy a package of four bulbs, and will use the four bulbs consecutively in the same lamp, then the total length of time you will get light from the bulbs in the package is

$$Y = \sum_{i=1}^{4} X_i.$$

From Theorem 5.4.5, Y again is a normal random variable with mean 4000 and variance $4(100)^2$; thus the standard deviation of Y is 200. The probability, then, that you would get at least 3900 hours of light from the package is

$$P(Y > 3900) = 1 - N_Z\left(\frac{3900 - 4000}{200}\right)$$

$$= .6915;$$

the probability that you get less than 4300 hours is

$$P(Y < 4300) = N_Z\left(\frac{4300 - 4000}{200}\right) = .9332.$$

Let us now make note of a theorem regarding the mean and the variance of a linear function of independent random variables. (The random variable $Y = \sum_{i=1}^{n} a_i X_i$ is called a linear function of $X_1, X_2, \ldots, X_n$ if $a_1, a_2, \ldots, a_n$ are any arbitrary set of constants.)

Theorem 5.4.6. $X_1, X_2, \ldots, X_n$ are independent random variables with means $\mu_1, \mu_2, \ldots, \mu_n$ and variances $\sigma_1^2, \sigma_2^2, \ldots, \sigma_n^2$, respectively. If $Y = \sum_{i=1}^{n} a_i X_i$, where the a_i's are arbitrary constants, then $\mu_Y = \sum_{i=1}^{n} a_i \mu_i$ and $\sigma_Y^2 = \sum_{i=1}^{n} a_i^2 \sigma_i^2$.

Proof:

$$\mu_Y = E[Y] = E\left[\sum_{i=1}^{n} a_i X_i\right]$$

$$= \sum_{i=1}^{n} E[a_i X_i]$$

$$= \sum_{i=1}^{n} a_i E[X_i]$$

$$= \sum_{i=1}^{n} a_i \mu_i.$$

(Note that this part of the theorem does not depend on the assumed independence.)

$$\sigma_Y^2 = E[(Y - \mu_Y)^2] = E\left[\left(\sum_{i=1}^{n} a_i X_i - \sum_{i=1}^{n} a_i \mu_i\right)^2\right]$$

$$= E\left[\left(\sum_{i=1}^{n} a_i (X_i - \mu_i)\right)^2\right]$$

$$= E\left[\sum_{i=1}^{n} a_i^2 (X_i - \mu_i)^2 + 2\sum_{\substack{i \ j \\ i<j}} a_i a_j (X_i - \mu_i)(X_j - \mu_j)\right]$$

$$= \sum_{i=1}^{n} a_i^2 E[(X_i - \mu_i)^2] + 2\sum_{\substack{i \ j \\ i<j}} a_i a_j E[(X_i - \mu_i)(X_j - \mu_j)]$$

$$= \sum_{i=1}^{n} a_i^2 \sigma_i^2. \qquad \blacktriangleleft$$

Every one of the terms of the double sum of cross products is zero since each is a constant times the covariance of 2 of the variables in $X_1, X_2, \ldots, X_n$. Each of these covariances is zero because of the independence of $X_1, X_2, \ldots, X_n$.

EXAMPLE 5.4.7. Suppose that $X_1, X_2, \ldots, X_n$ are independent, identically distributed random variables, each with mean μ and variance σ^2. Let $\bar{X}$ be the arithmetic average of these n random variables; that is,

$$\bar{X} = \frac{1}{n} \sum_{i=1}^{n} X_i \left(= \sum_{i=1}^{n} \frac{1}{n} X_i\right).$$

Thus $\bar{X}$ is a linear function of $X_1, X_2, \ldots, X_n$ with $a_i = 1/n$ for all i. Then, from Theorem 5.4.5,

$$\mu_{\bar{X}} = \sum_{i=1}^{n} \frac{1}{n} \mu = \mu$$

$$\sigma_{\bar{X}}^2 = \sum_{i=1}^{n} \left(\frac{1}{n}\right)^2 \sigma^2 = \frac{\sigma^2}{n}.$$

Thus the expected value of $\bar{X}$ is the same as the common expected value of the X_i's and the variance of $\bar{X}$ is $1/n$ times the common variance of the X's, which means that $\bar{X}$ is considerably less variable than the orignal X_i's. This fact has considerable impact on the theory of statistics.

EXERCISE 5.4

1. Suppose that you and a friend gamble for coffee on each of 20 successive days. Define:

$$X_i = 1, \quad \text{if you win on the } i\text{-th day}$$

$$= 0, \quad \text{if you lose on the } i\text{-th day},$$

   for $i = 1, 2, \ldots, 20$. Then $Y = \sum_{i=1}^{20} X_i$ is the total number of times you win in the 20 days. Assuming the game you are playing is fair (that is the probability of your winning is $\frac{1}{2}$ for each day) and the results are independent from day to day, compute the probability that $Y \geq 10$ and that $Y \geq 15$.

2. (a) If X_1, X_2, X_3 are independent Poisson random variables, each with parameter λs, write down their joint probability function $p_{X_1, X_2, X_3}(x_1, x_2, x_3)$.
   (b) If X_1, X_2 X_3 are independent Poisson random variables with parameters $\lambda_1 s$, $\lambda_2 s$, $\lambda_3 s$, respectively, write down their joint probability function

$$p_{X_1, X_2, X_3}(x_1, x_2, x_3).$$

3. The distance a football referee covers in 1 step is a normal random variable with mean $\mu = 1.05$ yards and standard deviation $\sigma = .1$ yard. Assume that this referee steps off a 5-yard penalty against your team (the distance covered by successive steps are independent). What is the probability that the total distance your team is penalized exceeds 5 yards?

4. If $X_1, X_2, \ldots, X_k$ are independent binomial random variables with parameters $n_1, p, n_2, p, \ldots, n_k, p$, respectively, show that $Y = \sum_{i=1}^{k} X_i$ is a binomial random variable with parameters $\sum_{i=1}^{k} n_i$ and p.

5. Show that if $X_1, X_2, \ldots, X_n$ are independent Poisson random variables with parameters $\lambda_1 s_1, \lambda_2 s_2, \ldots, \lambda_n s_n$, respectively, then $Y = \sum_{i=1}^{n} X_i$ is a Poisson random variable with parameter $\sum_{i=1}^{n} \lambda_i s_i$.

6. Suppose that $X_1, X_2, \ldots, X_n$ are independent, identically distributed, exponential random variables with parameter λ. Show that $Y = \sum_{i=1}^{n} X_i$ has moment generating function $m_Y(t) = \lambda^n / (\lambda - t)^n$.

7. X_1 and X_2 are independent, identically distributed random variables, each with mean μ and variance σ^2. Show that $V = X_1 - X_2$ has $\mu_V = 0$, $\sigma_V^2 = 2\sigma^2$. (Use Theorem 5.4.6.)

8. X_1 and X_2 are independent, identically distributed random variables, each with mean μ and variance σ^2. Define $Y = a_1 X_1 + a_2 X_2$. Show that if we restrict a_1 and a_2 to values such that $\mu_Y = \mu$, then the variance of Y is minimized with

$a_1 = a_2 = \frac{1}{2}$. (The condition that $\mu_Y = \mu$ implies $a_1 + a_2 = 1$; i.e., $a_2 = 1 - a_1$. Use this value for a_2 in σ_Y^2 and differentiate with respect to a_1.)

9. Generalize problem 8. If $X_1, X_2, \ldots, X_n$ are independent random variables, each with mean μ and variance σ^2, show that the variance of $Y = \sum_{i=1}^{n} a_i X_i$ is minimized with $a_i = 1/n$, $i = 1, 2, \ldots, n$, if we require that $\mu_Y = \mu$.

10. Given:

$$E[X_i] = \mu_i, \qquad i = 1, 2, \ldots, n$$
$$E[(X_i - \mu_i)^2] = \sigma_i^2, \qquad i = 1, 2, \ldots, n$$
$$\text{cov}\,(X_i, X_j) = \sigma_{ij}, \qquad i \neq j,$$

show that

$$\mu_Y = \sum_{i=1}^{n} a_i \mu_i$$
$$\sigma_Y^2 = \sum_{i=1}^{n} a_i^2 \sigma_i^2 + 2 \sum_{i<j} \sum a_i a_j \sigma_{ij}$$

where

$$Y = \sum_{i=1}^{n} a_i X_i.$$

11. Consider the four light bulbs discussed in Example 5.4.4.
    (a) What is the probability that bulbs numbered 1 and 2 each last more than 1100 hours, bulb 3 lasts less than 1100 hours and bulb 4 has a lifetime between 900 and 1100 hours?
    (b) What is the probability that two of the four bulbs last at least 1100 hours and the other two lifetimes lie between 900 and 1100 hours?

12. Patients with broken limbs arrive at a university hospital like events in a Poisson process at a rate of one per day.
    (a) What is the probability there will be exactly seven such arrivals in a seven-day period?
    (b) Given that there were exactly seven arrivals in a seven-day period, what is the probability that exactly one arrived on each of the seven days?

13. The manufacturer of a toy truck for children packs the trucks 24 to a carton to be sent to toy stores. The weight of an individual truck of this model is a normal random variable with mean 12 (ounces) and standard deviation .1. Let Y be the weight contained in one of these cartons.
    (a) What is the distribution for Y?
    (b) What is the probability a carton weighs between 18 and 19 pounds?

14. In Exercise 4 above, you were asked to show that $\sum X_i$ is a binomial random variable, if X_i is binomial with parameters n_i and p; in particular the parameter p had to be the same for all the X_i's. Suppose X_1 and X_2 are independent binomial random variables with parameters $n_1 = 3$, p_1 and $n_2 = 2$, p_2, respectively. Find the distribution for $X_1 + X_2$. Is it binomial in form?

15. An elevator has a plaque beside its door stating: "Capacity 6 persons, 990 pounds."
    (a) Assume that the weights of people using this elevator are selected from a normal distribution with $\mu = 140$ pounds, $\sigma = 30$ pounds. If six people enter the elevator, what is the probability their combined weight exceeds the maximum capacity figure of 990 pounds?
    (b) Answer the question posed in (a) if seven people crowd into the elavator.

16. The total length of time it takes to manufacture an item produced on an assembly line is the sum of the times required to complete each of the various stages. Assume the time required to complete stage i is a normal random variable X_i, $i = 1, 2, \ldots, n$, with parameters μ_i, σ_i.
    (a) Assuming the X_i's to be independent, what is the distribution for $Y = \sum X_i$?
    (b) Do you feel it is reasonable to assume the X_i's are independent, as in (a)?

5.5. The Chebychev Inequality and the Law of Large Numbers

The Chebychev inequality gives some insight into the fact that the standard deviation of a random variable is a rather natural unit for the probability law of a random variable. It was first derived by the Russian mathematician Chebychev in 1867. As we shall see, the inequality gives a bound on the probability that a random variable will be within k standard deviations of its mean. The result is presented as Theorem 5.5.1.

Theorem 5.5.1. (Chebychev) If X is a random variable with mean μ_X and standard deviation σ_X, then

$$P(|X - \mu_X| < k\sigma_X) \geq 1 - \frac{1}{k^2}$$

and

$$P(|X - \mu_X| \geq k\sigma_X) \leq \frac{1}{k^2}.$$

Proof: We shall go through the proof only for the discrete case. The same reasoning may be seen to hold in the continuous case and thus is true for any random variable (which has mean μ_X and standard deviation σ_X). Assume X is a discrete random variable with probability function $p_X(x)$, its mean is μ_X and its variance is σ_X^2; let $R_X = \{x : p_X(x) > 0\}$ be the range of X. For any constant k define

$$A = \{x : |x - \mu_X| \geq k\sigma_X\};$$

then the complement of A is

$$\bar{A} = \{x : |x - \mu_X| < k\sigma_X\}.$$

Now, by definition,

$$\sigma_X^2 = \sum_{x \in R_X} (x - \mu_X)^2 p_X(x)$$

$$= \sum_{x \in A \cap R_X} (x - \mu_X)^2 p_X(x) + \sum_{x \in \bar{A} \cap R_X} (x - \mu_X)^2 p_X(x)$$

$$\geq \sum_{x \in A \cap R_X} (x - \mu_X)^2 p_X(x),$$

since $\sum_{x \in \bar{A} \cap R_X} (x - \mu_X)^2 p_X(x) \geq 0$. For any $x \in A \cap R_X$, $|x - \mu_X| \geq k\sigma_X$ or $(x - \mu_X)^2 \geq k^2 \sigma_X^2$. Thus

$$\sum_{x \in A \cap R_X} (x - \mu_X)^2 p_X(x) \geq \sum_{x \in A \cap R_X} k^2 \sigma_X^2 p_X(x) = k^2 \sigma_X^2 \sum_{x \in A \cap R_X} p_X(x)$$

and we have

$$\sigma_X^2 \geq k^2 \sigma_X^2 \sum_{x \in A \cap R_X} p_X(x),$$

that is,

$$\frac{1}{k^2} \geq \sum_{x \in A \cap R_X} p_X(x).$$

But

$$\sum_{x \in A \cap R_X} p_X(x) = P(|X - \mu_X| \geq k\sigma_X)$$

and thus

$$P(|X - \mu_X| \geq k\sigma_X) \leq \frac{1}{k^2},$$

the second of the two statements given above. If we multiply this inequality by -1 (thus reversing the inequality) and then add 1 to both sides we have

$$1 - P(|X - \mu_X| \geq k\sigma_X) \geq 1 - \frac{1}{k^2}$$

or

$$P(|X - \mu_X| < k\sigma_X) \geq 1 - \frac{1}{k^2},$$

which is the original statement of the inequality given above.   ◄

Notice in particular that we make no distributional assumption about X when deriving the Chebychev inequality (other than the assumption that μ_X and σ_X^2 exist), and we are still able to place a bound on the probability that X is within k standard deviations of its mean. The following example compares some exact probabilities for the normal and exponential laws with the bound given by the Chebychev inequality.

EXAMPLE 5.5.1. Suppose that X is a normal random variable with mean μ_X and variance σ_X^2. Then

$$P[|X - \mu_X| < k\sigma_X] = P\left[\frac{|X - \mu_X|}{\sigma_X} < k\right]$$

$$= N_Z(k) - N_Z(-k)$$

$$= 2N_Z(k) - 1.$$

If X is an exponential random variable with parameter λ, then $\mu_X = \sigma_X = 1/\lambda$ and

$$P(|X - \mu_X| < k\sigma_X) = \int_{(1-k)/\lambda}^{(1+k)/\lambda} \lambda e^{-\lambda x} \, dx$$

$$= e^{-(1-k)} - e^{-(1+k)}, \qquad \text{for } 0 \le k \le 1$$

and

$$P(|X - \mu_X| < k\sigma_X) = \int_0^{(1+k)/\lambda} \lambda e^{-\lambda x} \, dx$$

$$= 1 - e^{-(1+k)}, \qquad \text{for } k > 1.$$

The following table compares these two exact probabilities with the bound provided by the Chebychev inequality for selected values of k. For these two

$$P(|X - \mu_X| < k\sigma_X)$$

k	Normal	Exponential	Chebychev Bound
1	.6826	.8647	0
1.5	.8664	.9179	.5556
2	.9546	.9502	.75
3	.9974	.9817	.8889
4	1.0000	.9933	.9334

probability laws the exact probability is considerably higher than the bound, especially for relatively small values of k. But then the exact probabilities are resting on the stronger assumption of knowing the exact probability law for X. Problem 2 in Exercise 5.5 shows a case in which the Chebychev bound is actually the exact probability; thus the bound is attained in some cases.

We should note that the Chebychev inequality says that

$$P(|X - \mu_X| < k\sigma_X) \ge 0$$

for any $k \le 1$, a statement we already knew to be true. Thus, for any $k \le 1$ the bound does not add anything useful for describing random variables. For $k \ge 1$, however, it is at times of use.

Let us now discuss the law of large numbers. If we were to toss a fair coin a large number of times, n, and let X_n be the number of heads that we observe, it seems that X_n/n should be close to $\frac{1}{2}$. In fact, it seems that the larger that n becomes, the closer this ratio should be to $\frac{1}{2}$. We cannot say with certainty that X_n/n will ever be $\frac{1}{2}$; we can, however, show that the probability it differs from $\frac{1}{2}$ by any small amount tends to zero as n increases. This is a special case of what is called the law of large numbers, presented as Theorem 5.5.2.

Theorem 5.5.2.   (Law of Large Numbers) Suppose that $X_1, X_2, \ldots,$ $X_k, \ldots$ is a sequence of independent, identically distributed random variables, each with mean μ and variance σ^2. Define the new sequence of $\bar{X}_i$ values by

$$\bar{X}_n = \frac{1}{n} \sum_{i=1}^{n} X_i, \qquad n = 1, 2, 3, \ldots.$$

(Thus, $\bar{X}_1 = X_1$, $\bar{X}_2 = \frac{1}{2}(X_1 + X_2)$, $\bar{X}_3 = \frac{1}{3}(X_1 + X_2 + X_3)$, etc.) Then,

$$\lim_{n \to \infty} P(|\bar{X}_n - \mu| > \varepsilon) = 0, \qquad \text{for any } \varepsilon > 0.$$

Proof: We saw in Example 5.4.7 that for any finite n

$$\mu_{\bar{X}_n} = E[\bar{X}_n] = \mu, \qquad \sigma^2_{\bar{X}_n} = E[(\bar{X}_n - \mu)^2] = \frac{\sigma^2}{n}.$$

By Chebychev's inequality,

$$P[|\bar{X}_n - \mu_{\bar{X}_n}| > k\sigma_{\bar{X}_n}] \leq \frac{1}{k^2}$$

for any $k > 0$. Then

$$P\left[|\bar{X}_n - \mu| > \frac{k\sigma}{\sqrt{n}}\right] \leq \frac{1}{k^2}.$$

Choose $k = \varepsilon\sqrt{n}/\sigma$ for any given $\varepsilon > 0$. Then

$$P[|\bar{X}_n - \mu| > \varepsilon] \leq \frac{\sigma^2}{\varepsilon^2 n}$$

and

$$\lim_{n \to \infty} P[|\bar{X}_n - \mu| > \varepsilon] \leq \lim_{n \to \infty} \frac{\sigma^2}{\varepsilon^2 n} = 0.$$

Since all of the terms in this sequence of probabilities must be nonnegative, we clearly have

$$\lim_{n \to \infty} P[|\bar{X}_n - \mu| > \varepsilon] = 0,$$

no matter how small ε may be.   ◀

It is important to realize that the law of large numbers gives the limit of a sequence of probability statements, not the limit of a sequence of random variables. It does not, for example, say that $\bar{X}_n$ is necessarily getting closer and closer to μ; it does say that the probability tends to zero that $\bar{X}_n$ differs from μ by more than ε, no matter how small ε may be. Let us consider the application of this theorem to the tossing of a fair coin, as mentioned earlier.

EXAMPLE 5.5.2.   Suppose that we are given a fair coin. Conceivably we could flip this coin any number of times that we like. In particular, we could define $X_i = 1$ if we get a head on the i-th flip and $X_i = 0$ if we get a tail on the i-th flip for $i = 1, 2, 3, \ldots$. Then we have defined a sequence of independent Bernoulli random variables, $X_1, X_2, \ldots$, each with parameter $p = \frac{1}{2}$. We can then define the sequence of proportions of heads that we observe:

$$\bar{X}_n = \frac{1}{n}\sum_{i=1}^{n} X_i, \quad \text{for } n = 1, 2, 3, \ldots.$$

In this case $\mu_{\bar{X}_n} = p = \frac{1}{2}$ and $\sigma_{\bar{X}}^2 = pq/n = 1/4n$. Then, by the law of large numbers, $\lim_{n\to\infty} P[|\bar{X}_n - \frac{1}{2}| > \varepsilon] = 0$; that is, the probability that the proportion of flips on which we observe a head differs from $\frac{1}{2}$ by more than ε tends to 0 as the number of flips increases indefinitely. We do not claim that the proportion itself will necessarily get closer and closer to $\frac{1}{2}$ because, in particular, we might get a head on every flip. The probability of this latter event happening is shrinking to zero, however, as $n \to \infty$.

EXAMPLE 5.5.3.   Suppose that we have a number n of vacuum tubes, all made by the same manufacturer and quite identical. If the time to failure, X_i, for the i-th tube is assumed to be an exponential random variable with parameter λ for each i, then the following is true. If we put all n of these tubes on test until they failed (independently) and we defined

$$\bar{X} = \frac{1}{n}\sum_{i=1}^{n} X_i,$$

then

$$\lim_{n\to\infty} P\left[\left|\bar{X} - \frac{1}{\lambda}\right| > \varepsilon\right] = 0.$$

Thus, if we tested a large number n of tubes and computed the arithmetic average $\bar{X}$ of their failure times (actually we could only compute $\bar{x}$, the observed value of $\bar{X}$), the probability is high that $|\bar{X} - 1/\lambda|$ is small. If we didn't know λ, it would seem plausible to use $1/\bar{x}$ as a guess of its value.

EXERCISE 5.5

1. If X is uniformly distributed on the interval $(0, 1)$, compare $P(|X - \mu_X| < k\sigma_X)$ with the values given by the Chebychev inequality for $k = 1\frac{1}{4}, 1\frac{1}{2}, 1\frac{3}{4}$, and 2.

2. For any value of $k \geq 1$, we can define a discrete random variable X to have probability function

$$p_X(x) = \frac{k^2 - 1}{k^2}, \qquad x = 0$$

$$= \frac{1}{2k^2}, \qquad x = -k, k$$

$$= 0, \qquad \text{otherwise.}$$

Compute μ_X and σ_X, and compare the exact probability $P(|X - \mu_X| < k\sigma_X)$ with the bound given by Chebychev's inequality. Note then that the bound given by Chebychev's inequality can not be improved for any $k \geq 1$.

3. Suppose that $X_1, X_2, X_3, \ldots$ are independent, identically distributed random variables each with probability function

$$p_X(x) = \tfrac{1}{2}, \qquad \text{for } x = -1, 1$$

$$= 0, \qquad \text{otherwise.}$$

Define the sequence of averages

$$\bar{X}_n = \frac{1}{n}\sum_{i=1}^{n} X_i$$

and show that the probability function for $\bar{X}_n$ is

$$p_{\bar{X}_n}(x) = \frac{\binom{n}{\frac{n}{2}(x+1)}}{2^n}, \qquad \text{for } x = \pm\frac{1}{n}, \pm\frac{3}{n}, \ldots, \pm\frac{n-2}{n}, \pm1,$$

$$= 0, \qquad \text{otherwise.}$$

(Assume that n is an odd number.)

4. Derive the mean and standard deviation for the probability function for $\bar{X}_n$ given in Exercise 3.

5. If you assume that $X_1, X_2, \ldots, X_n$ are independent, identically distributed, Poisson random variables, each with parameter λs (where n is quite large), what function of $X_1, X_2, \ldots, X_n$ might you guess would be close in value to λs? (See Example 5.5.3.)

6. If you assume that $X_1, X_2, \ldots, X_n$ (n large) are independent, identically distributed, geometric random variables with parameter p, what function of $X_1, X_2, \ldots, X_n$ would you guess should be close in value to p?

7. (a) Define the function $g(p) = p(1 - p)$ and show that g has a maximum at $p = \tfrac{1}{2}$. Thus, $g(p) \leq \tfrac{1}{4}$ for all p.
(b) If X is a binomial random variable with parameters n and p, then $\mu_X = np$, $\sigma_X^2 = npq$. Furthermore, if $Y = X/n$ then $\mu_Y = p$ and $\sigma_Y^2 = pq/n$; show that

$$P(|Y - p| < \delta) \geq 1 - pq/n\delta^2.$$

(c) For Y as defined in (b) show that

$$P(|Y - p| < \delta) \geq 1 - \tfrac{1}{4}\delta^2$$

(d) Suppose we wanted to find a value of n such that

$$P(|Y - p| < \delta) \geq .9,$$

where δ is some specified positive constant. Show that if $n \geq 2.5/\delta^2$, then the desired probability statement is satisfied.

8. A completely automated assembly line has been installed to manufacture a small part. Let p be the (unknown) proportion of defectives produced by this assembly line. How many items, n, should be examined from the line so that the probability is at least .9 that the sample proportion of defectives will differ from p by less than .05? By less than .01? (Use the technique outlined in Exercise 7. Since this technique is based on the Chebychev inequality, and not on a procedure tailor-made for the binomial case, the values of n derived here are considerably larger than necessary. In Section 5.6 we shall see a more realistic procedure for deriving a value for n.)

5.6. The Central Limit Theorem; Approximations

We have already seen the normal probability density function and several examples of its occurrence. Even if the normal random variable never occurred in practice, the normal distribution function would still be one of the most important of all probability laws because of its role in the central limit theorem. It can be shown that, under a wide range of conditions, the distribution function for a sum of random variables approaches the normal distribution function as the number of variables in the sum increases. One version of the central limit theorem is given as Theorem 5.6.1. (The assumptions regarding the random variables $X_1, X_2, X_3, \ldots$ can be weakened in several ways and still the conclusion follows.)

Theorem 5.6.1.   (Central Limit Theorem) $X_1, X_2, X_3, \ldots$ is a sequence of independent, identically distributed random variables, each with mean μ and variance σ^2. Define the sequence of random variables $Z_1, Z_2, Z_3, \ldots$ by

$$Z_n = \frac{\bar{X}_n - \mu}{\sigma/\sqrt{n}}, \qquad n = 1, 2, 3, \ldots,$$

where

$$\bar{X}_n = \frac{1}{n} \sum_{i=1}^{n} X_i.$$

Then, for all real t,

$$\lim_{n \to \infty} F_{Z_n}(t) = N_Z(t)$$

where N_Z is the standard normal distribution function.

Proof: The following heuristic argument is presented for the case in which the moment generating function, $m_X(t)$, exists for the random variables X_i in the original sequence. We shall show that $\lim_{n \to \infty} m_{Z_n}(t) = e^{t^2/2}$, the moment generating function of the standard normal distribution function. It can be shown that if the moment generating functions converge to a certain form (such as the above), the distribution functions of the Z_n converge to the corresponding distribution function. Now, for any n, we have seen that $E[\bar{X}_n] = \mu$, $E[(\bar{X}_n - \mu)^2] = \sigma^2/n$. Thus $E[Z_n] = 0$, $E[Z_n^2] = 1$, for any n. Now,

$$m_{Z_n}(t) = E[e^{tZ_n}]$$

$$= E[e^{t(\bar{X}_n - \mu)/(\sigma/\sqrt{n})}]$$

$$= E\left[\prod_{i=1}^{n} e^{t(X_i - \mu)/\sigma\sqrt{n}}\right]$$

$$= \prod_{i=1}^{n} E[e^{t(X_i - \mu)/\sigma\sqrt{n}}]$$

$$= \left[m_{(X-\mu)/\sigma}\left(\frac{t}{\sqrt{n}}\right)\right]^n$$

where $m_{(X-\mu)/\sigma}$ is the moment generating function for the standard form for X. Then

$$\log_e m_{Z_n}(t) = n \log_e \left[m_{(X-\mu)/\sigma}\left(\frac{t}{\sqrt{n}}\right)\right].$$

The standard form for X has mean 0 and variance 1, so

$$m_{(X-\mu)/\sigma}(t) = 1 + \frac{t^2}{2} + m_3 \frac{t^3}{3!} + m_4 \frac{t^4}{4!} + \cdots,$$

where m_3, m_4, etc., are the third, fourth, etc., moments of the standard form for X. Then

$$m_{(X-\mu)/\sigma}\left(\frac{t}{\sqrt{n}}\right) = 1 + \frac{t^2}{2n} + \frac{m_3}{3!}\left(\frac{t}{\sqrt{n}}\right)^3 + \frac{m_4}{4!}\left(\frac{t}{\sqrt{n}}\right)^4 + \cdots$$

$$= 1 + a(t)$$

and $\log_e m_{Z_n}(t) = n \log_e [1 + a(t)]$. It is shown in calculus courses that

$$\log_e (1 + b) = b - \frac{b^2}{2} + \frac{b^3}{3} - \frac{b^4}{4} + \cdots, \qquad \text{for } |b| < 1.$$

If we choose t close to zero, so $|a(t)| < 1$, then

$$\log_e m_{Z_n}(t) = [a(t) - \tfrac{1}{2}a^2(t) + \tfrac{1}{3}a^3(t) - \cdots].$$

By inspection

$$\lim_{n \to \infty} na(t) = \frac{t^2}{2}$$

$$\lim_{n \to \infty} na^k(t) = 0, \qquad k = 2, 3, 4, \ldots.$$

Thus

$$\lim_{n \to \infty} \log_e m_{Z_n}(t) = \frac{t^2}{2}$$

which implies that

$$\lim_{n \to \infty} m_{Z_n}(t) = e^{t^2/2},$$

the moment generating function for the standard normal random variable. Then

$$\lim_{n \to \infty} F_{Z_n}(t) = N_Z(t), \qquad \text{for all } t. \qquad \blacktriangleleft$$

Since $\lim_{n \to \infty} F_{Z_n}(t) = N_Z(t)$, we would expect that, for large n, $F_{Z_n}(t) \doteq N_Z(t)$ ($\doteq$ is meant to stand for approximate equality). But for any n,

$$F_{\bar{X}_n}(t) = F_{Z_n}\left(\frac{t - \mu}{\sigma/\sqrt{n}}\right),$$

and thus for large n

$$F_{\bar{X}_n}(t) \doteq N_Z\left(\frac{t - \mu}{\sigma/\sqrt{n}}\right).$$

That is, the distribution function of the arithmetic average of a large number of independent, identically distributed random variables is approximately equal to the standard normal distribution function (appropriately adjusted), regardless of the common distribution of the individual random variables averaged together. This fact is the basis of the normal approximation to a great number of probability laws. The accuracy of the approximation varies for different probability laws for the same value of n. It is not possible in general to say how large is large, that is, how large n should be for a given accuracy for a given probability law. The following examples illustrate applications of the normal approximation.

EXAMPLE 5.6.1. (Normal Approximation to the Binomial) We saw in Section 5.4. that if $X_1, X_2, \ldots, X_n$ are independent Bernoulli random variables, each with parameter p, then $Y = \sum_{j=1}^{n} X_j$ is a binomial random variable with parameters n and p. By the central limit theorem, the distribution function for Y/n tends to the normal form, and for large n we have

$$F_{Y/n}(t) \doteq N_Z\left(\frac{t - p}{\sqrt{pq/n}}\right)$$

or

$$F_Y(t) = F_{Y/n}\left(\frac{t}{n}\right) \doteq N_Z\left(\frac{t/n - p}{\sqrt{pq/n}}\right) = N_Z\left(\frac{t - np}{\sqrt{npq}}\right).$$

As an example of the use of this to approximate the binomial distribution function, suppose that we drop 49 coins onto a table and let Y be the number of coins that land heads up. Then Y is a binomial random variable with parameters 49 and $\frac{1}{2}$. The probability that at most 28 land heads up is

$$F_Y(28) \doteq N_Z\left(\frac{28 - 24\frac{1}{2}}{3\frac{1}{2}}\right) = .8413$$

(the exact value of this probability is .8736). The probability that between 20 and 25, inclusive, will show heads is

$$F_Y(25) - F_Y(19) \doteq N_Z\left(\frac{25 - 24\frac{1}{2}}{3\frac{1}{2}}\right) - N_Z\left(\frac{19 - 24\frac{1}{2}}{3\frac{1}{2}}\right)$$
$$= N_Z(.14) - N_Z(-1.57) = .4975$$

(the exact value is .5361).

In the above example the approximation was off by about .04 in both instances. We can materially improve the approximation of discrete probabilities by the continuous normal curve by realizing that if X is discrete, then

$$P(X = a) = P(a - \tfrac{1}{2} < X \leq a + \tfrac{1}{2}).$$

If, then, X is discrete and we want to approximate the probability that

$$a \leq X \leq b,$$

we should sum the approximations that X is equal to each of the integers from a to b, inclusive; that is,

$$P(a \leq X \leq b)$$

is, in terms of areas under a continuous density, approximated by the area between $a - \frac{1}{2}$ and $b + \frac{1}{2}$. This adjustment of $\frac{1}{2}$ unit on either side of the discrete interval is called the *continuity correction* for normal approximations.

Graphically, it is easy to see why the approximation is improved by the continuity correction. If we think of the exact discrete probabilities as being presented by a histogram, where the height of the bar centered at an integer is equal to the probability of occurrence of that integer, and the width of the bar is 1 unit, then the probability of any integer occurring is given by the area of the bar centered at that integer. Given a normal density centered the same way as the histogram, and with equal variability, the area of a bar then should be well approximated by the area under the normal density over the appropriate interval of length one. The sum of the areas of several successive bars should be well approximated by the area under the density

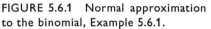

FIGURE 5.6.1   Normal approximation
to the binomial, Example 5.6.1.

FIGURE 5.6.2   Do not truncate normal
tail areas.

over the interval which is equal to the combined widths of the bars of interest. Figure 5.6.1 illustrates this reasoning for one of the probabilities computed in Example 5.6.1 above. The exact discrete probabilities are represented by the areas of the bars; the approximating normal density is also pictured. As the computation was performed in the example, we took the area under the normal density between 19 and 25, to approximate $20 \leq X \leq 25$; it is clear that we will get a better approximation if we use the normal area between $19\frac{1}{2}$ and $25\frac{1}{2}$.

Another point is illustrated in Figure 5.6.2. Suppose that we want the probability that X is no larger than 2. The exact probability is the sum of the areas of the bars labelled 0, 1, and 2. Rather than taking the area under the smooth curve between $-\frac{1}{2}$ and $2\frac{1}{2}$ to approximate this, we will in general get a closer value by taking the area under the smooth curve from $2\frac{1}{2}$ all the way back to $-\infty$. Thus, when approximating discrete tail probabilities, the normal approximation should not be truncated at the end of the last bar.

EXAMPLE 5.6.2.   (Binomial Approximation with Continuity Correction) Let us apply the continuity correction to the two probabilities computed in Example 5.6.1. As before, Y is binomial with parameters 49 and $\frac{1}{2}$.

$$F_Y(28) = F_Y(28\tfrac{1}{2})$$

$$\doteq N_Z\left(\frac{28\tfrac{1}{2} - 24\tfrac{1}{2}}{3\tfrac{1}{2}}\right) = .8729$$

$$P(20 \leq Y \leq 25) = F_Y(25\tfrac{1}{2}) - F_Y(19\tfrac{1}{2})$$

$$\doteq N_Z\left(\frac{25\tfrac{1}{2} - 24\tfrac{1}{2}}{3\tfrac{1}{2}}\right) - N_Z\left(\frac{19\tfrac{1}{2} - 24\tfrac{1}{2}}{3\tfrac{1}{2}}\right)$$

$$= .5377.$$

Comparison of these values with the exact probabilities given in Example 5.6.1 shows a considerable improvement in the accuracy of the approximation.

It can be shown that the normal approximation to the binomial is best for p close to $\frac{1}{2}$ for any n, and that as p gets further from $\frac{1}{2}$, n must be larger to get the same accuracy. Many different writers recommend various rules-of-thumb for deciding when the normal gives a good approximation to the binomial. Generally, if $9/(n + 9) \leq p \leq n/(n + 9)$ the approximation should be quite good.

EXAMPLE 5.6.3. (Normal Approximation to the Poisson) We saw in Section 5.4 that if $X_1, X_2, \ldots, X_n$ are independent Poisson random variables, each with parameter λ, then $Y = \sum_{i=1}^{n} X_i$ is a Poisson random variable with parameter $n\lambda$. Then, for large n,

$$F_Y(t) \doteq N_Z\left(\frac{t - n\lambda}{\sqrt{n\lambda}}\right),$$

giving the basis for approximating Poisson probabilities by the normal. (We shall find that again the continuity correction discussed for the binomial gives an improved approximation since the Poisson is also a discrete random variable.) Let us assume that a college hospital receives an average of 5 cases of tetanus per month and that it is reasonable to assume that incidences of tetanus among the students are independently distributed in time. Then, for a particular school year, let $X_1, X_2, \ldots, X_9$ be the number of cases that occur in the 9 individual months and assume each X_i is a Poisson random variable with parameter 5. The total number of cases for the year then is

$$Y = \sum_{i=1}^{9} X_i$$

and Y is also Poisson with parameter $9 \cdot 5 = 45$. Then, using the continuity correction, the probability that between 40 and 50 cases, inclusive, occur in this year is

$$F_Y(50\tfrac{1}{2}) - F_Y(39\tfrac{1}{2}) \doteq N_Z\left(\frac{50\tfrac{1}{2} - 45}{\sqrt{45}}\right) - N_Z\left(\frac{39\tfrac{1}{2} - 45}{\sqrt{45}}\right)$$

$$= N_Z(.82) - N_Z(-.82)$$

$$= .5878.$$

(The exact probability is .5879.) The probability that exactly 45 cases occur is

$$F_Y(45\tfrac{1}{2}) - F_Y(44\tfrac{1}{2}) \doteq N_Z\left(\frac{45\tfrac{1}{2} - 45}{\sqrt{45}}\right) - N_Z\left(\frac{44\tfrac{1}{2} - 45}{\sqrt{45}}\right)$$

$$= .0558.$$

(The exact probability is .0594.)

The normal approximation to the Poisson gets better the larger the value of $n\lambda$, the parameter of the sum. Thus, if λ, the common parameter of the individual variables making up the sum, is relatively small the number of variables in the sum must be relatively large to achieve a given degree of accuracy. If λ is quite large, then n can be correspondingly reduced.

The normal distribution can also be used to approximate probability statements about sums of continuous random variables. The following example is of this sort.

EXAMPLE 5.6.4. In making numerical computations, either by hand or within a computer, only a certain, finite number of significant digits can be retained at each step. If a fixed point arithmetic is used, then all computations completed will be rounded to the closest value having a fixed number of digits to the right of the decimal point. For example, suppose computations are carried to the closest 1/10,000-th; then all numbers involved will be restricted to having only 4 digits to the right of the decimal point and $3\frac{1}{3}$ will be written 3.3333, $\frac{2}{3}$ will be written 0.6667, π will be written 3.1416, etc. The most commonly used rule for the rounding is to round up if the digit in the first dropped position is 5 or more; otherwise the number is rounded down.

The difference between the number written in full as an infinite decimal and its rounded equivalent is called *rounding error.* For example, if computations are carried to the nearest 1/10,000-th, as discussed above, the rounding error in $3\frac{1}{3}$ is $(.333\cdots)10^{-4}$, the rounding error in $\frac{2}{3}$ is $-(.333\cdots)10^{-4}$, the rounding error in π is $-(.074\cdots)10^{-4}$, etc. The effects of rounding errors can be studied by assuming a distribution for such errors and then using such a distribution to make computations about the values of the rounding errrors.

Suppose, as discussed above, that computations are carried to the nearest 1/10,000-th. Then the smallest and largest rounding errors that could occur are $-(.499\cdots)10^{-4}$ and $(.499\cdots)10^{-4}$, respectively; furthermore, any value in between these two could occur as a rounding error. Lacking further knowledge of the particular errors to occur in a computation, it might seem reasonable to assume that the rounding errors are uniformly distributed between these two extremes. Granted this assumption, the central limit theorem can be used to make computations about the accumulation of rounding errors in a sum.

Specifically, assume that the rounding error involved in the computation of a number is a uniform random variable X on the interval $-(\frac{1}{2})10^{-k}$ to $(\frac{1}{2})10^{-k}$. Then, if n numbers are each computed and summed together, the rounding error of the sum is $Y = \sum_{i=1}^{n} X_i$, where each X_i has the assumed uniform distribution. If the X_i's are independent, Y is approximately normally distributed with mean $n\mu_X = 0$ and standard deviation $\sqrt{n}\sigma_X = 10^{-k}\sqrt{n/12}$. This distribution is useful in computing probability statements about the rounding error of the sum.

Suppose, as above, $k = 4$ and $n = 12$ numbers are summed together. Then approximately, the rounding error Y of the total is a normal random variable with mean 0 and $\sigma = 10^{-4}$. Thus

$$P(-10^{-4} < Y < 10^{-4}) \doteq N_Z(1) - N_Z(-1)$$

$$= .6836,$$

$$P(-(3)10^{-4} < Y < (3)10^{-4}) \doteq N_Z(3) - N_Z(-3)$$

$$= .9974.$$

The normal approximation to the binomial distribution can be used to find the number of trials, n, that should be performed if the sample proportion of successes is to lie within a given distance of p, with a specified probability. Assume we are going to perform n repeated, independent Bernoulli trials, each with probability p of success. Let X be the number of successes that will occur and then $Y = X/n$ is the proportion of successes in the n trials. How large should n be so that

$$P(|Y - p| \leq \delta) \geq \gamma?$$

We can answer this question approximately with the central limit theorem. As we have seen, if n is large then Y is approximately a normal random variable with mean p and variance pq/n. As we saw in exercise 5.5.7, $pq \leq \frac{1}{4}$, so the variance of Y is no larger than $\frac{1}{4}n$, no matter what the value for p. Then

$$P(|Y - p| \leq \delta) \doteq N_Z\left(\frac{\delta}{1/2\sqrt{n}}\right) - N_Z\left(-\frac{\delta}{1/2\sqrt{n}}\right)$$
$$= 2N_Z(2\delta\sqrt{n}) - 1.$$

Then, to have

$$2N_Z(2\delta\sqrt{n}) - 1 = \gamma$$

we require

$$N_Z(2\delta\sqrt{n}) = (\gamma + 1)/2$$

that is

$$2\delta\sqrt{n} = z_{(\gamma+1)/2}.$$

Thus, if we take

$$n = \frac{z^2_{(\gamma+1)/2}}{4\delta^2},$$

we will satisfy the probability statement (approximately). If it happens that this value is not an integer we will round up to the next larger integer.

EXAMPLE 5.6.5. Assume that a new machine has been purchased for the manufacture of a small part. Each part either is or is not defective. Assume there is a constant probability p that a part manufactured on this machine will be defective and that the defectives occur independently. How many parts should be examined if the proportion of defectives in those examined is not to differ from p by more than .01, with at least probability .95? Here we can directly apply the formula given above with $\delta = .01$, $\gamma = .95$, so $(\gamma + 1)/2 = .975$ and $z_{.975} = 1.96$. The number to be examined, then, should be at least

$$n = (1.96)^2/4(.01)^2 = 9,604.$$

If there is any reason to believe that p differs considerably from $\frac{1}{2}$, this number will be much too large (a smaller n will suffice), because the given formula for n replaced pq in the variance by $\frac{1}{4}$, the value it has with $p = \frac{1}{2}$. If we are sure $p \leq p_0$, say, then

$pq \leq p_0 q_0 < \frac{1}{4}$ where $q_0 = 1 - p_0$. The number of trials required then is

$$n = p_0 q_0 z^2_{(\lambda+1)/2}/\delta^2.$$

Suppose, in the example just quoted, we are willing to assume $p \leq .1$; then, keeping δ and γ unchanged, the number of trials we should perform is

$$n = (.1)(.9)(1.96)^2/(.01)^2 = 3457.44 \sim 3458,$$

which is much smaller than the original number.

EXERCISE 5.6

1. A pair of dice are rolled 180 times an hour (approximately) at a craps table in Las Vegas. What is the (approximate) probability that 25 or more rolls have a sum of 7 during the first hour? What is the (approximate) probability that between 700 and 750 rolls have a sum of 7 during 24 hours?

2. A lot of 1000 articles is to be accepted if a sample of 100 contains 3 or fewer defectives. If we assume that the process producing the article has a probability of .90 of making a nondefective article, what is the probability that the lot will be accepted?

3. A nursery man plants 115 cuttings of ivy in every flat he prepares. Assume that the probability that an individual cutting will develop roots is .9 and approximate the probability that the average number of rooted cuttings (per flat) in 50 flats is less than 100.

4. Experience has shown that the number of accidents to occur along a particular 10-mile stretch of 2-lane highway is a Poisson random variable with a mean of 2 per week. What is the (approximate) probability of there being less than 100 accidents on this stretch of road in a year?

5. The length of a continuous nylon filament that can be drawn without a break occurring is an exponential random variable with mean 5000 feet. What is the (approximate) probability that the average length of 100 filaments lies between 4750 feet and 5550 feet?

6. The probability that a person survives an attack of cholera (with good medical help) is assumed to be .4. What is the probability that at least half of 100 patients with cholera will survive?

7. The probability a person betting on black will win at roulette is 9/19, on a single spin. Assume a roulette wheel is spun 100 times in an hour. What is the (approximate) probability that a black number occurs at least 45 times?

8. Assume that a basketball player has a constant (unknown) probability p of sinking a free throw and that his attempts are independent. How many shots should he attempt, if the proportion of shots made is to lie within .1 of p, with probability at least .8?

9. If all numbers computed are rounded to the nearest 1/1000-th, what is the probability that a sum of six such numbers is accurate to the nearest 1/1000-th?

10. If all numbers computed are rounded to the nearest 1/1000-th, what is the probability that the average of six such numbers is accurate to the nearest 1/1000-th?

11. Two numbers are computed to the nearest $\frac{1}{10}$. Compute the exact probability that the rounding error of the sum is no larger than .03, assuming the rounding error for each is uniform on $(-.05, .05)$.

12. Assume that 10% of the tires produced by a manufacturer have surface defects. These tires are shipped in lots of 100.
    (a) What is the approximate probability a lot contains 8 or fewer tires with surface defects?
    (b) A large buyer orders 500 lots. What is the approximate probability that at least 140 lots each contain 8 or fewer tires with surface defects?

13. Incoming calls to an industrial switchboard arrive like events in a Poisson process at a rate of 120 per hour. What is the probability that between 110 and 125 calls arrive between 9 and 10 a.m. on any given day?

14. The negative binomial distribution can also be well approximated by a normal distribution for large r, since it is the sum of r independent geometric random variables with parameter p. (See Exercises 4.2.11 and 4.2.15.) Assume that 10% of the U.S. population have a malady and, for research purposes, it is desired to find 50 people with the malady. If people are examined at random until 50 are located, and X is the number examined then X is (at least approximately) a negative binomial random variable with parameter $r = 50$, $p = .1$. What is the (approximate) probability that at least 400 people must be examined to find 50 with the malady?

15. Assume in Exercise 14 that only half of those with the malady would be willing to take part in a medical experiment.
    (a) What is the distribution of Y, the number that must be examined to find 50 who would be willing to take part in an experiment?
    (b) What is the (approximate) probability that at least 800 must be examined to find 50 willing to participate in the experiment?

6

Sampling and Statistics

In this chapter we shall study some of the basic notions that are commonly used in applying probability theory to practical problems of inference. It is unfortunate, but perhaps necessary, that the application of statistics is still something of an art; the choice of the model to use and, thus, the appropriate statistical method to employ, is not always obvious. Indeed there are problems in which knowledgeable people cannot agree upon the "correct" model and method. Thus, it is not surprising perhaps that many students find statistics a confusing jumble of possibly conflicting ways of deriving an answer to a given practical problem.

Generally, it is much easier to compare two or more methods of answering a given question if the models used by the methods are clearly understood; the set of assumptions about the random variables occurring in the problem and their interrelationships, as well as the way in which they are assumed to represent practical reality, is called the model of the situation. Clearly, if two different models are assumed for the same problem it would not be surprising if they called for the use of different statistical methods and, conceivably, then led to different conclusions for the practical problem. Choosing the "correct" statistical method, thus, really comes down to the choice of the "correct" model of reality; as long as people are free to disagree about the correct method to improve society, they are still free to disagree about the "correct" model for certain aspects of reality. Thus, there will probably always be disagreement, and room for imagination and art, in the application of statistics.

In this chapter we shall study the basic idea of a population (or a population distribution) and the concept of sampling from that population. The inherent idea, of course, is to use known facts from the sample to make inferences about the population.

6.1. Populations and Samples

The word population is in common usage in everyday language. We are all familiar with such things as the population of the United States, the population of black bears on the North American continent, and the population of fish in Lake Superior. For any given point in time, each of these seems to be a well-defined entity, a set of individuals each of whom could be identified. Of course, as time passes, each of these populations changes, conceivably, through birth, death, and mobility. It is important to realize, then, in the contexts of such populations, that the character of the population may well vary with time, but for any given instant, such a population is in theory well defined.

Frequently, in applied problems, a specific aspect of the individuals in the population is the quantity to be studied. For example, we might want to estimate the average income of the population of the United States in 1970; then, rather than thinking of the population of the United States, as individual people, it is useful to consider only the incomes in 1970 of these individuals and, therefore, to consider the population of incomes in 1970; this, then, is a population of numbers, the numbers being the respective incomes of the individuals in the population. It might be of interest to a zoologist to estimate the average age, at a given time, of all the black bears on the North American continent or the average weight of all the fish in Lake Superior. Again, in these two examples, then, it is of interest to consider populations of numbers (ages and weights), rather than the populations of individual bears or fish.

In each of the three examples just discussed, the population of numbers of interest was derived from a concrete population of individuals which definitely exist in a real sense. Many practical problems are concerned with various aspects of such populations. However, if we were to restrict our attention to only this type of population we would be ignoring a very large body of important problems concerned with *conceptual populations*, ones which the mind can conceive of but which do not exist in the same concrete manner as those discussed above. For example, consider a specific process for manufacturing 60-watt light bulbs. Conceivably, any number of such bulbs could be produced in exactly the same way and each such light bulb would have some length of time that it would burn, without failure, in a given type of usage (in a home, or a factory, or on board a boat, etc.). We then can conceive of the population of lengths of time that this type of bulb would last in this usage, a conceptual population, and can pose problems that are of interest regarding this conceptual population. As a second example, consider a specific usage of a given drug in the treatment of a given ailment in a human being, say tetracycline and its effect on a given type of skin infection. It is possible to conceive of any number of individuals, each with the same skin infection, being treated in the same way with the same dosage of tetracycline. Each person receiving the treatment might be looked at as being cured (0), controlled from further spread (1), or not affected by the drug (2). Thus we could conceive of a population of numbers, each being 0 or 1 or 2, generated from such treatments; indeed, for any given dosage level of tetracycline we can conceive of such a population and we would probably be quite interested in how the proportions of 0's, 1's, and 2's were affected by the changes in the dosage.

Thus, our usage of the word *population* will be quite general. We may have in mind either a concrete population that really exists or a conceptual population which does not and never will exist. In either case we will almost exclusively be concerned with populations of numbers. Most statistical methods can be viewed as ways of making inferences about a population. Such inferences are made after the examination of a *sample* from the population. For example, we might be concerned with making a guess about the overall average value of all the numbers in the population (point estimation), or in giving an interval of values with a known probability of including the overall average (confidence interval), or in deciding whether a specific number is equal to the overall average (testing a hypothesis). In each of these cases we desire an answer based on only partial information, a sample from the population, rather than from a complete examination of the population itself.

A population can be either discrete or continuous, depending upon whether the set of numbers referred to is discrete or continuous. Continuous populations can occur in two ways. If we discuss the population of weights of people in the United States at a given time, then, since there is a finite number of people living there, the population of weights is truly discrete. However there is such a large number of distinct values that can occur as the weights of the individuals concerned that it is convenient to idealize the population and to think of the population of weights as being continuous; in that way continuous techniques can be applied which are considerably simpler to use than are their discrete counterparts. Conceptual populations of time, distance, etc., shall also be considered continuous; if we really take a sample from such a conceptual population, the values which we can observe are again truly necessarily discrete, because of the discreteness of the measuring instrument used. Again, however, the greater simplicity of continuous techniques makes it advantageous to consider the population, and sample values, as being continuous.

For any given population of numbers, there is a discrete probability function or continuous density function which describes the distribution of values in the population, depending upon whether the population is discrete or continuous. The random variable X which has the given probability or density function is called the *population random variable*. Population random variables can be either discrete or continuous; the range of the population random variable thus is the set of values in the population.

EXAMPLE 6.1.1. Consider the population generated by the treatment of skin infections using a given dosage of tetracycline, discussed above; if p_1 is the proportion of 0's, p_2 the proportion of 1's and p_3 the proportion of 2's, then the population random variable X is discrete with probability function

$$p_X(0) = p_1, \qquad p_X(1) = p_2, \qquad p_X(2) = p_3.$$

EXAMPLE 6.1.2. Assume that the distribution of weights of people in the United States is well approximated by a normal density with $\mu = 150$, $\sigma = 30$. (By this we mean the proportion of people having weights in the interval (a, b) is essentially the same as the area under the given normal density, for all possible (a, b).) Then the population random variable X is normal with $\mu_X = 150$, $\sigma_X = 30$. Note as well that the true range of the population is finite; surely the interval from 0 to 1000 pounds, say, should encompass all the weights in the population; yet the normal density we have assumed is actually positive over the whole real line. The use of the normal density in such a case represents a further idealization, which, for all practical purposes, should not seriously affect any desired manipulations. Note that if X is normal with $\mu_X = 150$, $\sigma_X = 30$, then $f_X(0)$ and $f_X(1000)$ are zero to many decimal places, although still strictly positive.

There are many ways in which a sample can be selected from a population. It could be selected quite purposefully; if we are going to select a sample of, say, 10 values from the population, we could quite deliberately attempt to find the 10 smallest values that exist in the population. Clearly, such a purposeful way of selecting the sample would not in general be of much use in making inferences about the population as a whole, although it would give us information about the smallest population values. Granted that we do in general want to make inferences about the population as a whole, we should find other methods of selecting the sample.

As a general rule, samples that are selected haphazardly or randomly from the population have great advantages for making inferences about the population. Let us now discuss an important definition.

DEFINITION 6.1.1.   A *random sample of size n of a random variable X* is a collection of n random variables, $X_1, X_2, \ldots, X_n$ such that

(i) $X_1, X_2, \ldots, X_n$ are independent.
(ii) $F_{X_i}(t) = F_X(t)$, for all t, $i = 1, 2, \ldots, n$.

In sampling at random from any finite population without replacement, we do not satisfy Definition 6.1.1. Suppose we have a finite population consisting of the numbers 1 through N and are going to select a sample of $n = 2$ of these values; thus, if X is the population random variable,

$$p_X(x) = \frac{1}{N}, \qquad x = 1, 2, \ldots, N.$$

There are $\binom{N}{2}$ different samples of size 2 that could be selected, and, if we choose any of the numbers in the population at random as our first sample value, and then choose at random any of the remaining $N - 1$ numbers as our second sample value, clearly each of the $\binom{N}{2}$ different possible pairs of values has the same probability of occurrence, and this would be called a *random sample* of size 2 *from the population*. However, if we let X_1 be the first value selected and X_2 the second, clearly the conditional probability function for X_2 depends upon the particular value X_1 takes on; thus X_1 and X_2 are not independent, although

$$F_{X_i}(t) = F_X(t) \qquad \text{for all } t, \qquad i = 1, 2.$$

Thus, X_1 and X_2 would not be a random sample of a random variable. If, on the other hand, the sample is selected with replacement, then X_1 and X_2, as defined above, would be a random sample of the population random variable X.

In essentially all applications to be discussed in this book we shall assume that we are dealing with a random sample of a population random variable X. Our object, of course, is to use the observed sample values to make inferences about various aspects of the underlying population distribution. Thus the sampling is assumed to have been done in such a manner that Definition 6.1.1 is satisfied.

EXAMPLE 6.1.3.   Assume that telephone calls are placed to a large industrial switchboard, between 8 a.m. and 5 p.m. on Monday through Friday, like events in a Poisson process with parameter $\lambda = 20$ per hour, and that the number of calls are independent from day to day. Then, if X_i, $i = 1, 2, \ldots, 5$, are the number of incoming calls on Monday through Friday of a given week, respectively, $X_1, X_2, \ldots,$ X_5 is a random sample of a Poisson random variable with parameter $9 \cdot 20 = 180$.

The population sampled is conceptual and corresponds to the population of business days this firm will be in existence, at the current level of activity. Such questions as the number of incoming lines that should be adequate for the company to install can be investigated by making inferences about this conceptual population.

EXAMPLE 6.1.4.   Every item coming off a production line is either defective or not. Assume that the probability a given item is defective is p and that the occurrences of defectives are independent. If we then select a random sample of n items from the output of this production line and define

$$X_i = 1 \quad \text{if } i\text{-th item is defective}$$
$$= 0 \quad \text{if not}$$

$i = 1, 2, \ldots, n$, then $X_1, X_2, \ldots, X_n$ is a random sample of a Bernoulli random variable X with parameter p. The sample can be thought of as being taken from the population of the (conceptually) infinite number of items to be produced by this production line. The sample would be of use in making inferences about p, the probability of occurrence of a defective item.

If $X_1, X_2, \ldots, X_n$ is a random sample of a random variable X, the random variables $X_1, X_2, \ldots, X_n$ are called the *elements* of the sample. Once the sample has in fact been taken, the observed values of $X_1, X_2, \ldots, X_n$ are known and available for any desired computations. Their observed values are used in making inferences about the population. Once their values have been observed, of course, they are known constants and are not random in any sense. Thus, any probability statements made refer to the random variables and may be thought of as being appropriate before the sampling is accomplished.

For example, suppose we take a random sample of size 1 from a normal population with unknown mean μ and unknown variance σ^2. Then, since the population density is symmetric about μ, the probability is $\frac{1}{2}$ that the value we will get in the sample will be less than μ (and the probability is also $\frac{1}{2}$ that it will exceed μ). Then, given we selected the single value and it turned out to equal 10, for example, either 10 does exceed μ or it does not; there is no randomness present. Unfortunately, granted μ is unknown, we do not know which of the two statements

1.   10 exceeds μ
2.   10 does not exceed μ

is true, but we can be assured that one or the other is. Granted this state of ignorance, in a subjective sense we may believe there is a 50% chance statement 1 is true, since before the sample was taken that was the probability that the value we had not yet observed would exceed μ. This line of reasoning may further be justified by realizing that $P(X > \mu) = .5$ implies that in 50% of all samples of size 1 the observed sample value will exceed μ; in the other 50% it will not. Thus, we may say that we have a 50-50 chance that our observed sample value of 10 came from either of these two possible types of sample. The proportion statement used is referring to results that would be observed over repeated samples of the same size from the population.

EXERCISE 6.1

1. It is known that a certain process makes light bulbs whose lifetimes are normally distributed with average lifetime of 200 hours and standard deviation of 20 hours. If we randomly select 3 bulbs made by this process, what is the probability that all 3 lifetimes exceed 195 hours?

2. Assume that vacuum tubes of a certain type have lifetimes that are exponentially distributed with parameter λ. If we take a random sample of n of these tubes and let X_i = length of life of i-th tube, $i = 1, 2, \ldots, n$, what is the joint density function of the sample?

3. Suppose that X is uniformly distributed on the interval from zero to 1. If we take a random sample of 5 observations of X, what is the joint density function for the sample?

4. Assume that weights of college students are normally distributed. If we take a random sample of 20 students and record the weight of each, what is the joint density function for the sample weights? As a function of μ and σ, what is the probability that the total of the weights exceeds 4000 pounds?

5. It is known that scores on a certain aptitude test are normally distributed with mean μ and variance σ^2. If we select 10 people at random to take this test, what is the joint density function for their scores? What is the probability that the average of the 10 scores is less than μ?

6. In Exercise 5, assume that $\mu = 50$, $\sigma^2 = 25$, and that anyone who scores more than 60 on the test is certain to make a good pilot. What is the probability that there are no good pilots in the group of 10 people taking the test? What is the probability that there is one or more good pilots in the group of 10?

7. An urn contains 4 white and 1 red balls. We are going to draw balls, with replacement, from the urn until we get the red ball. Let X be the number of draws required. If we take a random sample of 5 observations of X, what is the joint probability function of the sample?

8. Assume that a certain college football team has a constant probability p of winning every game it plays and that all games it plays are independent. In any year this team plays exactly 10 games. Let X_1, X_2, X_3, X_4, X_5 be a random sample of 5 successive years of this team's performance (i.e., X_i is the number of games won in the i-th year, $i = 1, 2, \ldots, 5$). What is the joint probability function for the sample?

9. Assume that all burglars have a constant probability p of being caught while doing a job and let X be the number of jobs done until the burglar is first caught.

If we take a random sample of 10 burglars, what is the joint probability function of the number of jobs each accomplished before first being caught?

10. Suppose a small town has a total of 500 residents, every one of whom is listed by name in the town directory. A television program rating service would like to select a sample of 10 residents to query about their viewing habits. It is decided to pick one name at random from the first 50 listed and then to select each 50-th name thereafter. (Thus, if person number 31 is selected, so are those numbered 81, 131, etc.)
    (a) How many different samples could occur, with this procedure?
    (b) Would you say this procedure gives a random sample from the population?

11. Assume $X_1, X_2, \ldots, X_{10}$ is a random sample of a random variable.
    (a) Assume that we split the sample of size 10 into two parts, the first part consisting of the first five sample values selected and the second part consisting of the remaining five values. Are the two parts independent?
    (b) Again the sample of 10 is split into two parts, the first part consisting of the five smallest values (from the 10). Are the two parts independent?

12. Fifteen independent, repeated Bernoulli trials are performed and 12 successes are observed. (Thus, X, the number of successes in 15 trials, is binomial, $n = 15$, and the observed value of X was 12.)
    (a) What is the probability of observing $X = 12$, as a function of p?
    (b) What value of p maximizes the probability in (a)?
    (c) If p is unknown, would you say that the value derived in (b) is a good guess for its unknown value?

13. Assume X is a standard normal random variable and that X_1, X_2, X_3, X_4 is a random sample of X; define

$$\bar{X} = \sum_{i=1}^{4} X_i/4. \quad \text{What is the probability that } |\bar{X}| > \tfrac{1}{2}?$$

6.2. Statistics and Distributions

In this section we shall see the technical definition of the word *statistic*, some commonly used statistics, and the distributions over repeated samples of some of these statistics.

DEFINITION 6.2.1. A *statistic* is any function of the elements of a random sample which does not depend on unknown parameters.

Statistics, then, are quantities whose observed values can be computed, once the sample has been taken. They form the bases of all statistical methods of making inferences about populations. Let us define some of the most commonly used statistics.

DEFINITION 6.2.2.   Given $X_1, X_2, \ldots, X_n$ is a random sample of a random variable X

(i) The k-th *sample moment* is

$$M_k = \frac{1}{n} \sum_{i=1}^{n} X_i^k, \qquad k = 1, 2, \ldots$$

(ii) The *sample mean*, $\bar{X}$, is the first sample moment:

$$\bar{X} = \frac{1}{n} \sum_{i=1}^{n} X_i.$$

(iii) The *sample variance* is

$$S^2 = \frac{1}{n-1} \sum_{i=1}^{n} (X_i - \bar{X})^2$$

and the *sample standard deviation* is

$$S = \sqrt{S^2}.$$

Notice first of all that each sample moment (no matter what the value of k) then is a statistic and is itself a random variable. Once we actually take the sample and observe the values of $X_1, X_2, \ldots, X_n$, we can then compute the observed value of M_k for any k. This observed value of M_k will be denoted by $\hat{m}_k$; we use the lower case m to be consistent with our notation of capital letters for random variables and lower case letters for their observed values. The carat ($\hat{}$) is placed over the lower case m to distinguish this observed sample value from the like numbered moment of X (the notation used has been $m_k = E[X^k]$). Clearly, the observed value, $\hat{m}_2$ say, is not necessarily the same number as $m_2 = E[X^2]$. Just as the moments of the random variable measure certain aspects of the probability law of the random variable, we shall see that the sample moments measure certain aspects of the sample.

EXAMPLE 6.2.1.   Assume that heights of college students are normally distributed with mean μ and variance σ^2. We take a random sample of 5 students, measure their heights, and find them to be 70, 74, 66, 69 and 72 (measured in inches). Once we have actually done the sampling and observed the sample values that occurred, we are able to compute the observed values of any sample moments in which we might be interested. Thus

$$\hat{m}_1 = \tfrac{1}{5}(70 + 74 + 66 + 69 + 72) = 70.2$$
$$\hat{m}_2 = \tfrac{1}{5}[(70)^2 + (74)^2 + (66)^2 + (69)^2 + (72)^2] = 4935.4$$

and

$$\hat{m}_3 = \tfrac{1}{5}[(70)^3 + (74)^3 + (66)^3 + (69)^3 + (72)^3] = 347{,}495.4.$$

Since $\bar{X} = M_1$, the observed value of the sample mean is $\bar{x} = \hat{m}_1 = 70.2$. **The** observed value of the sample variance is

$$s^2 = \tfrac{1}{4} \sum (x_i - \bar{x})^2$$
$$= \tfrac{1}{4}[(70 - 70.2)^2 + (74 - 70.2)^2 + \cdots + (72 - 70.2)^2]$$
$$= \tfrac{1}{4}[36.8] = 9.2$$

and the observed value of the sample standard deviation is

$$s = \sqrt{9.2} = 3.03.$$

Rather than computing the sample variance directly from the definition, Theorem 6.2.1 gives a computational formula which involves less numerical effort.

Theorem 6.2.1.   If S^2 is the sample variance then

$$S^2 = \frac{1}{n-1} \left(\sum X_i^2 - n\bar{X}^2 \right)$$

Proof: Note that

$$\sum_{i=1}^{n} (X_i - \bar{X})^2 = \sum_{i=1}^{n} (X_i^2 - 2X_i\bar{X} + \bar{X}^2)$$
$$= \sum X_i^2 - 2\bar{X} \sum X_i + n\bar{X}^2.$$

Since $n\bar{X} = \sum X_i$

$$\sum (X_i - \bar{X})^2 = \sum X_i^2 - n\bar{X}^2,$$

from which the result follows directly.   ◀

Note as well that

$$n\bar{X}^2 = n \left(\frac{1}{n} \sum X_i \right)^2 = \frac{\left(\sum X_i \right)^2}{n}$$

so that an alternative computational formula may be written

$$S^2 = \frac{1}{n-1} \left(\sum X_i^2 - \frac{\left(\sum X_i \right)^2}{n} \right)$$

and, since $nM_2 = \sum X_i^2$,

$$S^2 = \frac{n}{n-1} [M_2 - M_1^2];$$

thus the sample variance may be evaluated directly from the first two sample moments in much the same way that σ_X^2 is determined by the first two moments for X.

EXAMPLE 6.2.2.   A group of 20 students took an examination graded from zero to 100. The 20 scores which they achieved were as follows: 99, 87, 94, 86, 89,

79, 84, 90, 86, 86, 81, 98, 85, 91, 87, 89, 95, 88, 85 and 94. Looking at these numbers as though they are the observed values of a random sample of size 20 of a random variable X, we find

$$\sum x_i = 1773$$

$$\sum x_i^2 = 157{,}703$$

and thus the observed value of the sample mean is

$$\bar{x} = 1773/20 = 88.65$$

while the observed value of the sample variance is

$$s^2 = \tfrac{1}{19}[\sum x_i^2 - 20(\bar{x})^2]$$

$$= \tfrac{1}{19}[157{,}703 - 20(88.65)^2]$$

$$= \frac{526.55}{19} = 27.71;$$

we also have $s = \sqrt{27.71} = 5.26$.

The order statistics of a sample can be used for many purposes; in a sense they are somewhat analogous to percentiles of a random variable discussed in Chapter 3. The definition of the order statistics follows.

DEFINITION 6.2.3.   Given a random sample $X_1, X_2, \ldots, X_n$ of a random variable X, rank the elements in order of increasing numerical magnitude, yielding $X_{(1)}, X_{(2)}, \ldots, X_{(n)}$, where $X_{(1)}$ is the smallest and $X_{(n)}$ the largest of $X_1, X_2, \ldots, X_n$. $X_{(i)}$ is called the i-th *order statistic*, $i = 1, 2, \ldots, n$. ($X_{(1)}$ is also referred to as the *minimum* sample value and $X_{(n)}$ as the *maximum*.)

Let us immediately look at an example of computations of order statistics

EXAMPLE 6.2.3.   The ranked sample of heights given in Example 6.2.1 is 66, 69, 70, 72, and 74. Thus, for this sample, $x_{(1)} = 66, x_{(5)} = 74$. For the sample of 20 scores given in Example 6.2.2, we find that $x_{(1)} = 79, x_{(20)} = 99, x_{(19)} = 98, x_{(4)} = 85$, etc.

It will be recalled that the median value m_e of a random variable has the property that

$$P(X \le m_e) \ge \tfrac{1}{2};$$

that is, half the values of X exceed m_e and half do not. This same idea can be used to define the median of a sample as the middle value in the ranked sample; if n is odd there is such a middle value. If n is even we shall take the average of the two middle-ranked values to be the median. The following definition defines both the sample median and the sample range.

DEFINITION 6.2.4.   Given a random sample $X_1, X_2, \ldots, X_n$ of X, the *sample median* is defined to be

$$M_0 = X_{([n+1]/2)}, \text{ if } n \text{ is odd}$$

$$= \tfrac{1}{2}[X_{(n/2)} + X_{([n+2]/2)}] \text{ if } n \text{ is even};$$

the *sample range* is

$$R_a = X_{(n)} - X_{(1)}.$$

The sample median and the sample range are alternative ways of measuring the middle and the variability, respectively, of a sample (alternative to the sample mean and the sample standard deviation). The following example illustrates their computation.

EXAMPLE 6.2.4.   For the ranked heights (first given in Example 6.2.1) we find $m_0 = 70$, $r_a = 8$. For the exam scores (first given in Example 6.2.2) we find $m_0 = 87.5$, $r_a = 20$.

It should be kept in mind that the statistics we have discussed, or any others, are random variables; thus the particular values they will equal vary from sample to sample. The *sampling distribution* of a statistic describes the way in which the statistic will vary from one sample to another (of the same size). Thus, such sampling distributions are useful in studying the behavior of various statistics and can be used to evaluate probability statements involving the statistic. We shall now examine some facts concerning the sampling distribution of the k-th sample moment, which are true no matter what the specific population distribution may be.

Theorem 6.2.2.   Assume $X_1, X_2, \ldots, X_n$ is a random sample of X and let M_k be the k-th sample moment. Then

(i) $E[M_k] = m_k$, and

(ii)   Variance $[M_k] = \dfrac{1}{n}[m_{2k} - m_k^2]$,   for $k = 1, 2, 3, \ldots$.

Proof: Let us first derive the mean of the sampling distribution for M_k. We find

$$E[M_k] = E\left[\frac{1}{n}\sum_{i=1}^{n} X_i^k\right]$$

$$= \frac{1}{n}\sum_{i=1}^{n} E[X_i^k]$$

$$= \frac{1}{n}\sum_{i=1}^{n} m_k = m_k,$$

since $E[X_i^k] = m_k$, the k-th moment of X, for each element of the sample. To get the variance of M_k, we first will evaluate $E[M_k^2]$; thus

$$E[M_k^2] = E\left[\left(\frac{1}{n}\sum_{i=1}^{n} X_i^k\right)^2\right]$$

$$= E\left[\frac{1}{n^2}\left(\sum_{i=1}^{n} X_i^{2k} + \sum_{\substack{i=1 \\ i \neq j}}^{n}\sum_{j=1}^{n} X_i^k X_j^k\right)\right]$$

$$= \frac{1}{n^2}\sum_{i=1}^{n} E[X_i^{2k}] + \frac{1}{n^2}\sum_{\substack{i=1 \\ i \neq j}}^{n}\sum_{j=1}^{n} E[X_i^k]E[X_j^k]$$

since the elements of the sample are independent. Thus,

$$E[M_k^2] = \frac{1}{n^2}[nm_{2k}] + \frac{1}{n^2}[n(n-1)m_k^2]$$

$$= \frac{1}{n}m_{2k} + \frac{n-1}{n}m_k^2$$

and

$$\text{Variance }[M_k] = \frac{1}{n}m_{2k} + \frac{n-1}{n}m_k^2 - [E(M_k)]^2$$

$$= \frac{1}{n}(m_{2k} - m_k^2),$$

as was to be proved. ◀

Notice, then, that no matter what population we may have selected our sample from, the expected value of M_k is m_k. This says that if we were to repeatedly select random samples of X, each of size n, the average value of M_k over an infinite number of repetitions is m_k, the k-th moment of X. We shall have available only a single sample and the value of M_k may or may not be particularly close to m_k; if m_k is unknown, of course, we can not know how close the observed value of M_k is to m_k for our given sample.

The fact that the variance of M_k about its mean value is $\frac{1}{n}(m_{2k} - m_k^2)$ enables us to get some feeling for how variable M_k is about m_k, from sample to sample, and thus to have some idea how likely M_k is to be within given distances of m_k. The fact that the variance of M_k is an inverse function of n, the sample size, is very important for statistical methods; the larger the sample size n the less variable M_k is about m_k from sample to sample. Notice, then, as we have seen previously, $E[\bar{X}] = m_1 = \mu$ and $\sigma_{\bar{X}}^2 = \frac{1}{n}(m_2 - m_1^2) = \sigma_X^2/n$ and the sample mean is less variable about the population mean than are individual population values. The higher order sample moments also have variances that decrease with n.

Although its proof is beyond the level of this text, the following theorem proves quite useful in deriving sampling distributions for various statistics of interest.

Theorem 6.2.3. Assume $X_1, X_2, \ldots, X_n$ are independent random variables and define n new random variables, $Y_1, Y_2, \ldots, Y_n$, by

$$Y_i = g_i(X_i), \qquad i = 1, 2, \ldots, n,$$

where $g_1, g_2, \ldots g_n$ are any one-dimensional functions possibly involving known (or unknown) constants. Then $Y_1, Y_2, \ldots, Y_n$ are independent random variables.

The results given in Theorem 6.2.3, together with the central limit theorem, enable us to say that the sampling distribution for M_k is approximately normal in form, for sufficiently large n, with mean and variance given in Theorem 6.2.2. This is because, granted $X_1, X_2, \ldots, X_n$ is a random sample of X, then $\dfrac{1}{n} X_1^k, \dfrac{1}{n} X_2^k, \ldots, \dfrac{1}{n} X_n^k$ are independent (these are the random variables Y_i of Theorem 6.2.3,); notice then that M_k is the sum of independent, identically distributed random variables and from Theorem 5.6.1 the distribution of M_k approaches that of a normal random variable as

$$n \to \infty.$$

EXAMPLE 6.2.5. Suppose n dice are tossed onto a table and let $X_1, X_2, \ldots, X_n$ be the numbers uppermost. It is reasonable to assume that $X_1, X_2, \ldots, X_n$ is a random sample of a discrete variable X with probability function

$$p_X(x) = 1/6, \qquad x = 1, 2, \ldots, 6.$$

It is easily verified that $E[X] = m_1 = 3.5$, $E[X^2] = \frac{91}{6}$, $E[X^4] = \frac{2275}{6}$. Then the sample mean $\bar{X}$ is approximately a normal random variable with mean 3.5 and variance $35/12n$ $(= \sigma_X^2/n)$. The second sample moment, M_2, is approximately a normal random variable with mean $\frac{91}{6}$ and variance $\dfrac{1}{n} \left[\dfrac{2275}{6} - \left(\dfrac{91}{6} \right)^2 \right] = 5369/36n$. For any given (large) n, these normal distributions can be used to approximate the values of probability statements about M_1 or about M_2.

If we have a random sample of a normal random variable, we have seen that $\bar{X}$ will be normal with mean μ and variance σ^2/n, where μ and σ^2 are the parameters of the population random variable. As we shall see, it will be useful to know the distribution of the square of a standard normal random variable, as well as the distribution for the sum of the squares of independent standard normal random variables. In Section 4.5 we saw that if X is a normal random variable with parameters μ and σ^2, then $(X - \mu)/\sigma$ is $N(0, 1)$ and

$$W = \frac{(X - \mu)^2}{\sigma^2}$$

has density

$$f_W(w) = \frac{w^{-1/2}}{\sqrt{2\pi}} e^{-w/2}, \qquad w > 0;$$

W is called a χ^2 random variable with 1 degree of freedom. It can be shown that its moment generating function is

$$m_W(t) = \frac{1}{(1 - 2t)^{1/2}}.$$

The evaluation of this moment generating function involves the *gamma function*, which is defined by

$$\Gamma(k) = \int_0^\infty t^{k-1} e^{-t} \, dt, \qquad k > 0;$$

integration by parts shows that

$$\Gamma(k) = (k - 1)\Gamma(k - 1).$$

Thus, if k is a positive integer,

$$\Gamma(k) = (k - 1)(k - 2) \cdots 2 \cdot 1 \Gamma(1).$$

But

$$\Gamma(1) = \int_0^\infty t^{1-1} e^{-t} \, dt = \int_0^\infty e^{-t} \, dt = 1,$$

so, for $k = 1, 2, 3, \ldots$, we find

$$\Gamma(k) = (k - 1)!.$$

It is also true that, for $k = 1, 2, 3, \ldots$,

$$\Gamma(k + \tfrac{1}{2}) = (k - \tfrac{1}{2})(k - \tfrac{3}{2}) \cdots \tfrac{3}{2} \cdot \tfrac{1}{2} \Gamma(\tfrac{1}{2})$$

and it can be shown that

$$\Gamma(\tfrac{1}{2}) = \int_0^\infty t^{-1/2} e^{-t} \, dt = \sqrt{\pi}.$$

Thus, if k is a positive integer

$$\Gamma(k + \tfrac{1}{2}) = \frac{(2k - 1)(2k - 3) \cdots (3)(1)\sqrt{\pi}}{2^k}.$$

Notice, then, that

$$\frac{t^{k-1}}{\Gamma(k)} e^{-t}, \qquad t > 0$$

will integrate to 1 and is a density function, for any $k > 0$. If we let $y = 2t$ and thus $dy = 2dt$ this function becomes

$$\frac{y^{k-1}}{2^k \Gamma(k)} e^{-y/2}, \qquad y > 0;$$

replacing k by $d/2$ this density can be written

$$f_Y(y) = \frac{y^{(d/2)-1}}{2^{d/2} \Gamma(d/2)} e^{-y/2}, \qquad y > 0.$$

Notice that if $d = 1$ this is exactly the χ^2, one degree of freedom, density given above (the square of a single $N(0, 1)$ variable). A random variable Y having this density is called a χ^2 random variable with d degrees of freedom, its only parameter. Its moment generating function is

$$m_Y(t) = \frac{1}{(1 - 2t)^{d/2}}.$$

Suppose, then, that $W_1, W_2, \ldots, W_n$ are independent χ^2 random variables, each with one degree of freedom; then $Y = \sum_{i=1}^{n} W_i$ has moment generating function

$$m_Y(t) = \prod_{i=1}^{n} m_{W_i}(t)$$

$$= \prod_{i=1}^{n} \frac{1}{(1 - 2t)^{1/2}}$$

$$= \frac{1}{(1 - 2t)^{n/2}}$$

and thus Y is a χ^2 random variable with $d = n$ degrees of freedom (n measures the number of independent, single degree of freedom, χ^2 random variables that have been added together). We thus have proved the following theorem.

Theorem 6.2.4. Assume $X_1, X_2, \ldots, X_n$ is a random sample of a normal random variable X with mean μ and variance σ^2. Then

$$Y = \sum_{i=1}^{n} \frac{(X_i - \mu)^2}{\sigma^2}$$

is a χ^2 random variable with n degrees of freedom.

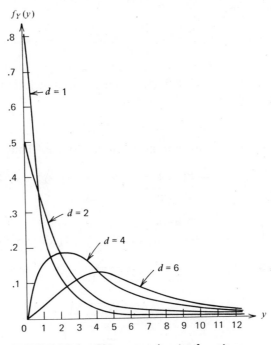

FIGURE 6.2.1   Chi-square density functions.

This χ^2 density function will occur in several inference problems which we shall study. Figure 6.2.1 gives some graphs of the χ^2 density function for varying degrees of freedom. Note that if the number of degrees of freedom d exceeds 2 the density function has a maximum value at $d - 2$. By differentiating the moment generating function, it is easy to see that the mean of a χ^2 random variable is d, its degrees of freedom, and that its variance is $2d$. It can be shown that as the number of degrees of freedom, d, gets larger and larger, the χ^2 density function gets more and more symmetric. In fact, its limit, as $d \to \infty$, is the normal density function.

Table 4 in the Appendix gives selected percentiles of χ^2 distributions for degrees of freedom $d = 1, 2, \ldots, 30$. If Y is a χ^2 random variable with $d = 5$ degrees of freedom

$$P(Y \leq 1.61) = .10$$

$$P(Y \leq 11.1) = .95;$$

if $d = 18$

$$P(Y \leq 10.9) = .10$$

$$P(Y \leq 28.9) = .95.$$

We shall define χ_α^2 by

$$P(Y \leq \chi_\alpha^2) = \alpha;$$

note then that the value for χ_α^2 will depend on d, the degrees of freedom. As discussed above, if $d = 5$, then $\chi_{.1}^2 = 1.61$, $\chi_{.95}^2 = 11.1$, whereas, if $d = 18$, $\chi_{.1}^2 = 10.9$, $\chi_{.95}^2 = 28.9$. If Y is a χ^2 random variable with degrees of freedom $d > 30$ it can be shown that

$$\sqrt{2Y} - \sqrt{2d - 1}$$

is approximately a standard normal random variable. Thus, for $d > 30$,

$$\alpha = P(Y \leq \chi_\alpha^2) = P(\sqrt{2Y} - \sqrt{2d - 1} \leq \sqrt{2\chi_\alpha^2} - \sqrt{2d - 1})$$
$$\doteq N_Z(\sqrt{2\chi_\alpha^2} - \sqrt{2d - 1})$$
$$= N_Z(z_\alpha),$$

where z_α is the 100α-th percentile of the $N(0, 1)$ distribution. Thus, for $d > 30$,

$$\sqrt{2\chi_\alpha^2} - \sqrt{2d - 1} \doteq z_\alpha$$

so

$$\chi_\alpha^2 \doteq \tfrac{1}{2}(z_\alpha + \sqrt{2d - 1})^2.$$

If $d = 51$, then,

$$\chi_{.1}^2 \doteq (z_{.1} + \sqrt{101})^2$$
$$= \tfrac{1}{2}(-1.28 + 10.05)^2$$
$$= 38.46,$$

while

$$\chi_{.95}^2 \doteq \tfrac{1}{2}(z_{.95} + \sqrt{101})^2$$
$$= \tfrac{1}{2}(1.64 + 10.05)^2$$
$$= 68.33.$$

EXAMPLE 6.2.6. Suppose that we select a random sample of size 10 of a normal random variable with parameters μ and σ^2. Then $\sum_{i=1}^{10} \dfrac{(X_i - \mu)^2}{\sigma^2}$ is a χ^2 random variable with $d = 10$ degrees of freedom; from Table 4, with $d = 10$, we find $\chi_{.05}^2 = 3.94$, $\chi_{.95}^2 = 18.3$ and thus

$$P\left(3.94 < \sum_{i=1}^{10} \frac{(X_i - \mu)^2}{\sigma^2} < 18.3\right) = .90.$$

This means that in 90% of all random samples of size 10 of a normal random variable X, $\sum_{i=1}^{10} \dfrac{(X_i - \mu)^2}{\sigma^2}$ will lie between 3.94 and 18.3; in 10% of such samples it will lie outside this interval. Notice as well that $\bar{X} = \tfrac{1}{10} \sum_{i=1}^{10} X_i$ is normal with mean μ and variance $\sigma^2/10$; thus

$$\frac{\sqrt{10}}{\sigma} (\bar{X} - \mu) \quad \text{is } N(0, 1)$$

and $\dfrac{10(\bar{X} - \mu)^2}{\sigma^2}$ is a χ^2 random variable with one degree of freedom. With $d = 1$, we find from Table 4 in the Appendix that $\chi^2_{.9} = 2.71$; thus

$$P\left(\frac{10}{\sigma^2} (\bar{X} - \mu)^2 \leq 2.71\right) = .90$$

or

$$P(\,(\bar{X} - \mu)^2 \leq .271\sigma^2) = .90.$$

This says that in 90% of the random samples of size 10 of a normal random variable, $(\bar{X} - \mu)^2$ will be no larger than $.271\sigma^2$; in 10% of such samples the inequality is reversed. Thus, if we happened to know that $\sigma^2 = 16$, say, the probability is .9 that $(\bar{X} - \mu)^2 \leq 4.34$, or that $|\bar{X} - \mu| \leq 2.08$.

Another important result regarding samples of a normal random variable is given without proof in the following theorem.

Theorem 6.2.5. Assume $X_1, X_2, \ldots, X_n$ is a random sample of a normal random variable. Then

$$\frac{\sum (X_i - \bar{X})^2}{\sigma^2}$$

is a χ^2 random variable with $d = n - 1$ degrees of freedom and S^2 and $\bar{X}$ are independent random variables.

It is rather remarkable that S^2 and $\bar{X}$, both functions of the same sample random variables end up being independent random variables (when X itself is normal). This fact has some rather important implications, as we shall see.

EXAMPLE 6.2.7. The fact that $\sum (X_i - \bar{X})^2/\sigma^2$ is a χ^2 random variable can be used to make probability statements about how close the sample variance may be to the population variance. Suppose we have a random sample of size $n = 15$ of a normal random variable. Then, from Table 4, we find, with $d = n - 1 = 14$ degrees of freedom, $\chi^2_{.05} = 6.57$, $\chi^2_{.95} = 23.7$.

Thus, the probability is .9 that

$$6.57 < \frac{\sum (X_i - \bar{X})^2}{\sigma^2} < 23.7$$

But this inequality is equivalent to

$$\frac{6.57}{14} \sigma^2 < S^2 < \frac{23.7}{14} \sigma^2;$$

thus, we can say

$$P(.47\sigma^2 < S^2 < 1.69\sigma^2) = .9.$$

That is, in 90% of the samples of size 15 of a normal random variable, S^2 will lie between $.47\sigma^2$ and $1.69\sigma^2$.

Let us close this section with a theorem about the distribution of the order statistics for a random sample of a random variable.

Theorem 6.2.6. Let $X_1, X_2, \ldots, X_n$ be a random sample of a random variable X and let $X_{(1)}, X_{(2)}, \ldots, X_{(n)}$ be the order statistics of the sample. Then the distribution function for $X_{(j)}$ is, for any real t, given by

$$F_{X_{(j)}}(t) = \sum_{k=j}^{n} \binom{n}{k} [F_X(t)]^k [1 - F_X(t)]^{n-k}$$

where $F_X(t)$ is the distribution function of the population random variable.

Proof: The j-th order statistic, $X_{(j)}$, is less than or equal to t if and only if at least j of the sample random variables are less than or equal to t. Each sample random variable has the same probability, $F_X(t)$, of not exceeding t and the sample random variables are independent. Thus, the number of sample random variables which do not exceed t is a binomial random variable with parameters n and $p = F_X(t)$, so

$$F_{X_{(j)}}(t) = \sum_{k=j}^{n} \binom{n}{k} [F_X(t)]^k [1 - F_X(t)]^{n-k},$$

as was to be proved.   ◀

The distribution functions for the sample minimum, $X_{(1)}$, and sample maximum, $X_{(n)}$, are especially simple. Note that

$$F_{X_{(1)}}(t) = \sum_{k=1}^{n} \binom{n}{k} [F_X(t)]^k [1 - F_X(t)]^{n-k}$$

$$= 1 - [1 - F_X(t)]^n,$$

$$F_{X_{(n)}}(t) = \sum_{k=n}^{n} \binom{n}{k} [F_X(t)]^k [1 - F_X(t)]^{n-k}$$

$$= [F_X(t)]^n.$$

If X is a continuous random variable so are each of the order statistics; the density function for the sample minimum is

$$f_{X_{(1)}}(t) = \frac{d}{dt} F_{X_{(1)}}(t)$$

$$= \frac{d}{dt} \{1 - [1 - F_X(t)]^n\}$$

$$= n f_X(t) [1 - F_X(t)]^{n-1}$$

and the density function for the sample maximum is

$$f_{X_{(n)}}(t) = \frac{d}{dt} F_{X_{(n)}}(t)$$

$$= \frac{d}{dt} [F_X(t)]^n$$

$$= n f_X(t)[F_X(t)]^{n-1}.$$

EXAMPLE 6.2.8. Suppose we have a random sample of size n of an exponential random variable X, with parameter λ. Thus $F_X(t) = 1 - e^{-\lambda t}$, $t > 0$. Then the distribution and density functions for $X_{(1)}$ are, respectively,

$$F_{X_{(1)}}(t) = 1 - [1 - (1 - e^{-\lambda t})]^n$$

$$= 1 - e^{-n\lambda t}$$

$$f_{X_{(1)}}(t) = n\lambda e^{-n\lambda t},$$

so the minimum value, in a sample of an exponential random variable, again is an exponential random variable with parameter $n\lambda$. Similarly, the distribution function and density function for the sample maximum are, respectively,

$$F_{X_{(n)}}(t) = [1 - e^{-\lambda t}]^n$$

$$f_{X_{(n)}}(t) = n\lambda e^{-\lambda t}[1 - e^{-\lambda t}]^{n-1},$$

so the sample maximum does not have an exponential distribution.

EXERCISE 6.2

1. A builder of sports cars took 6 identical cars of one of his models out to a track to see how the top speed attainable varied from car to car. The maximum speeds observed for the 6 were 118, 121, 120, 110, 125 and 116 (measured in miles per hour). Assume that these numbers are the observed values of a random sample of a random variable and compute the observed values of the first 3 sample moments, of $\bar{X}$, of S^2, of M_0, of $X_{(4)}$ and of R_a.

2. A college sprinter competed in 8 100-yard dash races during spring; the times recorded for his races were 10.2, 9.6, 9.8, 9.8, 10.0, 9.6, 9.7 and 9.6 (in seconds). Assuming that these numbers are the observed values of a random sample of a random variable, compute $\bar{x}$, s^2 and the observed values of the order statistics.

3. A child-development study included, among other things, observing the age at which babies are first able to sit up alone. The ages at which 5 of the babies first sat up alone were 155, 162, 150, 180 and 171 (days). Compute $\bar{x}$, s^2, s, m_0 and r_a.

4. Four sons from the same family all attended the same college. Their grade-point averages, at graduation, were 3.63, 3.10, 2.75 and 3.22 (4.0 is an A). Compute $\bar{x}$, s, m_0 and r_a.

5. Five students enrolled in the same 6-month typing course. Upon completion the numbers of words per minute which they could type were 55, 70, 72, 38 and 46. Compute $\bar{x}$, s^2, m_0 and the observed values of the order statistics.

6. A fair coin is flipped 5 times. What is the exact probability that the sample proportion of heads is within .05 of $\frac{1}{2}$? That it is within .15 of $\frac{1}{2}$?

7. What is the probability that the largest value of a sample of 10 of a uniformly distributed random variable on $(0, 1)$ exceeds .9? What is the probability that it is less than $\frac{1}{2}$?

8. For the sample mentioned in Exercise 7, what is the probability that the smallest value is less than $\frac{1}{2}$? Greater than .9?

9. Given that $X_1, X_2, \ldots, X_n$ is a random sample of a Bernoulli random variable with parameter p, show that

$$P(X_{(n)} = 1) = 1 - (1 - p)^n$$
$$P(X_{(1)} = 1) = p^n$$

10. Let $X_1, X_2, \ldots, X_n$ be a random sample of a normal random variable with parameters μ and σ^2. Derive the density functions for $X_{(1)}$ and $X_{(n)}$. Are either of these normally distributed?

11. Given that $X_1, X_2, \ldots, X_n$ is a random sample of a continuous random variable X, evaluate $E[F_X(X_{(n)})]$ and $E[F_X(X_{(1)})]$.

$$\left(\text{Hint: } \int n[F_X(t)]^n f_X(t)\, dt = \frac{n}{n+1} [F_X(t)]^{n+1}. \right)$$

12. Assume $X_1, X_2, \ldots, X_n$ is a random sample of a normal random variable with unknown mean μ and unknown variance σ^2. Define $\bar{X} = \frac{1}{n}\sum X_i$.

   (a) What is the distribution of $\sqrt{n}(\bar{X} - \mu)/\sigma$?
   (b) Is $\sqrt{n}(\bar{X} - \mu)/\sigma$ a statistic?

13. In Example 6.2.5, assume $n = 12$.
   (a) Evaluate (approximately) $P(40 \leq \sum X_i \leq 45)$.
   (b) Evaluate (approximately) $P(170 \leq \sum X_i^2 \leq 190)$.

14. Compute $E[S^2]$ and $E[S]$ for a random sample of a normal random variable. (Hint: $(n-1)S^2/\sigma^2$ is a χ^2 random variable.)

15. Assume the probability an assembly line turns out a defective is p and that defectives occur independently. n items from the assembly line are examined and classified as being defective or nondefective. Define X_i to be 1 if the i-th item is defective and define $X_i = 0$, otherwise, $i = 1, 2, \ldots, n$.
   (a) $X_1, X_2, \ldots, X_n$ is a random sample of a random variable X. What is the distribution for X?
   (b) Evaluate m_1, m_2, m_3, m_4, the first four moments of X.
   (c) What are the approximate sampling distributions for $\bar{X}$ and m_2?

16. Assume that the time you arrive at a stop light on your way to work is a uniform random variable on the interval from 0 to 10, and that you drive to work 5 times in a week.
    (a) What is the distribution function for your minimum weekly arrival time, $X_{(1)}$, at the light?
    (b) Evaluate $E[X_{(1)}]$.

17. In Exercise 16, what is the distribution function for your maximum weekly arrival time, $X_{(5)}$, at the light? Evaluate $E[X_{(5)}]$.

18. In Exercise 16, what is the distribution function for your median arrival time, $X_{(3)}$, at the light? Evaluate $E[X_{(3)}]$ and compare with $\frac{1}{2}(E[X_{(1)}] + (E[X_{(5)}])$.

19. Assume $X_1, X_2, \ldots, X_{10}$ is a random sample of a standard normal random variable. Evaluate $P(2.56 < \sum_{i=1}^{10} X_i^2 < 18.3)$.

20. In Exercise 19, evaluate

$$P(\bar{X} < .271),$$

where

$$\bar{X} = \frac{1}{10} \sum X_i.$$

21. In Exercise 19, evaluate

$$P(S^2 < 1.88),$$

where S^2 is the sample variance.

22. Show that if W_1 is a χ^2 random variable with d_1 degrees of freedom and W_2 is an independent χ^2 random variable with d_2 degrees of freedom, then $W_1 + W_2$ is again a χ^2 random variable with $d_1 + d_2$ degrees of freedom.

23. Verify that the mean of a χ^2 random variable is d and its variance is $2d$, where d is its degrees of freedom.

24. Assume that the length of life, X, of a vacuum tube in a home radio is an exponential random variable with parameter λ. If $X_1, X_2, \ldots, X_n$ is a random sample of X, what is the sampling distribution of $2\lambda n\bar{X}$?

25. Assume $X_1, X_2, \ldots X_n$ is a random sample of a continuous random variable X with density $f_X(t)$ and distribution function $F_X(t)$. Show that the density function for the j-th order statistic is

$$f_{X_{(j)}}(t) = \frac{n!}{(j-1)!\,(n-j)!} \, f_X(t)[F_X(t)]^{j-1}[1 - F_X(t)]^{n-j}.$$

26. Use the result in Exercise 25 to find the density of the sample median of a sample of size n of a uniform $(0, 1)$ random variable, where n is an odd number.

7

Statistical Inference: Estimation

We now begin our study of problems of statistical inference. The first of these inference problems which we shall study is that of estimation of parameters. Suppose that we are quite sure a random variable X has a normal probability law but we do not know the parameters of the probability law (μ and σ^2). Thus our assumption of a normal probability law has picked out a shape for the density of X but we do not know which of the infinite number of possible values for μ and σ^2 seem appropriate. In this situation we might decide to take a sample of n observations of X and ask what function (statistic) of the sample values would give a good guess of the unknown values of μ or of σ^2.

Or, again, an analysis of the situation might make it reasonable to assume that a random variable X has a Poisson probability function, but we do not know the value of λ (which would uniquely specify the probability law). Given a random sample of n observations of X, what function of the sample values would be a good guess for the unknown value of λ?

Many similar problems could be posed; each is called an estimation problem because we are asking what manipulation we might make of the observed values in a sample to get a good guess, or estimate, of the value of an unknown parameter or parameters. In this chapter we shall study two general methods of generating estimators of parameters, criteria frequently employed for comparing estimators, and then methods of interval estimation.

A word may be in order at this point regarding the notation that we shall use. As has been mentioned, in estimation problems we are trying to find a statistic (function of the elements of a random sample) which would be a good guess to make for the value of an unknown parameter. This statistic will be called the *estimator* of the unknown parameter; since the estimator is a statistic it is a random variable and will be denoted by a capital letter [frequently with a tilde ($\sim$) or caret ($\wedge$) over it to indicate the rationale used in deriving it]. Thus, if our unknown parameter is denoted by γ, the estimator of γ will be denoted by $\tilde{\Gamma}$ (if the method of moments was used) or by $\hat{\Gamma}$ (if maximum likelihood was used). The observed value of the estimator, once we have the observed values of the random variables, is called the *estimate* of the parameter and will be denoted by $\tilde{\gamma}$ or $\hat{\gamma}$.

7.1.   The Method of Moments and Maximum Likelihood

The oldest general method for generating estimates of unknown parameters, given a sample, is called the method of moments. It was first proposed by Karl Pearson in 1894 and was used extensively for many years. It is a simple method to use, in most cases, and generates "reasonable" estimators. However, it does not share the universal good points of the maximum likelihood method which we shall also study.

Briefly, the rationale behind the method of moments (and the method itself) proceeds as follows. We are given a random variable X which has a distribution function F_X; this distribution function F_X is indexed or specified by an unknown parameter γ. Generally, the first moment of X (its mean) will depend in some relatively simple way upon γ, say $\mu_X = g(\gamma)$ (g may be the identity function). Given a sample of n values of X, we can define the first sample moment $\bar{X}$; the method of moments then specifies that we equate $\bar{X}$ to $g(\tilde{\Gamma})$ and solve for $\tilde{\Gamma}$. The resulting value of $\tilde{\Gamma}$ is called the method of moments estimator of γ.

EXAMPLE 7.1.1.   Suppose that X is uniformly distributed over the interval $(0, \gamma)$. Thus the distribution function (and density function) of X is specified, once we know γ. The mean of X is

$$\int_0^{\gamma} x \frac{1}{\gamma}\, dx = \frac{\gamma}{2} \, ;$$

thus $g(\gamma) = \gamma/2$. Given a random sample of n observations of X, we then set up the equation

$$\bar{X} = g(\tilde{\Gamma}) = \frac{\tilde{\Gamma}}{2}$$

whose solution immediately is $\tilde{\Gamma} = 2\bar{X}$.

Our method of moments estimator of γ then is twice the sample mean; if our sample values consist of .5, .6, .1, 1.3, .9, 1.6, .7, .9, 1.0 we find that $\bar{x} = 7.6/9 = .84$ and $\tilde{\gamma} = 1.68$ (the estimate based on this sample). An obvious defect of this estimator of γ is that it does not necessarily give rise to an estimate that generates an interval containing all of the sample values. Obviously, each sample value must lie between zero and γ, if our initial assumption is correct; then it would seem foolish to use as a guess of the value of γ a number which did not exceed each of the sample values. For example, if we had taken a sample of 3 values of X (for this same example) and found them to be 4, 6, and 50, our estimate would be $2(20) = 40$, yet we know that γ cannot be smaller than 50. Thus, we see it is possible for this rather intuitively appealing procedure to give unsatisfactory results. Nonetheless, it is still generally simple to use and most frequently gives "good" estimators.

EXAMPLE 7.1.2.   Suppose we assume the number of traffic fatalities in San Francisco per day to be a Poisson random variable X with parameter λ (which is unknown to us). Then the mean of X is λ and thus $g(\lambda) = \lambda$; the method of moments estimator of λ, given a random sample, then is simply $g(\tilde{\Lambda}) = \tilde{\Lambda} = \bar{X}$. Suppose that we took a sample of 15 days and found the number of fatalities to be 0, 0, 2, 0, 0, 1, 0, 0, 0, 0, 0, 0, 0, 0, 0. Then $\bar{x} = \frac{3}{15} = .2$ and $\tilde{\lambda} = .2$.

As we have seen, many probability laws depend on two or more unknown parameters. Suppose then that F_X is indexed by γ and λ (that is, values must be specified for both γ and λ to specify F_X completely). Then, generally, both of the first two moments of X will be functions of γ and λ; i.e.,

$$\mu_X = g(\gamma, \lambda)$$
$$E(X^2) = h(\gamma, \lambda).$$

The method of moments then specifies that we should set the first two sample moments equal to g and h, respectively, and solve the resulting equations simultaneously for $\tilde{\Gamma}$ and $\tilde{\Lambda}$, the method of moments estimators. (If F_X contains more than two unknown parameters, then a correspondingly larger number of sample moments is used and an equal number of equations must be solved simultaneously; the procedure is identical to the two-parameter case which we now illustrate.)

EXAMPLE 7.1.3.   Assume that X, the distance that a discus will be tossed by a particular person, is normally distributed with unknown mean μ and unknown variance σ^2. Given a random sample of n tosses this person has made, the method of moments estimators of μ and σ^2 are determined as follows. From our study of the normal probability law, we know that μ is the mean of X and that $E[X^2] = \sigma^2 + \mu^2$.

Then $g(\mu, \sigma^2) = \mu$, $h(\mu, \sigma^2) = \sigma^2 + \mu^2$, and the equations whose solutions determine the method of moments estimators are

$$\bar{X} = \tilde{M}$$

$$M_2 = \tilde{\Sigma}^2 + \tilde{M}^2.$$

(M is capital μ, Σ is capital σ.)

The solutions obviously are

$$\tilde{M} = \bar{X}$$

$$\tilde{\Sigma}^2 = M_2 - \bar{X}^2 = \frac{n-1}{n} S^2.$$

EXAMPLE 7.1.4.   Let us now assume that X is uniformly distributed on (a, b) where neither a nor b is known. We recall that

$$\mu = g(a, b) = \frac{b+a}{2}$$

$$E[X^2] = h(a, b) = \frac{b^2 + ab + a^2}{3};$$

then given a random sample of n observations of X the equations which determine the method of moments estimators of a and b are

$$\bar{X} = \frac{\tilde{B} + \tilde{A}}{2}$$

$$M_2 = \frac{\tilde{B}^2 + \tilde{A}\tilde{B} + \tilde{A}^2}{3}.$$

Solving the first equation for $\tilde{A}$ and substituting this in the second results in the solutions:

$$\tilde{B} = \bar{X} + \sqrt{3(M_2 - \bar{X}^2)}$$

$$\tilde{A} = \bar{X} - \sqrt{3(M_2 - \bar{X}^2)}.$$

If we assume that the 9 sample values given in Example 7.1.1 are our observed values for the sample of this random variable, we find that $\bar{x} = .84$, $m_2 = .89$, $\sqrt{3(m_2 - \bar{x}^2)} = .744$, and thus $\tilde{b} = 1.584$ and $\tilde{a} = .096$. Note that the interval $(\tilde{a}, \tilde{b})$ does not include all of the sample values. If we had a sample of 3 whose values were 4, 6, and 50, then $\bar{x} = 20$, $m_2 = 850.67$, $\sqrt{3(m_2 - \bar{x}^2)} = 36.76$, and $\tilde{b} = 56.76$, $\tilde{a} = -16.76$. Notice that this interval does include all of the sample values.

Let us now investigate the *maximum likelihood* method of estimating unknown parameters. While this procedure was employed by the great mathematician C. F. Gauss 170 years ago for isolated problems, it was described as a generally appropriate method, and extensively used, by Sir Ronald A. Fisher. It has proved a very powerful technique and is extensively used today. As we shall see, for some problems both the method of moments and maximum likelihood lead to exactly the same estimators, and for others they do not; when the two methods do not agree the maximum likelihood estimators are generally to be preferred.

The maximum likelihood method may be described as follows. Suppose that X is a random variable with distribution function F_X indexed by, or dependent on, an unknown parameter λ. Fisher proposed that the likelihood function $L(\lambda)$ be defined as the joint density function of the sample if X is continuous or as the joint probability function of the sample if X is discrete; in either case it should be evaluated at the observed sample values $x_1, x_2, \ldots,$ x_n. $L(\lambda)$ then is a function of the single unknown parameter λ. Choosing that value of λ, say $\hat{\lambda}$, that maximizes $L(\lambda)$ yields the maximum likelihood estimate of λ; since $L(\lambda)$ has been evaluated at the observed sample values the maximizing value $\hat{\lambda}$ will generally be a function of the observed sample values. Let us formally define the likelihood function of the sample and the maximum likelihood estimator.

DEFINITION 7.1.1. (1) Let X be a discrete random variable whose probability function p_X depends on an unknown parameter λ. Let $X_1, X_2, \ldots, X_n$ be a random sample of X and let $x_1, x_2, \ldots, x_n$ be the observed sample values. The *likelihood function* of the sample is defined to be

$$L(\lambda) = p_X(x_1)p_X(x_2) \cdots p_X(x_n).$$

(2) Let X be a continuous random variable whose density function f_X depends upon an unknown parameter λ. Let $X_1, X_2, \ldots, X_n$ be a random sample of X and let $x_1, x_2, \ldots, x_n$ be the observed sample values. The *likelihood function* of the sample is defined to be

$$L(\lambda) = f_X(x_1)f_X(x_2) \cdots f_X(x_n).$$

DEFINITION 7.1.2. Given the likelihood function of the sample $L(\lambda)$, let $\hat{\lambda} = g(x_1, x_2, \ldots, x_n)$ be the maximizing value of $L(\lambda)$; i.e., $L(\hat{\lambda}) = \max_{\lambda} L(\lambda)$. The *maximum likelihood estimator* of λ then is

$$\hat{\Lambda} = g(X_1, X_2, \ldots, X_n).$$

Notice that if X is discrete, $L(\lambda)$ gives the probability of observing the sample results, as a function of λ; if X is continuous then $L(\lambda)$ is proportional to the probability of observing a sample outcome in a neighborhood of the values actually observed.

By basing our estimator on the value of λ that maximizes the likelihood function, we have in a sense taken that value of λ that maximizes the probability of occurrence of the sample results, an intuitively appealing thing to do. If we use the estimate $\hat{\lambda}$ as our guess of the unknown value of λ (our estimate of λ), we have chosen that value which maximizes the probability of what we observed in our sample.

Before looking at some examples, let us take note of a calculus theorem which is almost always of use in finding $\hat{\lambda}$, the maximizing value of λ. Generally, since $L(\lambda)$ is either a product of probability functions or density functions, it will always be positive (for the range of possible values of λ). Thus, $K(\lambda) = \log_e L(\lambda)$ can always be defined; the theorem referred to merely states that the value of λ that maximizes K also maximizes L. It is almost always easier to maximize K than L.

EXAMPLE 7.1.5.   Let X be a Bernoulli random variable with unknown parameter p and let $X_1, X_2, \ldots, X_n$ be a random sample of X. Then

$$p_X(x) = p^x(1 - p)^{1-x}, \qquad x = 0, 1$$

and the likelihood function of the sample is

$$L(p) = \prod_1^n p_X(x_i) = \prod_1^n p^{x_i}(1 - p)^{1-x_i}$$
$$= p^{\Sigma x_i}(1 - p)^{n-\Sigma x_i}.$$

Then we define

$$K(p) = \log L(p) = \sum x_i \log p + (n - \sum x_i) \log (1 - p).$$

K is a continuous function of p and, thus, if there exists a value of p such that

$$\frac{dK}{dp} = 0, \qquad \frac{d^2K}{dp^2} < 0,$$

then this value makes K a maximum.   Now

$$\frac{dK}{dp} = \frac{\sum x_i}{p} - \frac{n - \sum x_i}{1 - p};$$

setting this equal to zero and solving for $\hat{p}$ (the extreme point) yields

$$\hat{p} = \frac{\sum x_i}{n} = g(x_1, x_2, \ldots, x_n).$$

We note that

$$\frac{d^2K}{dp^2} = -\frac{\sum x_i}{p^2} - \frac{n - \sum x_i}{(1 - p)^2}$$

which is negative for any value of p, so $\hat{p}$ does correspond to a maximum value of K (and L). The maximum likelihood estimator of p then is

$$\hat{P} = g(X_1, X_2, \ldots, X_n) = \frac{\sum X_i}{n} = \bar{X}.$$

EXAMPLE 7.1.6.   Suppose that X is a normal random variable whose mean μ is unknown and whose variance is equal to 1. If $X_1, X_2, \ldots, X_n$ is a random sample of X, then the likelihood function of the sample is

$$L(\mu) = \prod_{i=1}^{n} f_X(x_i) = \prod_{1}^{n} \frac{1}{\sqrt{2\pi}} e^{-(x_i-\mu)^2/2}$$

$$= \left(\frac{1}{2\pi}\right)^{n/2} e^{-\Sigma(x_i-\mu)^2/2};$$

its logarithm is

$$K(\mu) = \log L(\mu) = -\frac{n}{2}\log 2\pi - \sum \frac{(x_i - \mu)^2}{2}$$

and

$$\frac{dK}{d\mu} = \sum (x_i - \mu) = \sum x_i - n\mu$$

$$\frac{d^2K}{d\mu^2} = -n < 0.$$

Thus

$$\hat{\mu} = \frac{1}{n}\sum x_i = \bar{x}$$

maximizes K (and L) and the maximum likelihood estimator is

$$\hat{M} = \frac{1}{n}\sum X_i = \bar{X}.$$

(This is also the value that the method of moments yields in this case.)

In each of the two examples discussed so far for maximum likelihood estimation we were able to take the derivative of the log of the likelihood function and equate it to zero to find the estimator. If we were to plot the likelihood function as a function of the unknown parameter in these two cases, we would find it had a graph like Figure 7.1.1. The likelihood function in these cases has a unique maximum value and the derivative is zero at that point. It is possible, however, for the likelihood function to have one of the shapes pictured in Figure 7.1.2; as illustrated in 7.1.2a, there may be two

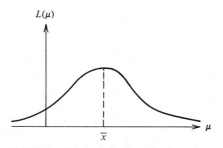

FIGURE 7.1.1 Likelihood function, Examples 7.1.5 and 7.1.6.

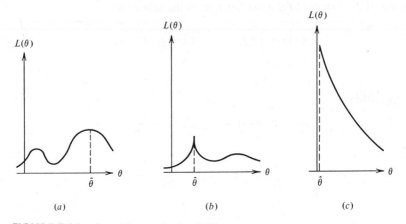

FIGURE 7.1.2   Possible graphs for **L**(θ).

points at which the derivative is zero. The maximum likelihood estimator is the value corresponding to the absolute maximum value, as marked. For the shapes given in 7.1.2*b* and *c*, the derivative is not zero at the maximum value. In these cases other techniques must be relied upon to find the estimator. Example 7.1.7 gives an example in which the likelihood function has a shape like that illustrated in Figure 7.1.2*c*.

EXAMPLE 7.1.7.   Suppose that X is a uniform random variable on the interval $(0, \gamma)$ where γ is an unknown positive parameter. Then

$$f_X(x) = \frac{1}{\gamma}, \qquad \text{for } 0 < x < \gamma$$

and the likelihood function for a random sample of n is

$$L(\gamma) = \frac{1}{\gamma^n}.$$

The graph of this function is given in Figure 7.1.2*c*. We note that the slope is not zero anywhere, so there would be no point in taking the derivative of L (or of K) and equating it to zero. However, the likelihood increases in value the closer that γ is to zero; thus $L(\gamma)$ is maximized by setting γ equal to the smallest value it could possibly equal. Clearly, γ can be no smaller than the largest value that occurs in our sample (since each element of the sample is assumed to have the same density as X). Thus $\hat{\gamma} = x_{(n)}$, and the maximum likelihood estimator is $\hat{\Gamma} = X_{(n)}$, the maximum value in the sample. Note that this estimator would not be subject to the criticism levelled against the method of moments estimator for this parameter in Example 7.1.1.

If the random variable X has a distribution function indexed by two parameters γ and λ, for example, then the likelihood function is defined in exactly the same way but it is now a function of two arguments, λ and γ, rather than one. The maximum likelihood estimators of the two unknown parameters are specified by the two coordinates $\hat{\gamma}$ and $\hat{\lambda}$ (each of which will be functions of the observed sample values) corresponding to the maximum value of $L(\gamma, \lambda)$.

If F_X is indexed by more than two unknown parameters, the likelihood function L is a function of more than two parameters. The maximum likelihood estimators are again specified simply by the coordinates, no matter how many there may be, of the maximum value of L. Let us consider a specific example.

EXAMPLE 7.1.8. Suppose we assume that the weights of Valencia oranges are normally distributed with mean μ and variance σ^2, both of which are unknown. If we were to take a random sample of n oranges of this type, we could use the method of maximum likelihood to estimate μ and σ^2 as follows. The likelihood function for the sample is a function of the two arguments, μ and σ^2, and is

$$L(\mu, \sigma^2) = \prod_{i=1}^{n} \frac{1}{\sigma\sqrt{2\pi}} e^{-(x_i-\mu)^2/2\sigma^2}$$

$$= \frac{1}{(2\pi\sigma^2)^{n/2}} e^{-(1/2)\Sigma(x_i-\mu)^2/\sigma^2}.$$

Then, since L is a continuous function of μ and σ^2, we might expect to find the maximizing values, $\hat{\mu}$ and $\hat{\sigma}^2$, by setting the two first partial derivatives of L (or of $K = \log L$) equal to zero and solving. Thus,

$$K(\mu, \sigma^2) = -\frac{n}{2}\log 2\pi - \frac{n}{2}\log \sigma^2 - \frac{1}{2\sigma^2}\sum(x_i - \mu)^2$$

and

$$\frac{\partial K}{\partial \mu} = \frac{1}{\sigma^2}\sum(x_i - \mu)$$

$$\frac{\partial K}{\partial \sigma^2} = -\frac{n}{2\sigma^2} + \frac{1}{2(\sigma^2)^2}\sum(x_i - \mu)^2.$$

The equations to be solved for the estimates then are

$$\frac{1}{\hat{\sigma}^2}\sum(x_i - \hat{\mu}) = 0$$

$$-\frac{n}{2\hat{\sigma}^2} + \frac{1}{2(\hat{\sigma}^2)^2}\sum(x_i - \hat{\mu})^2 = 0.$$

The first equation clearly has as its solution

$$\hat{\mu} = \frac{1}{n}\sum x_i = \bar{x}.$$

Substituting this in the second equation and solving for $\hat{\sigma}^2$ yields

$$\hat{\sigma}^2 = \frac{1}{n}\sum (x_i - \bar{x})^2.$$

It can be shown that these two quantities do correspond to the maximum value of L. The maximum likelihood estimators of μ and σ^2 then are

$$\hat{M} = \bar{X}$$

$$\hat{\Sigma}^2 = \frac{1}{n}\sum (X_i - \bar{X})^2 = \frac{n-1}{n} S^2.$$

These are the same estimators that the method of moments yields for this problem.

EXERCISE 7.1

1. Assume that the amount of rainfall recorded at a certain station on a given date (for example, September 30) is uniformly distributed on the interval $(0, b)$. If a sample of 10 years' records shows that the following amounts were recorded on that date, compute the estimate $\tilde{b}$ of b: 0, 0, .7, 1, .1, 0, .2, .5, 0, .6. (Measurements are recorded in inches.)

2. Assume that the amount of growth in height of 1-foot tall Monterey pine trees in a year is a normal random variable with unknown mean and unknown variance. The growths of 5 trees were recorded as: 3 feet, 5 feet, 2 feet, 1.5 feet, 3.5 feet. Compute the method of moments estimates for μ and σ^2.

3. The number of cars arriving at a supermarket parking lot per hour is assumed to be a Poisson random variable with parameter λ. The following numbers of cars were observed arriving during the hour from 9 to 10 a.m. on 6 successive days: 50, 47, 82, 91, 46, 64. Compute $\tilde{\lambda}$.

4. X is a geometric random variable with parameter p. Given a random sample of n observations of X, what is the method of moments estimator of p?

5. Richard is allowed to shoot a basketball from the free-throw line until he makes a basket. He does this 5 times, with the number of shots necessary being 5, 1, 7, 4, and 9, respectively. Compute the method of moments estimate of p, the probability he will make a basket, shooting from the free-throw line. (Assume that the number of shots necessary is a geometric random variable.)

6. The number of hours an electron tube will work is assumed to be an exponential random variable with parameter λ. Given a sample of n lifetimes for tubes of this sort, compute the method of moments estimator for λ.

7. X is a Bernoulli random variable with parameter p. Given a random sample of n observations of X, compute the method of moments estimator for p.

8. X is a binomial random variable with parameters n (known) and p. Given a random sample of N observations of X, compute the method of moments estimator for p.

9. Suppose that X is a normal random variable with mean $\mu = 5$ and unknown variance σ^2. What is the method of moments estimator for σ^2, based on a random sample of n observations of X?

10. If X is assumed to be uniformly distributed on the interval $(b - \frac{1}{4}, b + 5)$, what is the method of moments estimator for b based on a random sample of n observations?

11. X is a binomial random variable with parameters n and p, both unknown. Given a random sample of N observations of X, compute the method of moments estimators of n and p.

12. Suppose that X is a normal random variable with mean $\mu = 10$ and variance σ^2 unknown. What is the maximum likelihood estimator for σ^2, based on a random sample of n observations of X?

13. Suppose that you buy a dozen Valencia oranges, weigh each of them, and find

$$\sum_{i=1}^{12} x_i = 180, \qquad \sum_{i=1}^{12} x_i^2 = 2799$$

(in ounces). Compute the maximum likelihood estimates of μ and σ^2, assuming that the 12 oranges you have purchased are a random sample of all Valencia oranges (and their weights are normally distributed).

14. Suppose that X is a Poisson random variable with parameter λ. Given a random sample of n observations of X, what is the maximum likelihood estimator for λ?

15. Assume that the number of new car sales, per day, that a particular dealer makes is a Poisson random variable with parameter λ. Given that in 20 days the total number of sales he made was 30 cars, what is the maximum likelihood estimate of λ?

16. If X is a geometric random variable with parameter p, what is the maximum likelihood estimator for p based on a random sample of n observations of X?

17. Suppose that X were uniformly distributed on the interval $b - \frac{1}{2}$ to $b + \frac{1}{2}$. What is the maximum likelihood estimator for b based on a random sample of n observations of X?

18. What is the maximum likelihood estimator for the parameter λ of an exponential random variable for a sample of size n?

19. A random sample of 30 electron tubes was placed in operation and the time to failure for each was recorded. If you assume that the recorded times were such that $\Sigma \, x_i = 32{,}916$ hours, what is the maximum likelihood estimate for the parameter of the exponential distribution of tube lives?

20. Suppose that X is a binomial random variable with parameters n (known) and p. What is the maximum likelihood estimator for p based on a random sample of N observations of X?

21. A used car salesman is willing to assume the number of sales he makes, per day, is a Poisson random variable with parameter λ. Over the past 30 days, he made 0 sales on 20 days and one or more sales on each of the remaining 10 days. What is the method of moments estimate for λ?

22. Answer Exercise 21 using the maximum likelihood method.

23. Assume X is a normal random variable with mean μ and variance μ^2. Given a random sample of n observations of X, what is the method of moments estimator for μ?

24. As in 23, assume X is normal with mean μ and variance μ^2. What is the maximum likelihood estimator for μ, given a random sample of n observations of X?

25. Assume we have a random sample of size n_1 of a normal random variable X, whose parameters are μ_1, σ_1^2. We also have an independent random sample of size n_2 of a normal random variable Y with parameters μ_2, σ_2^2.
    (a) What are the method of moments estimates for μ_1, σ_1^2, μ_2, σ_2^2?
    (b) What are the maximum likelihood estimators for μ_1, σ_1^2, μ_2, σ_2^2?

26. Assume $\mu_1 = \mu_2 = \mu$ and $\sigma_1^2 = \sigma_2^2 = \sigma^2$ in 25 and find the method of moments and maximum likelihood estimators for μ and σ^2.

27. Assume $\sigma_1^2 = \sigma_2^2 = \sigma^2$ in 25 and find the method of moments and maximum likelihood estimators for μ_1, μ_2 and σ^2.

28. Assume $X_1, X_2, \ldots, X_n$ is a random sample of a uniform random variable on the interval $(-b, b)$ and find the method of moments and maximum likelihood estimators for b. This is one of those cases in which the equation defining the method of moments estimator is not $E[X] = \bar{X}$.

29. Assume that X is a continuous random variable with density

$$f_X(x) = e^{\alpha - x}, \, x > \alpha.$$

(This is an exponential with displaced origin.)
(a) Find the method of moments estimator for α.
(b) Find the maximum likelihood estimator for α.

7.2. Properties of Estimators

In the last section we have seen two different, generally applicable methods for constructing estimators of unknown parameters. In many cases the two methods generate the same estimator, but in many important problems they do not lead to the same estimator. When we are faced with the choice of two or more estimators for the same parameter, it becomes important to develop criteria for comparing them; if we were able to develop some scale of "goodness" of estimators, then we would always want to use the estimator that was best for the given problem. Unfortunately there is no universally appropriate scale of goodness which can be used in comparing estimators of the same parameter.

As we have seen, the estimator of an unknown parameter is a statistic, a random variable whose value can be observed on the basis of a sample. Since an estimator $\hat{\Gamma}$ is always a random variable, the particular value it takes on varies from one sample to another; thus we certainly would not expect to have the estimate $\hat{\gamma}$ equal to the true unknown value γ for every sample we take. If we are considering two estimators, $\tilde{\Gamma}$ and $\hat{\Gamma}$, for the same parameter γ, we can in theory derive the probability laws of $\tilde{\Gamma}$ and of $\hat{\Gamma}$ from the probability law of the random sample. Comparison of $\tilde{\Gamma}$ and $\hat{\Gamma}$ then reduces to a comparison of their two respective probability laws in some way. If, for example, $\tilde{\Gamma}$ were uniformly distributed from $\gamma - \frac{1}{4}$ to $\gamma + \frac{1}{4}$ while $\hat{\Gamma}$ was uniformly distributed on the interval from $\gamma - \frac{1}{16}$ to $\gamma + \frac{1}{16}$, we would obviously prefer $\hat{\Gamma}$ as an estimator of γ (and thus would use $\hat{\gamma}$ rather than $\tilde{\gamma}$ as our estimate for any particular sample). Unfortunately the comparison of probability laws of estimators is generally not this straightforward; in most cases the probability density functions (or probability functions) of various estimators are of widely divergent types and there is then a variety of ways in which they could be compared. Perhaps the most obvious property to investigate of the probability law of an estimator Γ would be its mean value. As the following definition states, if the mean value of Γ is γ, then Γ is unbiased; otherwise it is a biased estimator.

DEFINITION 7.2.1. An estimator Γ of an unknown parameter γ is *unbiased* if $E[\Gamma] = \gamma$, for all values of γ.

Thus an unbiased estimator is a random variable whose expected value is the parameter being estimated; if we were to take samples of size n repeatedly and for each compute the observed value of Γ (the estimate for that sample outcome), then the average of these observed values would be γ, the parameter being estimated.

EXAMPLE 7.2.1. We saw in Chapter 6, Section 2, that $E[\bar{X}] = \mu$, no matter what the probability law from which we were sampling. Thus $\bar{X}$ is an unbiased estimator of μ in a normal density, of the parameter p for a Bernoulli random variable, and of the parameter λ for a Poisson random variable (since in each of these cases the parameter mentioned is the mean of the random variable being sampled).

The property of unbiasedness, while generally desirable for an estimator, is not the only criterion that should be employed when comparing estimators. In all cases we can find a large number of unbiased estimators of a parameter; thus we should have at least one additional criterion for judging which is best among all unbiased estimators.

If an estimator is unbiased, then the average of the values it takes on over an infinite number of samples is the parameter estimated; however, this statement does not imply that the estimator has very high probability of lying close to the unknown parameter for any given sample. Clearly, if Γ_1 and Γ_2 are two different unbiased estimators of γ for the same sample, and Γ_1 had probability .9 of being within 1 unit of γ while Γ_2 had probability .5 of being within 1 unit of γ, then we would prefer to use the observed value of Γ_1 as our estimate rather than the observed value of Γ_2. This additional idea of choosing the unbiased estimator that has high probability of lying close to the parameter leads one to consider the variances of the unbiased estimators. The unbiased estimator that has smaller variance (about the true value of the parameter) then would have the larger probability of being close to the unknown parameter. Generally in the literature this comparison of variances of estimators is described as comparisons of efficiency of estimators (the more efficient estimator has smaller variance). Accordingly, we have the following definition.

DEFINITION 7.2.2. If Γ_1 and Γ_2 are both unbiased estimators of γ for the same sample, then Γ_1 is *more efficient* than Γ_2 if

$$V(\Gamma_1) < V(\Gamma_2)$$

($V(\Gamma_1)$ is the variance of Γ_1).

The following example compares two unbiased estimators.

EXAMPLE 7.2.2. Suppose we have a random sample of size 2 of an exponential random variable X with unknown parameter $1/\theta$; thus the population random variable has density

$$f_X(x) = \frac{1}{\theta} e^{-x/\theta}, \quad x > 0$$

and $\mu_X = \theta$, $\sigma_X^2 = \theta^2$. As we know in general $\bar{X} = (X_1 + X_2)/2$ has mean $\mu_{\bar{X}} = \mu_X = \theta$ and variance $\sigma_{\bar{X}}^2 = \sigma_X^2/2 = \theta^2/2$. Let us consider the geometric mean of X_1 and X_2 as an alterative estimator; note

$$E[\sqrt{X_1 X_2}] = E[\sqrt{X_1}]E[\sqrt{X_2}]$$

$$= \{E[\sqrt{X}]\}^2$$

$$= \left\{ \int_0^\infty x^{1/2} \frac{1}{\theta} e^{-x/\theta} \, dx \right\}^2$$

$$= \left\{ \sqrt{\theta} \int_0^\infty u^{1/2} e^{-u} \, du \right\}^2$$

$$= \theta \left\{ \Gamma\left(\frac{3}{2}\right) \right\}^2 = \theta \frac{\pi}{4}.$$

So, if we define $\Theta = \dfrac{4}{\pi}\sqrt{X_1 X_2}$, then Θ is also an unbiased estimator for θ. To evaluate its variance we find

$$E[\Theta^2] = \frac{16}{\pi^2} E[X_1 X_2] = \frac{16}{\pi^2}\theta^2.$$

Thus its variance is

$$\sigma_\Theta^2 = \theta^2(16/\pi^2 - 1) = .621\theta^2.$$

Notice then that $\theta^2/2 < .621\theta^2$ so $\bar{X}$ is a more efficient estimator of θ.

DEFINITION 7.2.3. Γ is the *best linear unbiased estimator* for γ if:
   (1) Γ is a linear function of $X_1, X_2, \ldots, X_n$,
   (2) $E[\Gamma] = \gamma$ for all values of γ, and
   (3) among all estimators Γ^* satisfying (1) and (2), $V[\Gamma] \le V[\Gamma^*]$.

A best linear unbiased estimator then is one that has both the property of unbiasedness and the smallest possible variance within a certain class of estimators. (Thus it could also be called most efficient within the class of linear unbiased estimators.)

Theorem 7.2.1. X is a random variable with mean μ and variance σ^2. If $X_1, X_2, \ldots, X_n$ is a random sample of X, $\bar{X}$ is the best linear unbiased estimator of μ.

Proof: Let M be a linear unbiased estimator of μ. Then

$$\mathbf{M} = \sum_{i=1}^{n} a_i X_i + b$$

and

$$E[\mathbf{M}] = \mu \sum a_i + b.$$

Since M must be unbiased, we must have $\mu\sum a_i + b = \mu$, no matter what value μ may have. This implies then that $b = 0$ and $\sum a_i = 1$. The variance of M is

$$V(\mathbf{M}) = V(\sum a_i X_i)$$
$$= \sigma^2 \sum a_i^2.$$

Finding the best linear unbiased estimator then reduces to finding $a_1, a_2, \ldots,$ a_n such that $\sum a_i = 1$ and $\sum a_i^2$ is as small as possible. Now, $\sum a_i = 1$ implies that

$$a_1 = 1 - \sum_{2}^{n} a_i$$

and thus we want to minimize

$$Q = \left[1 - \sum_{2}^{n} a_i\right]^2 + \sum_{2}^{n} a_i^2.$$

This minimization is accomplished by setting

$$\frac{\partial Q}{\partial a_2}, \frac{\partial Q}{\partial a_3}, \dots, \frac{\partial Q}{\partial a_n}$$

each equal to zero and solving the resulting equations for $a_2^*, a_3^*, \dots, a_n^*$ (then $a_1^* = 1 - \sum_2^n a_i^*$). We find

$$\frac{\partial Q}{\partial a_j} = 2\left[1 - \sum_2^n a_i\right](-1) + 2a_j, \qquad j = 2, 3, \dots, n$$

and thus

$$a_j^* = \left[1 - \sum_2^n a_i^*\right] = a_1^*, \qquad j = 2, 3, \dots, n.$$

But then

$$\sum_1^n a_i^* = na_1^* = 1$$

and thus

$$a_j^* = \frac{1}{n}, \qquad j = 1, 2, \dots, n$$

are the values that minimize $\sum_1^n a_i^2$ subject to the condition that $\sum_1^n a_i = 1$. Thus,

$$M = \sum \frac{1}{n} X_i = \bar{X}$$

is the linear unbiased estimator of μ with the smallest variance. It can, in fact, be shown that $\bar{X}$ has the smallest variance among all unbiased estimators of μ, regardless of whether they are linear functions of $X_1, X_2, \dots, X_n$. ◀

Note that the preceding theorem holds true without assuming a specific form for the distribution function for X (other than the fact that μ and σ^2 are assumed to exist). Thus the sample mean is the best linear unbiased estimator for p in the Bernoulli probability function, for λ in the Poisson probability function, and for μ in the normal density function.

Let us make note of one further property which estimators may possess, that of consistency. The subscript on Θ denotes the sample size.

DEFINITION 7.2.4. Θ_n is a *consistent* estimator of θ if

$$\lim_{n \to \infty} P[|\Theta_n - \theta| > \varepsilon] = 0$$

or, equivalently, if

$$\lim_{n \to \infty} P[|\Theta_n - \theta| < \varepsilon] = 1, \qquad \text{for any } \varepsilon > 0$$

(n is the sample size).

Note first of all that an estimator Θ_n is consistent if a sequence of probabilities (evaluated from the probability or density function of Θ_n) converges to zero (or 1) as the sample size increases. Thus consistency, strictly speaking, has to do only with the limiting behavior of an estimator as the sample size increases without limit and does not imply that the observed value of Θ_n is necessarily close to θ for any specific size of sample n. The consistency of an estimator is called an *asymptotic* or large sample *property* of the estimator; presumably if n is fixed at some "large" value, then

$$P[|\Theta_n - \theta| > \varepsilon]$$

is quite small. What sample size is "large" for one estimator is not necessarily large for another in this sense. In particular, if we are only going to have a relatively small sample available, and only one at that, it would seem quite immaterial whether we use a consistent estimator or not.

The following theorem, which we shall not prove, gives a relatively easy check on whether a particular estimator is consistent.

Theorem 7.2.2. If

$$\lim_{n \to \infty} E[\Theta_n] = \theta$$

and

$$\lim_{n \to \infty} V(\Theta_n) = 0,$$

then Θ_n is a consistent estimator of θ (n in the sample size).

Notice that if Θ_n is a random variable whose expected value gets closer and closer to θ as n increases, and whose variance shrinks to zero as n increases, it will be a consistent estimator of θ.

EXAMPLE 7.2.3. Suppose that X is a random variable with mean μ and variance σ^2. If $X_1, X_2, \ldots, X_n$ is a random sample of X, then we know that

$$E[\bar{X}] = \mu, \qquad V(\bar{X}) = \frac{\sigma^2}{n}.$$

Thus

$$\lim_{n \to \infty} E[\bar{X}] = \mu, \qquad \lim_{n \to \infty} V(\bar{X}) = \lim_{n \to \infty} \frac{\sigma^2}{n} = 0$$

and $\bar{X}$ is a consistent estimator of μ.

Sufficient statistics, and estimators based on them, have many desirable properties. Roughly speaking, if we can find a statistic which summarizes all the information about an unknown parameter, then that statistic is called a *sufficient* statistic for the unknown parameter. This idea is made more exact in the following definition.

DEFINITION 7.2.5.   Given a random sample of a random variable X, whose distribution depends on an unknown parameter θ, the statistic $Y = g(X_1, X_2, \ldots, X_n)$ is a *sufficient statistic* for θ if and only if the conditional distribution of *any* other statistic $h(X_1, X_2, \ldots, X_n)$, given $Y = y$, does not depend on θ.

Notice, then, that in a sense all the sample information concerning θ has been condensed into the statistic Y, if Y is sufficient for θ; this is reflected in the fact that the conditional distribution for any other statistic does not involve θ, once we observe $Y = y$. Since their conditional distributions do not involve θ the particular values they take on cannot contain any more information about θ, beyond the given value of the sufficient statistic. A complete general discussion of sufficient statistics, and their properties involve mathematical results beyond those assumed in this text. However, it can be shown that $Y = \sum_{i=1}^{n} X_i$ is a sufficient statistic for the parameter p of a Bernoulli random variable, for any sample size n. The basic reasoning behind this statement is well illustrated by the special case with $n = 2$, treated in the following example.

EXAMPLE 7.2.4.   Assume that an 1899 Indian head penny has (unknown) probability p of turning up heads when flipped in an unbiased manner. The coin is to be flipped 2 times. Let X_1 be the 1 if heads occur on the first flip, and $X_1 = 0$, otherwise; similarly, $X_2 = 1$ if the second flip results in a head and $X_2 = 0$, otherwise. Then X_1, X_2 is a random sample of a Bernoulli random variable X with probability function

$$p_X(x) = p^x (1 - p)^{1-x}, \quad x = 0, 1.$$

Define $Y = X_1 + X_2 = g(X_1, X_2)$; then the marginal probability function for Y is

$$p_Y(y) = \binom{2}{y} p^y (1 - p)^{2-y}, \quad y = 0, 1, 2,$$

(since Y is binomial with parameters $n = 2$ and p). Then, let $V = h(X_1, X_2)$ be any other statistic. We can see then that the joint probability function for Y and V is positive only at $(0, h(0, 0))$, $(1, h(1, 0))$, $(1, h(0, 1))$, $(2, h(1, 1))$ in the (y, v) plane and it has values $(1 - p)^2$, $p(1 - p)$, $(1 - p)p$, p^2, respectively, at these points. Notice, then, if we are given $Y = 0$, the conditional probability function for V is

$$p_{V|Y}(v \mid 0) = \frac{(1 - p)^2}{\binom{2}{0} p^0 (1 - p)^{2-0}} = 1, \quad \text{for } v = h(0, 0);$$

furthermore, given $Y = 1$, the conditional probability function for V is

$$p_{V|Y}(v \mid 1) = \frac{p(1 - p)}{\binom{2}{1} p(1 - p)} = \tfrac{1}{2}, \quad \text{for } v = h(0, 1), h(1, 0).$$

and given $Y = 2$, the conditional probability function for V is

$$p_{V|Y}(v \mid 2) = \frac{p^2}{\binom{2}{2}p^2(1-p)^0} = 1, \quad \text{for } v = h(1, 1).$$

Since the conditional probability function for V does not involve p, no matter what the particular function V may be, $Y = X_1 + X_2$ is a sufficient statistic for p.

The Fisher-Neyman factorization criterion, stated without proof in the following theorem, generally provides a fairly easy way of finding the sufficient statistic for an unknown parameter (if one exists).

Theorem 7.2.3. Assume $X_1, X_2, \ldots, X_n$ is a random sample of a random variable whose distribution depends on an unknown parameter θ. Let $Y = g(X_1, X_2, \ldots, X_n)$ be a statistic. Then Y is a sufficient statistic for θ if

$$\prod_{i=1}^{n} p_X(x_i) = G(g(x_1, x_2, \ldots, x_n), \theta)H(x_1, x_2, \ldots, x_n),$$

if X is discrete, or if

$$\prod_{i=1}^{n} f_X(x_i) = G(g(x_1, x_2, \ldots, x_n), \theta)H(x_1, x_2, \ldots, x_n),$$

if X is continuous; in either case H does not involve θ and may depend on $x_1, x_2, \ldots, x_n$.

Let us immediately apply this theorem to a sample of size n of a Bernoulli random variable.

EXAMPLE 7.2.5.   Assume that $X_1, X_2, \ldots, X_n$ is a random sample of a Bernoulli random variable with parameter p. Then

$$\prod_{i=1}^{n} p_X(x_i) = \prod_{i=1}^{n} [p^{x_i}(1-p)^{1-x_i}]$$

$$= p^{\Sigma x_i}(1-p)^{n-\Sigma x_i}$$

$$= G(\textstyle\sum x_i, p)H(x_1, x_2, \ldots, x_n)$$

where $H(x_1, x_2, \ldots, x_n) = 1$. Thus $Y = \sum_{i=1}^{n} X_i$ is a sufficient statistic for p.

The reason sufficient statistics are of interest in estimation problems, of course, is that if we can find a sufficient statistic for a parameter then an estimator based on that statistic should in a sense contain everything that the sample has to offer regarding the value of the unknown parameter. If we base our guess, or estimate, of the unknown parameter on the observed value of the sufficient statistic, we should expect to have used the full amount of information available about θ in the sample. In the case of estimating the parameter p for the Bernoulli random variable, the unbiased estimator of p based on the sufficient statistic is $\hat{P} = \bar{X}$ (note that both the method of moments and maximum likelihood yield this estimator for this case).

It can be shown that if Θ is the unbiased estimator for θ, based on the sufficient statistic, then Θ has the smallest variance among all unbiased estimators for θ. This is a major reason for basing estimators on the sufficient statistics.

EXAMPLE 7.2.6.   Assume X is a normal random variable with mean 0 and unknown variance σ^2; $X_1, X_2, \ldots, X_n$ is a random sample of X.   Then

$$\prod_{i=1}^{n} f_{X_i}(x_i) = \prod_{i=1}^{n} \left(\frac{1}{\sigma \sqrt{2\pi}} \right) e^{-x_i^2/2\sigma^2}$$

$$= \left(\frac{1}{\sigma} \right)^n e^{-\Sigma x_i^2/2\sigma^2} \cdot \left(\frac{1}{2\pi} \right)^{n/2}$$

$$= G(\textstyle\sum x_i^2, \sigma^2) \, H(x_1, x_2, \ldots, x_n),$$

where $H(x_1, x_2, \ldots, x_n) = \left(\dfrac{1}{2\pi} \right)^{n/2}$.   Thus $\displaystyle\sum_{i=1}^{n} X_i^2$ is the sufficient statistic for σ^2 for this case.   Since

$$E\left[\sum_{i=1}^{n} X_i^2 \right] = \sum_{i=1}^{n} E[X_i^2] = n\sigma^2$$

notice that $\dfrac{1}{n} \sum X_i^2$ is an unbiased estimator for σ^2 and, using the result mentioned above, has the smallest possible variance among all such unbiased estimators.

Let us now make note of some general results regarding maximum likelihood estimators of parameters.   The first of these is generally called the *invariant property* of maximum likelihood estimators.   Suppose $\hat{\Gamma}$ is the maximum likelihood estimator of a parameter γ.   If for some reason it is desired to estimate a function of γ, say $\theta = h(\gamma)$, rather than γ itself, we could revert to the likelihood function $L(\gamma)$ and express it in terms of θ rather than γ (by substituting the values of θ corresponding to those of γ through the function $\theta = h(\gamma)$).   Then the maximum likelihood estimator of θ, $\hat{\Theta}$, would be determined by the value that maximizes L.   The invariant property of maximum likelihood estimators says that this maximizing value $\hat{\theta}$ of L will yield the estimator $\hat{\Theta} = h(\hat{\Gamma})$; thus there is no need actually to express L in terms of θ to get the estimator if we already have $\hat{\Gamma}$.

EXAMPLE 7.2.7.   Suppose that the time to failure X of a certain type of electron tube is an exponential random variable with parameter θ.   Thus, $E[X] = 1/\theta$.   If we take a random sample of n observations of X, the maximum likelihood estimator of θ is

$$\hat{\Theta} = \frac{1}{\bar{X}}$$

By the invariance principle the maximum likelihood estimator of $E[X] = 1/\theta$ is $1/\hat{\Theta} = \bar{X}$, the sample mean.

Another result that has some practical significance is that if a sufficient statistic, Y, exists for an unknown θ, based on a random sample of a random variable, then the maximum likelihood estimator of θ will be a function of Y. Thus the maximum likelihood estimator uses all the sample information about θ. Additionally, maximum likelihood estimators are generally consistent and, for large sample sizes, approximately normally distributed, as is given without proof in the following theorem.

Theorem 7.2.4. If X is a random variable with probability function p_X or density function f_X which depends on an unknown parameter γ and $\hat{\Gamma}$ is the maximum likelihood estimator based on a random sample of size n of X, then $\hat{\Gamma}$ is approximately normally distributed for large n with mean γ and variance

$$\left\{ nE\left[\frac{d}{d\gamma} \log_e p_X(X)\right]^2 \right\}^{-1}$$

or

$$\left\{ nE\left[\frac{d}{d\gamma} \log_e f_X(X)\right]^2 \right\}^{-1},$$

depending on whether X is discrete or continuous.

This result again is one where "large" must remain undefined; for some probability laws $n = 50$ would undoubtedly be large enough for sufficient accuracy while for others the sample size might have to be 500 to attain the same accuracy.

EXAMPLE 7.2.8. Suppose that $X_1, X_2, \ldots, X_n$ is a random sample of a geometric random variable with parameter p. Then the maximum likelihood estimator of p is

$$\hat{P} = \frac{1}{\bar{X}}$$

For large n, then, Theorem 7.2.3 says that $1/\bar{X}$ is approximately normally distributed with mean p and variance $p^2(1 - p)/n$. This large sample variance is derived as follows: Since X is a geometric random variable,

$$p_X(X) = (1 - p)^{X-1}p$$

$$\frac{d}{dp} \log_e p_X(X) = \frac{1}{1-p}\left[\frac{1}{p} - X\right]$$

$$E\left[\frac{d}{dp} \log_e p_X(X)\right]^2 = \frac{1}{(1-p)^2} E\left[\left(X - \frac{1}{p}\right)^2\right]$$

$$= \frac{1}{(1-p)^2} \frac{1-p}{p^2} = \frac{1}{p^2(1-p)}$$

and thus

$$\left\{nE\left[\frac{d}{dp} \log_e p_X(X)\right]^2\right\}^{-1} = \frac{p^2(1-p)}{n}.$$

EXERCISE 7.2

1. Suppose that $X_1, X_2, \ldots, X_6$ is a random sample of a normal random variable with mean μ and variance σ^2. Determine C such that

$$C[(X_1 - X_2)^2 + (X_3 - X_4)^2 + (X_5 - X_6)^2]$$

   is an unbiased estimator of σ^2.

2. Compute the variance of the estimator of σ^2 given in problem 1. Is it more efficient than the sample variance S^2?

3. Suppose that the time it takes a distance runner to run a mile is a normal random variable with mean μ and variance σ^2. Suppose he will run n_1 mile races in May and n_2 mile races in June. Let $\bar{X}_1$ be the mean value of his times in May and $\bar{X}_2$ be the mean value of his times in June. Then $a\bar{X}_1 + (1 - a)\bar{X}_2$ will be an unbiased estimator of μ for any value of a. Find the value of a that minimizes the variance of $a\bar{X}_1 + (1 - a)\bar{X}_2$, assuming $\bar{X}_1$ and $\bar{X}_2$ are independent.

4. $X_1, X_2, \ldots, X_n$ is a random sample of a normal random variable with mean μ and variance σ^2 (n is an even number). Is

$$\frac{1}{n} \sum_{i=1}^{n/2} (X_{2i} - X_{2i-1})^2$$

   a consistent estimator for σ^2?

5. On the basis of his experience, an automobile dealer feels that the number of new car sales he makes per working day is a Poisson random variable with parameter λ. He examines his records for the past year (which consisted of 310 working days) and finds that the total number of cars he sold was 279. Compute the maximum likelihood estimate of the probability that he will sell no cars on his next working day.

6. Assume that the lengths of lives of American males are normally distributed with mean μ and variance σ^2. A random sample of 10,000 American males'

mortality histories yielded the estimate $\bar{x} = 72.1$ years, $s^2 = 144$ years^2. Compute the maximum likelihood estimate of the probability that an American male will live to be 50 or older and the estimate that he will not live to be 90.

7. Derive the asymptotic density function for the maximum likelihood estimator of the parameter λ of the exponential random variable.

8. Assume that X is uniform on the interval $(0, \gamma)$. Based on a random sample of n observations, the maximum likelihood estimator is $\hat{\Gamma} = \max(X_1, X_2, \ldots, X_n)$ while the method of moments estimator is $\tilde{\Gamma} = 2\bar{X}$. Compute $E[\hat{\Gamma}]$ and $E[\tilde{\Gamma}]$ and the variances of the two estimators. Which would you prefer? Are they consistent?

9. Suppose that T is an exponential random variable with parameter λ. Given a random sample of n observations of T, construct an unbiased estimator of $1/\lambda$, using the minimum sample value. Would you prefer this estimator or $\bar{T}$ for estimating $1/\lambda$?

10. In Example 7.2.6 it was shown that the minimum variance unbiased estimator for σ^2 was $\dfrac{1}{n} \sum X_i^2$ (for the situation assumed there). Evaluate the variance of this estimator.

11. Suppose that Θ_1 and Θ_2 are unbiased independent estimators of an unknown parameter θ, with known variances σ_1^2 and σ_2^2 respectively.
    (a) Show that $\Theta = (1 - a)\Theta_1 + a\,\Theta_2$ is also unbiased, for any value of a.
    (b) Find the value of a which minimizes σ_Θ^2.

12. Assume that the time to failure of an electron tube is a random variable X with density $\lambda e^{-\lambda x}$, $x > 0$. Find the minimum variance unbiased estimator of $E[X]$.

13. Derive the asymptotic distribution of the estimator in Exercise 12.

14. During the calendar year 1970 there were 40 fatal traffic accidents at one intersection in the city of São Paulo, Brazil; during the calendar year 1971 there were 42 fatal accidents at the same intersection. Assume that fatal accidents occur like events in a Poisson process with unknown parameter λ, per week, during these two calendar years.
    (a) Let $\hat{\lambda}_1$ be the estimate of λ from 1970 and $\hat{\lambda}_2$ the estimate from 1971. Evaluate $\hat{\lambda}_1$ and $\hat{\lambda}_2$.
    (b) Using the result of Exercise 11, combine these two estimates.
    (c) Compare the answer in (b) with the estimate gotten by simply combining the data for the two years to estimate λ.

15. Another measure of variability used to describe the behavior of estimators is the *mean square error* (*MSE*). If Λ is an estimator for λ, then by definition

$$MSE(\Lambda) = E[(\Lambda - \lambda)^2].$$

Note that if Λ is unbiased, $MSE(\Lambda) = \sigma_\Lambda^2$. In any case, show that

$$MSE(\Lambda) = \sigma_\Lambda^2 + (E[\Lambda] - \lambda)^2.$$

The difference, $b_\Lambda = E[\Lambda] - \lambda$, is called the bias of Λ in estimating λ. Thus, note

$$MSE(\Lambda) = \sigma_\Lambda^2 + b_\Lambda^2.$$

16. For the case discussed in Example 7.2.2, show that the mean square error of the geometric mean, $\sqrt{X_1 X_2}$, is smaller than the variance of $\bar{X}$ in estimating θ.

17. Can you find a biased estimate using the sufficient statistic, $X_1 + X_2$, whose mean square error is smaller than that of $\sqrt{X_1 X_2}$, for the case discussed in Example 7.2.2 and Exercise 16 above?

7.3.   Confidence Intervals

So far in this chapter we have considered the problem of finding: (1) a point estimator Γ of an unknown parameter γ, given a random sample of observations of a random variable, and (2) criteria for comparing two or more estimators of the same parameter. By point estimator we are simply referring to the fact that, after the sampling has been done and the observed value of the estimator computed, our end product is this single number, a point on the real line, which we feel is a good guess for the unknown true value of the parameter. If we are using a good estimator, according to some criteria, then we feel that the estimate we give should probably be close to the unknown true value. However, the single number itself does not include any indication of how high the probability might be that the estimator has taken on a value close to the unknown true value. The methods of confidence intervals are meant to give both an idea of the actual numerical value the parameter may have and also an indication of how confident we may be, on the basis of the sample, that we have given a correct indication of the possible numerical value of the parameter. Let us now define a confidence interval for an unknown parameter.

DEFINITION 7.3.1.   Suppose that X is a random variable whose probability law depends on an unknown parameter θ. Given a random sample of X, $X_1, X_2, \ldots, X_n$, the two statistics L_1 and L_2 form a 100 $(1 - \alpha)\%$ *confidence interval* for θ if

$$P(L_1 \leq \theta \leq L_2) \geq 1 - \alpha,$$

no matter what the unknown value of θ. (Either of L_1 or L_2 may be constants, including $-\infty$ or ∞.)

Note first of all that the interval (L_1, L_2) might be termed a *random interval*, since at least one of its end points is a random variable and hence will vary from one sample to another. Once we have taken the sample, we can compute the observed values for L_1 and L_2 and then can say that we are

$100(1 - \alpha)\%$ confident that the interval bracketed by their observed values contains the unknown value of θ.

EXAMPLE 7.3.1.   Let us first examine the concept of a random interval. It seems reasonable that the final digits of telephone numbers might take on the values $0, 1, 2, \ldots, 9$ with equal probabilities. That is, if X is the final digit of a randomly selected telephone number, let us assume that its probability function is

$$p_X(x) = \tfrac{1}{10}, \qquad x = 0, 1, 2, \ldots, 9.$$

The mean of this distribution is $\mu_X = 4.5$. Notice that $P(2.25 \le X \le 9) = .7$ and the event $\{2.25 \le X \le 9\}$ is equivalent to $\left\{\dfrac{X}{2} \le \mu_X \le 2X\right\}$, where $\mu_X = 4.5$. Thus, we can also say

$$P\left(\frac{X}{2} \le \mu_X \le 2X\right) = .7$$

and the interval $(X/2, 2X)$ forms a 70% confidence interval for μ_X. Select 100 last digits from your telephone book, randomly if possible, and for each compute $(x_i/2, 2x_i)$. How many of your intervals include $\mu_X = 4.5$?

The following example gives a more realistic example of a confidence interval.

EXAMPLE 7.3.2.   Suppose that the number of ounces of beer that a machine puts into a bottle is a normal random variable with unknown mean μ and known standard deviation of $\tfrac{1}{2}$ ounce. If we randomly select n bottles filled by this machine and let $X_1, X_2, \ldots, X_n$, respectively, be the number of ounces they contain, then $\bar{X}$ is a normal random variable with mean μ and standard deviation $1/2\sqrt{n}$. From a standard normal table we know that

$$P(-1.96 \le Z \le 1.96) = .95,$$

where Z is a standard normal random variable. But, for our sample,

$$\frac{\bar{X} - \mu}{1/2\sqrt{n}}$$

is a standard normal random variable and thus

$$P\left(-1.96 \le \frac{\bar{X} - \mu}{1/2\sqrt{n}} \le 1.96\right) = .95.$$

The inequality,

$$-1.96 \le \frac{\bar{X} - \mu}{1/2\sqrt{n}} \le 1.96,$$

is equivalent to

$$\bar{X} - 1.96 \frac{1}{2\sqrt{n}} \le \mu \le \bar{X} + 1.96 \frac{1}{2\sqrt{n}},$$

since any time the first statement is true so is the second and vice versa; thus their probabilities must be equal and we have

$$P\left(\bar{X} - 1.96\,\frac{1}{2\sqrt{n}} \le \mu \le \bar{X} + 1.96\,\frac{1}{2\sqrt{n}}\right) = .95.$$

The two statistics

$$L_1 = \bar{X} - 1.96\,\frac{1}{2\sqrt{n}}$$

$$L_2 = \bar{X} + 1.96\,\frac{1}{2\sqrt{n}}$$

then form a 95 % confidence interval for μ. If we actually took a sample of 75 bottles filled by this machine and found $\bar{x} = 12.1$, the observed values of L_1 and L_2, the 95 % confidence interval for μ, would be 11.99 and 12.21, respectively. We then say that we are 95 % confident that the interval (11.99, 12.21) includes μ.

It is important to note the sort of statement that we can make regarding the observed values of a confidence interval. If we were repeatedly to take samples of 75 bottles from the output of this machine, and compute the observed values of L_1 and L_2 for each, we would find that 95 % of these observed intervals actually did include μ while 5 % did not. Given the results of a single sample, such as that quoted above, we do not know whether the particular interval we have computed brackets μ or not. However, 95 % of the intervals computed in this fashion bracket μ, which is the basis for saying we are 95 % confident that we have bracketed μ, with our single sample.

The method used in Example 7.3.2 to generate a confidence interval for the mean of a normal distribution, with σ^2 known, illustrates a general type of procedure which is useful in getting confidence intervals for unknown parameters. Notice that the procedure consisted of finding a random variable ($\sqrt{n}(\bar{X} - \mu)/\sigma$ in that case, where σ was known) whose distribution was completely known (standard normal). Thus it was possible to make a probability statement about the value of the random variable and then, since the random variable involved an unknown parameter (μ), the inequality for the random variable could be transformed into an inequality for the unknown parameter; since the two inequalities were equivalent, the probabilities attached to them are identical. The same sort of manipulation may be done to derive a confidence interval for the variance (or standard deviation) of a normal random variable. Theorem 7.3.1 gives this result.

Theorem 7.3.1. If $X_1, X_2, \ldots, X_n$ is a random sample of a normal random variable with unknown mean μ and variance σ^2, then

$$L_1 = \frac{\sum (X_i - \bar{X})^2}{\chi^2_{1-\alpha/2}}$$

and

$$L_2 = \frac{\sum (X_i - \bar{X})^2}{\chi^2_{\alpha/2}}$$

form a $100(1 - \alpha)\%$ confidence interval for σ^2. ($\chi^2_{\alpha/2}$ and $\chi^2_{1-\alpha/2}$ are the 100 $\alpha/2$ and $100(1 - \alpha/2)$ percentiles, respectively, of the χ^2 distribution with $n - 1$ degrees of freedom.) $\sqrt{L_1}$ and $\sqrt{L_2}$ form a $100(1 - \alpha)\%$ confidence interval for σ.

Proof: From Theorem 6.2.5, we know that

$$\frac{\sum (X_i - \bar{X})^2}{\sigma^2}$$

is a χ^2 random variable with $n - 1$ degrees of freedom. Thus

$$P\left[\chi^2_{\alpha/2} \leq \frac{\sum (X_i - \bar{X})^2}{\sigma^2} \leq \chi^2_{1-\alpha/2}\right] = 1 - \alpha,$$

where $\chi^2_{\alpha/2}$ and $\chi^2_{1-\alpha/2}$ are the $100\alpha/2$ and $100(1 - \alpha/2)$ percentiles of the χ^2 distribution with $n - 1$ degrees of freedom. The inequality

$$\chi^2_{\alpha/2} \leq \frac{\sum (X_i - \bar{X})^2}{\sigma^2} \leq \chi^2_{1-\alpha/2}$$

is equivalent to

$$\frac{\sum (X_i - \bar{X})^2}{\chi^2_{1-\alpha/2}} \leq \sigma^2 \leq \frac{\sum (X_i - \bar{X})^2}{\chi^2_{\alpha/2}},$$

so $L_1 = \sum(X_i - \bar{X})^2/\chi^2_{1-\alpha/2}$ and $L_2 = \sum(X_i - \bar{X})^2/\chi^2_{\alpha/2}$ form a $100(1 - \alpha)\%$ confidence interval for σ^2. Taking square roots we find $\sqrt{L_1}$ and $\sqrt{L_2}$ form a $100(1 - \alpha)\%$ confidence interval for σ. ◀

EXAMPLE 7.3.3. Assume that a machine is used to clip metal stock into pieces (which may be used to make bolts, nuts, face plates, etc.) and that the natural variability of the machine causes the length of a clipped piece to be a normal random variable with unknown mean μ and unknown variance σ^2. If the variance σ^2 appears to be excessively large, then the machines will have to be reset. Assume that we have selected a random sample of 25 pieces clipped by this machine and find that $\sum x_i = 101.1, \sum x_i^2 = 412.75$ (x_i's are measured in inches). Then $\sum (x_i - \bar{x})^2 = $ 3.90 and a 90% confidence interval for σ^2 would extend from $\frac{3.90}{36.4} = .11$ to $\frac{3.90}{13.8} = $ .28. If this range of values makes it appear that σ^2 is excessively large—the machine is producing too large a proportion of pieces too short and too long—we would want to stop production and reset the machine.

The confidence interval given for μ in Example 7.3.2 is unrealistic since in general we will not know the values of either μ or σ^2, in sampling from a normal population. However, by using what is called the t-distribution, it is possible to construct confidence intervals for the mean of a normal random variable when its variance is unknown. Theorem 7.3.2, which we shall not prove, shows how the T random variable occurs. (We shall use a capital letter to denote the random variable, as usual, but due to common usage shall refer to the distribution as the t-distribution.)

Theorem 7.3.2. If Z is a standard normal random variable and V is a χ^2 random variable with n degrees of freedom, then if Z and V are independent,

$$T = \frac{Z}{\sqrt{V/n}}$$

has density

$$f_T(t) = \frac{\Gamma((n+1)/2)}{\Gamma(n/2)\sqrt{n\pi}}\left[1 + \frac{t^2}{n}\right]^{-(n+1)/2}.$$

This is known as the t-distribution with n degrees of freedom.

Figure 7.3.1 gives a picture of the t-density function, for $n = 5$ degrees of freedom, as well as a plot of a standard normal density. Note that the two are extremely similar in shape but that the normal includes more density close to zero. It can be shown that as n, the degrees of freedom, becomes large without limit the t-distribution approaches the standard normal density function.

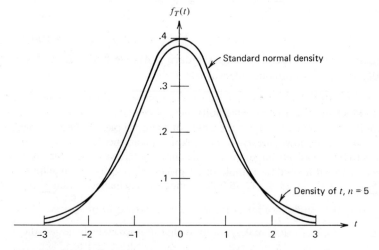

FIGURE 7.3.1   Comparison of normal and t densities, $n = 5$.

Table 5 in the Appendix presents selected percentiles of the t-distribution. The $100(1 - \gamma)$ percentile, $t_{1-\gamma}$, is the number such that $F_T(t_{1-\gamma}) = 1 - \gamma$. We see then, for a t random variable with 5 degrees of freedom, that $t_{.95} = 2.015$, $t_{.99} = 3.365$. For a t random variable with 20 degrees of freedom, $t_{.90} = 1.325$, $t_{.975} = 2.086$.

The following two theorems show the way which the t-distribution can be used to construct a confidence interval for the mean of a normal distribution.

Theorem 7.3.3. Suppose that X is a normal random variable with mean μ and variance σ^2. Let $X_1, X_2, \ldots, X_n$ be a random sample of n observations of X. Then

$$\frac{\overline{X} - \mu}{S/\sqrt{n}}$$

has the t-distribution with $n - 1$ degrees of freedom. (S is the sample standard deviation.)

Proof: We know that

$$\frac{\overline{X} - \mu}{\sigma/\sqrt{n}}$$

is a standard normal random variable, that

$$\frac{(n - 1)S^2}{\sigma^2}$$

is a χ^2 random variable with $n - 1$ degrees of freedom, and that $\overline{X}$ and S^2 are independent random variables (Theorem 6.2.5). Then, by Theorem 7.3.2,

$$\frac{\overline{X} - \mu}{\sigma/\sqrt{n}} \Big/ \sqrt{\frac{(n - 1)S^2}{\sigma^2(n - 1)}} = \frac{\overline{X} - \mu}{S/\sqrt{n}}$$

has the t-distribution with $n - 1$ degrees of freedom (the number of degrees of freedom of the χ^2 random variable in the denominator). ◄

Theorem 7.3.4. Suppose that X is a normal random variable with mean μ and variance σ^2, both of which are unknown. Let $X_1, X_2, \ldots, X_n$ be a random sample of X; then

$$L_1 = \overline{X} - t_{1-\alpha/2} \frac{S}{\sqrt{n}}$$

and

$$L_2 = \overline{X} + t_{1-\alpha/2} \frac{S}{\sqrt{n}}$$

constitute a $100(1 - \alpha)\%$ confidence interval for μ ($t_{1-\alpha/2}$ is the $100(1 - \alpha/2)$ percentile of the t-distribution with $n - 1$ degrees of freedom).

Proof: We saw in Theorem 7.3.3 that

$$\frac{\bar{X} - \mu}{S/\sqrt{n}}$$

is a t random variable with $n - 1$ degrees of freedom. Then

$$P\left[-t_{1-\alpha/2} \leq \frac{\bar{X} - \mu}{S/\sqrt{n}} \leq t_{1-\alpha/2}\right] = 1 - \alpha,$$

where $t_{1-\alpha/2}$ is the $100(1 - \alpha/2)$ percentile of the t-distribution with $n - 1$ degrees of freedom. But

$$-t_{1-\alpha/2} \leq \frac{\bar{X} - \mu}{S/\sqrt{n}} \leq t_{1-\alpha/2}$$

is equivalent to

$$\bar{X} - t_{1-\alpha/2}\frac{S}{\sqrt{n}} \leq \mu \leq \bar{X} + t_{1-\alpha/2}\frac{S}{\sqrt{n}}.$$

Thus,

$$P[L_1 \leq \mu \leq L_2] = 1 - \alpha$$

where

$$L_1 = \bar{X} - t_{1-\alpha/2}\frac{S}{\sqrt{n}}$$

$$L_2 = \bar{X} + t_{1-\alpha/2}\frac{S}{\sqrt{n}}.$$ ◄

This confidence interval for μ, with σ^2 unknown, is interpreted like any other. If we take our sample and find the observed values of the 95% confidence interval to be 1 and 2, for example, then we say we are 95% confident that the interval $(1, 2)$ includes μ. We justify this statement by the fact that in 95% of all possible samples of the given size the observed values of L_1 and L_2 will bracket the unknown value of μ; in 5% of such samples the observed values will not bracket μ.

EXAMPLE 7.3.4.   On a bright sunny day in May, a track man put the shot 6 times; the distances thrown were 58 feet, 69 feet, 62 feet, 55 feet, 64 feet, and 65 feet. We might assume that the distance he will put the shot on a day such as this is a normal random variable X with unknown mean μ and unknown variance σ^2. If we look at these 6 values as being the observed values of a random sample of size 6 of the random variable X, we can compute the observed value of a 90% confidence interval for μ, the unknown average distance he would put the shot, as follows.

We find

$$\sum_{i=1}^{6} x_i = 373, \qquad \sum_{i=1}^{6} x_i^2 = 23{,}315$$

and thus $\bar{x} = 62.2$ feet, $s = 5.04$; then the observed value of the 90% confidence interval is from $62.2 - 2.015(2.06) = 58.1$ feet to $62.2 + (2.015)(2.06) = 66.3$ feet. We are 90% sure that the interval (58.1 feet, 66.3 feet) brackets the unknown value of μ.

The parameter λ of an exponential random variable is one for which it is possible to find a random variable, which is a function of λ, but whose distribution contains no unknown parameters. The following theorem gives this result.

Theorem 7.3.5. Assume $X_1, X_2, \ldots, X_n$ is a random sample of an exponential random variable with parameter λ. Then

$$L_1 = \frac{\chi_{\alpha/2}^2}{2n\bar{X}} \quad \text{and} \quad L_2 = \frac{\chi_{1-\alpha/2}^2}{2n\bar{X}}$$

form a $100(1 - \alpha)\%$ confidence interval for λ.

Proof: If $X_1, X_2, \ldots, X_n$ are independent exponential random variables with parameter λ, then the moment generating function for their sum is

$$m_{\Sigma X_i}(t) = \prod_{i=1}^{n} \frac{1}{(1 - t/\lambda)} = \frac{1}{(1 - t/\lambda)^n}.$$

The moment generating function of $2\lambda n\bar{X} = 2\lambda\sum X_i$ is

$$m_{2\lambda\Sigma X_i}(t) = m_{\Sigma X_i}(2\lambda t)$$

$$= \frac{1}{(1 - 2t)^n}$$

which is the moment generating function of a χ^2 random variable with $2n$ degrees of freedom (see Section 6.2 and Exercise 6.2.25). Thus $2\lambda n\bar{X}$ is a χ^2 random variable with $2n$ degrees of freedom and with probability $1 - \alpha$

$$\chi_{\alpha/2}^2 \leq 2\lambda n\bar{X} \leq \chi_{1-\alpha/2}^2;$$

but this inequality is equivalent to

$$\frac{\chi_{\alpha/2}^2}{2n\bar{X}} \leq \lambda \leq \frac{\chi_{1-\alpha/2}^2}{2n\bar{X}}$$

and the result is established. ◄

EXAMPLE 7.3.5.   Assume that the time to failure of an electron tube is an exponential random variable with parameter λ. If 25 of these tubes are put on test and the sum of their times to failure is

$$\sum x_i = 25,242 \text{ hours},$$

we can construct a 95 % (two-sided) confidence interval for λ by first calculating $\chi^2_{.025}$ and $\chi^2_{.975}$ for a χ^2 distribution with $2n = 50$ degrees of freedom. We recall from Section 6.2 that

$$\chi^2_\alpha \doteq \frac{1}{2} (z_\alpha + \sqrt{2d - 1})^2$$

where d is the degrees of freedom. Thus, for $d = 50$

$$\chi^2_{.975} \doteq \tfrac{1}{2}(z_{.975} + \sqrt{99})^2$$
$$= 70.92$$

$$\chi^2_{..025} \doteq \tfrac{1}{2}(z_{.025} + \sqrt{99})^2$$
$$= 31.92$$

so the 95 % confidence interval for λ goes from $\dfrac{31.92}{50484} = .00063$ to $\dfrac{70.92}{50484} = .00140$.

Frequently, *one-sided confidence intervals* are of interest. In such cases L_1 or L_2 is set equal to a constant. The following example illustrates such a case.

EXAMPLE 7.3.6.   The *reliability* of a piece of equipment for a period of time t is defined to be the probability that the equipment will perform satisfactorily for at least a period of time of length t (thus its time to failure exceeds t). That is, if X is the time to failure for the item, its reliability for a period of time t is by definition

$$R(t) = P(X > t), \quad \text{for } t > 0.$$

If the time to failure for the item is an exponential random variable, the reliability function is especially simple

$$R(t) = 1 - F_X(t) = e^{-\lambda t}.$$

Based on a sample of failure times of an exponential random variable it is possible to derive a confidence interval and then to translate this into a confidence interval for $R(t)$.

Note that

$$\lambda \leq a$$

is equivalent to

$$R(t) = e^{-\lambda t} \geq e^{-at};$$

thus to get a lower confidence interval for $R(t)$, we need an upper confidence interval for λ. Using the data given in Example 7.3.4, we are 97.5 % confident that

λ (for that tube) is no larger than .00140. Thus we are also 97.5 % confident that $R(t) \geq e^{-.0014t}$, for this tube. We are 97.5 % sure that $R(100) \geq e^{-.14} = .869$.

To this point we have only considered the problem of constructing various confidence intervals for the parameters of certain continuous random variables. Let us now discuss the construction of a $100(1 - \alpha) \%$ confidence interval for p, the parameter of a Bernoulli random variable. If $X_1, X_2, \ldots,$ X_n is a random sample of a Bernoulli random variable with parameter p, we know that $\bar{X}$ is approximately normally distributed with mean p and variance $p(1 - p)/n$ for n large. Thus

$$P\left[\frac{|\bar{X} - p|}{\sqrt{p(1 - p)/n}} \leq z_{1-\alpha/2}\right] \doteq 1 - \alpha$$

where $z_{1-\alpha/2}$ is the $100(1 - \alpha/2)$ percentile of the standard normal distribution function. This inequality is equivalent to

$$(\bar{X} - p)^2 \leq z^2 \frac{p(1 - p)}{n}$$

where the subscript has been dropped from $z_{1-\alpha/2}$. Placing all terms on the left-hand side of the inequality, we have

$$p^2\left(1 + \frac{z^2}{n}\right) - p\left(2\bar{X} + \frac{z^2}{n}\right) + \bar{X}^2 \leq 0;$$

note that the left-hand side is simply the equation of a parabola, such as appears in Figure 7.3.2. We thus want to find the two numbers, p_1 and p_2,

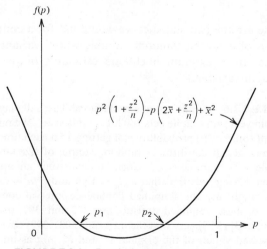

FIGURE 7.3.2   Parabola determining p_1 and p_2.

defining the interval for which the parabola is negative. These two numbers, for a given sample value of $\bar{X}$, will be the observed value of the confidence interval for p. Using the formula for the roots of a quadratic equation we find

$$P_1 = \frac{\bar{X} + z^2/2n - (z/\sqrt{n})\sqrt{\bar{X}(1 - \bar{X}) + z^2/4n}}{1 + z^2/n}$$

$$P_2 = \frac{\bar{X} + z^2/2n + (z/\sqrt{n})\sqrt{\bar{X}(1 - \bar{X}) + z^2/4n}}{1 + z^2/n}.$$

Assuming that n is large enough for the normal approximation to be sufficiently accurate, we have

$$P(P_1 \leq p \leq P_2) = 1 - \alpha;$$

thus the two statistics, P_1 and P_2, form an approximate $100(1 - \alpha)\%$ confidence interval for p. Especially if n is quite large compared to z^2, we might just as well set the terms in z^2/n equal to zero; the two limits then become

$$P_1 = \bar{X} - \frac{z}{\sqrt{n}}\sqrt{\bar{X}(1 - X)}$$

$$P_2 = \bar{X} + \frac{z}{\sqrt{n}}\sqrt{\bar{X}(1 - \bar{X})}.$$

Note that these are the two statistics we would use for a confidence interval for the mean μ of a normal random variable whose variance is $\bar{X}(1 - \bar{X})$. (This is in fact the maximum likelihood estimator of the variance of a Bernoulli random variable.)

EXAMPLE 7.3.6.   Suppose that a professional baseball player has 400 times at bat in a certain year and that he gets 140 hits in this year. We might assume that he has a constant (unknown) probability p of getting a hit each time he is at bat and that his 400 times at bat constitute a random sample of size 400 of a Bernoulli random variable with parameter p. Then, to construct an approximate 95% confidence interval for p, we find that $z_{.975} = 1.96$ and $z^2/n = .01$. Since z^2/n is quite small, we might use the simplified confidence interval computations. The observed values of these simplified statistics are .303 and .397, respectively. Thus we are (approximately) 95% sure that the interval $(.303, .397)$ covers the true value of p. (The observed values of the original P_1 and P_2, not assuming z^2/n is zero, are .305 and .398, respectively.)

EXERCISE 7.3

1. Assume that the top speed attainable by a certain design racing car is a normal random variable X with unknown mean μ and standard deviation σ. A random sample of 10 cars, built according to this design, was selected and each car was tested. The sum of squares of the top speeds attained (in miles per hour) were

$$\sum_{i=1}^{10} x_i = 1652, \qquad \sum_{i=1}^{10} x_i^2 = 273,765.$$

Compute the observed value of a 95 % confidence interval for μ.

2. Find a statistic L_2 such that

$$P(\sigma^2 \leq L_2) = 1 - \alpha,$$

given a random sample of size n of a normal random variable with mean μ and variance σ^2.

3. The quality control group of an industrial plant routinely samples the production line for a given item and, daily, computes a 90 % confidence interval for the mean length of the parts produced that day. After 100 days they have computed 100 90 % confidence intervals.
   (a) Let X be the (unknown) number of these intervals which did in fact cover the unknown mean length of parts produced that day. What is the distribution for X?
   (b) Compute the (approximate) probability that exactly 90 of the 100 intervals covered the true mean.

4. Assume that $X_1, X_2, \ldots, X_{n_1}$ is a random sample of a normal random variable X with mean μ_X and variance σ^2. $Y_1, Y_2, \ldots, Y_{n_2}$ is an independent random sample of a normal random variable Y with mean μ_Y and variance σ^2. (Note that the two variances are equal.) Derive a $100(1 - \alpha)$ % confidence interval for the difference $\mu_X - \mu_Y$. (Hint: Let S_X^2 be the sample variance of the X_i's and let S_Y^2 be the sample variance of the Y_i's. Consider the distribution of

$$[(n_1 - 1)S_X^2 + (n_2 - 1)S_Y^2]/\sigma^2;$$

it is independent of $\bar{X} - \bar{Y}$.)

5. Two different feeding programs for fattening beef cattle are to be compared. $n_1 = 8$ cattle are used on the first program for a month; the sum and the sum of squares of the gains of these 8 cattle were $\sum x_i = 416, \sum x_i^2 = 21,807$. $n_2 = 10$ cattle are used on the second program for the same period; the sum and the sum of squares of the gains of these 10 cattle were $\sum y_i = 468$, $\sum y_i^2 = 22,172$. Use the procedure from Exercise 4 to construct a 95 % confidence interval for the difference in mean gains, $\mu_X - \mu_Y$.

6. Assume as in Example 7.3.5 that the time to failure of an electron tube is an exponential random variable with parameter λ. A sample of $n = 20$ tubes is placed on test and the sum of their times to failure was 20,169 hours.
   (a) What is the maximum likelihood estimator for $R(1000)$, the reliability of this tube for 1000 hours?
   (b) Construct a 90 % lower confidence limit for $R(1000)$.

7. (a) What is the distribution of the length of the confidence interval for σ^2 given in Theorem 7.3.1?
   (b) For a sample of size $n = 20$, what is the probability this length is no larger than σ^2?

8. (a) What is the distribution of the length of the confidence interval for the mean of a normal random variable, given in Theorem 7.3.4?
   (b) For a sample of size $n = 20$, what is the probability the length of this interval is less than σ?

9. How large should n be for the length of the confidence interval discussed in Example 7.3.2 to be $\frac{1}{2}$ ounce? $\frac{1}{4}$ ounce?

10. Assume that the time to failure (breakage) of a 30-foot length of clothes line is a normal random variable with $\sigma = 1$ year.
    (a) What is the reliability function for this clothes line for t years? (Refer to Example 7.3.5.)
    (b) Describe how you could compute a confidence interval for $R(t)$, given a random sample of n values from this distribution.
    (c) Assume $n = 10$ pieces of this clothes line were tested and that the sum of their times to failure was 111. Compute a 95 % two-sided confidence interval for $R(10)$.

11. Assume that a random variable X is uniformly distributed on the interval $(0, b)$. Using the maximum value of a sample of size n of X, construct a confidence for b. (Hint: Look at the distribution for $X_{(n)}/b$.)

12. Suppose we take a random sample of size n of a normal random variable with unknown mean μ and whose variance is known to be $\sigma^2 = 1$. Then, either the procedure given in Example 7.3.2 could be used to construct a confidence interval for μ, or the procedure given in Theorem 7.3.4 could be used. Under what circumstances will the latter procedure give an interval of shorter length? Which of the two procedures is generally to be recommended, when σ^2 is known?

13. Assuming that we have a large sample of a Poisson random variable X with parameter λ, use the normal approximation to the distribution of $\bar{X}$ to derive an approximate $100(1 - \alpha)$ % confidence interval for λ similar to the binomial derivation discussed above.

14. X is a normal random variable with $\mu = 0$ and σ^2 unknown. Given a random sample of n observations of X, derive a $100(1 - \alpha)$ % confidence interval for σ^2 which is different than the one derived in Theorem 7.3.1. Which of the two is better (and in what sense)?

8

Statistical Inference: Tests of Hypotheses

In this chapter we shall study the classical theory of tests of hypotheses. A hypothesis, first of all, is simply a statement about the probability law of a random variable (which can be sampled). A test of a hypothesis, then, amounts to sampling the random variable whose probability law is referred to in the hypothesis and, on the basis of the sample, deciding to accept or reject the stated hypothesis.

In general, if our sample results seem consistent with the hypothesis, we would like to accept the stated hypothesis; if however the sample results do not seem consistent with the stated hypothesis, we would like to reject it. Usually there are many different ways of making the decision to accept or reject a hypothesis on the basis of a sample; we shall examine some of the methodology which has been developed for judging the best way (in a sense) of making the decision.

8.1. Hypotheses and Tests

As was just mentioned, for our purposes a hypothesis will always be a statement about the probability law of a random variable. There are two distinct sorts of statements which can be made about the probability law of a random variable. The first of these is to assume that the form of the distribution function F_X for X is known and the hypothesis then merely refers to the value or values of the parameters of F_X. The second type of hypothesis is a statement regarding the form of F_X itself. An example of the first type of hypothesis would be a case in which we know, or are willing to assume, that a random variable X is binomial and the hypothesis is a statement regarding the value of p. An example of the second type would be a case in which our hypothesis is that X is a binomial random variable. For reasons that will become apparent as we proceed, it is advantageous to treat these two types of hypotheses separately. At first we shall concern ourselves with only hypotheses of the first type, those in which the form of the density or probability function is assumed known and we are interested in accepting or rejecting statements about the values of the parameters.

If our hypothesis gives exact values for all unknown parameters of the assumed probability law, we shall call it a *simple* hypothesis; otherwise the hypothesis is *composite*. The following example illustrates this difference.

EXAMPLE 8.1.1. Suppose that a random variable X is binomial with parameter $n = 20$, and we hypothesize that $p = \frac{1}{4}$ or $p = \frac{1}{2}$ or $p = .9$. Each of these is an example of a simple hypothesis. However, the statements

$$0 \le p \le .1, \qquad p < \tfrac{1}{4}, \qquad .49 < p < .51$$

are each composite hypotheses. Or, if Y is a normal random variable, $\mu = 2$, $\sigma = 1$ is a simple hypothesis while $\mu = 2$ is a composite hypothesis if σ is unknown (since σ is unspecified).

As has been mentioned, we shall decide to accept or reject a hypothesis on the basis of a random sample of the random variable concerned. Thus we are, in effect, partitioning our sample space of all possible outcomes for the experiment into two disjoint pieces, those outcomes for which we decide to accept the hypothesis versus those outcomes for which we decide to reject the hypothesis. We would like to find that way of partitioning the sample space which is good in some sense.

DEFINITION 8.1.1. A *test* of a hypothesis is a partitioning of the sample space into two parts, called the *rejection* region (or *critical* region) and the *acceptance* region.

If we compare two different ways of partitioning the sample space, then we say we are comparing two tests (of the same hypothesis).

Tests are generally compared on the basis of the probabilities of errors which might be committed. A little reflection will reveal that there are two types of errors which might be committed. We might reject the hypothesis when it is true or we might accept the hypothesis when it is false.

DEFINITION 8.1.2. If we reject the hypothesis when it is true, we commit a *type I error*. If we accept the hypothesis when it is false, we commit a *type II error*.

Then, if we could find a test for a hypothesis which simultaneously minimized the probabilities of both of these types of errors, it would clearly be the test we would want to use. Unfortunately, it is not in general possible to do this, as a simple example will illustrate.

EXAMPLE 8.1.2.   Suppose that we have a brand new coin and wish to test the hypothesis that the probability p of getting a head when this coin is flipped is equal to $\frac{1}{2}$. We might decide to flip the coin 100 times, in which case we would have a random sample of 100 observations of a Bernoulli random variable. We could decide to accept the hypothesis no matter what happens when the coin is flipped 100 times (thus the rejection region is null and the whole sample space is the acceptance region). This partition yields P(type I error) equal to zero, clearly the smallest possible value it might have. Or we might decide to reject the hypothesis no matter what happens on the 100 flips (now the acceptance region is null), in which case P(type II error) equals zero. Thus, we could always get a test which makes the probability of one of the two errors equal to zero.

The way in which different tests are compared, therefore, is to fix the probability of type I error at α (.05 and .01 are common values) and then choose the test that gives the smallest probability of type II error. If we recall that accepting the hypothesis when it is false is called a type II error, it is evident that it is necessary to assume some value of the parameter not specified by the hypothesis before the probability of a type II error can be calculated. That is, we must specify what we assume the true value of the parameter to be before we can do the calculation; we must state what it is we do accept as the value of the parameter. To make this comparison of probabilities of type II errors more concrete (really to specify those values of the parameter which should be relevant to the comparison of tests), we shall always state two hypotheses, one of which is to be accepted on the basis of the sample. The hypothesis which we want to test is called the *null* hypothesis; the range of parameter values pertinent to the comparison of probabilities of type II errors specifies the *alternative* hypothesis. We shall use H_0 to indicate the null hypothesis and H_1 to indicate the alternative hypothesis. Thus, in Example 8.1.2 above, we would write $H_0: p = \frac{1}{2}$ and, depending on which alternatives we might want to consider, we would write $H_1: p \neq \frac{1}{2}$ or $H_1: p > \frac{1}{2}$ or $H_1: p = \frac{3}{4}$, etc.

EXAMPLE 8.1.3.   Suppose that the lengths of lives of light bulbs made by a certain company are, through past experience, assumed to be normally distributed with $\mu = 1500$ hours and $\sigma = 100$ hours. Someone in the research division of this company invents a new process for making bulbs which preliminary research shows may have a mean lifetime of more than 1500 hours with a standard deviation of 100 hours (again it would appear plausible that the distribution of lifetimes for bulbs made the new way is normal in form). It is apparent that the new process would cost the same amount per bulb once installed. The company feels that if this new method does produce bulbs with longer average lifetimes, they should adopt it since it would lead to increased consumer goodwill, possibly a bigger share of the light bulb market, etc.

It is assumed, then, that lifetimes of bulbs made the new way are normally distributed with unknown mean μ and known standard deviation of 100 hours. The company wants to test $H_0: \mu = 1500$ versus $H_1: \mu > 1500$. A group of 100 bulbs are made the new way (and it is assumed that their lifetimes constitute a random sample of size 100 from the distribution of all possible lifetimes for this process). It is decided that if the sample mean exceeds 1525 hours, then H_0 would be rejected (H_1 accepted) and the new process would be adopted. Otherwise H_0 would be accepted and they would continue to make bulbs in the old way. Thus the partition of the sample space is into those outcomes where $\bar{X} > 1525$ (rejection region) versus those for which $\bar{X} \le 1525$ (acceptance region). The probability of a type I error using this test then is

$$P(\bar{X} > 1525 \mid \mu = 1500) = P\left(\frac{\bar{X} - 1500}{10} > \frac{1525 - 1500}{10} \,\middle|\, \mu = 1500\right)$$

$$= P(Z > 2.5) = .0062,$$

where Z is a standard normal variable.

In the preceding example, H_0 specified that the new process was the same as the old (in terms of average lifetime), while H_1 then consisted of the possible pertinent alternative values for μ. Note that rejecting H_0 when it is true (type I error) then would consist of adopting the new process when in fact it was no better than the old; this would be an error which would cause the company needless expense. We noted that its probability of occurrence is quite small.

As we shall see, the usual methods for testing hypotheses allow easy control of the probability of type I error. Thus, in practical examples, H_0 and H_1 are selected so that the more important or more serious error is the type I; its probability of occurrence can then be easily made as small as one would like. In Example 8.1.3 it was deemed more important not to adopt the new process when it was no better than the old rather than to fail to adopt it when in fact it was better.

We made no mention of the probability of type II error in Example 8.1.3. Since H_1 is composite, the probability of type II error varies with the particular value of μ specified by H_1 that we want to assume is true. The *operating characteristic* (*OC*) curve describes how the probability of type II error varies with μ.

DEFINITION 8.1.3.   When testing hypotheses about a parameter μ, the probability of accepting H_0 is a function of μ, denoted by $C(\mu)$, and is called the *OC* curve of the test. $Q(\mu) = 1 - C(\mu)$ is called the *power function* of the test.

FIGURE 8.1.1 Operating characteristic curve, Example 8.1.4.

EXAMPLE 8.1.4.   In Example 8.1.3, H_0 is accepted if $\overline{X} \leq 1525$. Thus

$$C(\mu) = P(\overline{X} \leq 1525)$$

$$= P\left(\frac{\overline{X} - \mu}{10} \leq \frac{1525 - \mu}{10}\right)$$

$$= N_Z\left(\frac{1525 - \mu}{10}\right),$$

where N_Z is the standard normal distribution function. Since

$$\frac{d}{d\mu} N_Z\left(\frac{1525 - \mu}{10}\right) = -\frac{1}{10} n_Z\left(\frac{1525 - \mu}{10}\right),$$

we know that $C(\mu)$ is a monotonically decreasing function of μ. Figure 8.1.1 gives a sketch of this curve. The power function for the test then is

$$Q(\mu) = 1 - C(\mu)$$

$$= 1 - N_Z\left(\frac{1525 - \mu}{10}\right)$$

and is sketched in Figure 8.1.2. Since the power curve, for a given value of μ specified by H_1, gives the probability of making no error we would clearly prefer the test that has the biggest power, preferably for all μ specified by H_1.

EXAMPLE 8.1.5.   Joe and Hugh are in the habit of flipping coins daily to see which of them will buy their coffee. Suppose that Joe is of a suspicious nature and would like to test formally the hypothesis that his probability of winning, p, is $\frac{1}{2}$ in playing this game versus the alternative that $p < \frac{1}{2}$. He decides to keep track of the outcomes the next 10 times that he and Hugh gamble and to reject $H_0: p = \frac{1}{2}$

FIGURE 8.1.2   Power functions, Example 8.1.4.

if he does not win at least 2 times out of the 10. Then, formally, he is defining

$$X_i = 1, \quad \text{if he wins on the } i\text{-th game}$$
$$= 0, \quad \text{if he loses on the } i\text{-th game},$$

for $i = 1, 2, \ldots, 10$; thus $X_1, X_2, \ldots, X_{10}$ is a random sample of 10 observations of a Bernoulli random variable X with parameter p. His critical region then is defined by

$$\sum_{i=1}^{10} X_i \le 1.$$

$$Y = \sum_{i=1}^{10} X_i$$

is a binomial random variable with parameters 10 and p; his probability of type I error then is

$$P[Y \le 1 \,|\, p = \tfrac{1}{2}] = (\tfrac{1}{2})^{10} + \frac{10}{2^{10}} \doteq .01.$$

The OC curve for this test then is

$$C(p) = 1 - P[Y \le 1 \,|\, p]$$
$$= 1 - p^{10} - 10p^9(1 - p)$$

and we find, for example, that

$$C(.4) = .95, \quad C(.3) = .85, \quad C(.2) = .63, \quad C(.1) = .27.$$

EXERCISE 8.1

1. Suppose that X is a Poisson random variable with parameter λ. We take a sample $X_1, X_2, \ldots, X_n$ of X and want to test $H_0: \lambda = 2$ versus $H_1: \lambda > 2$. What do you feel would be a reasonable critical region?

2. Suppose that X is a normal random variable with unknown mean μ and variance 9. If we take a random sample of 16 observations of X and want to test $H_0: \mu = 2$, show that each of the following critical regions is of size .05 (that is, the probability of type I error is .05 for each).

(a) $\bar{X} > 3.23$ (b) $\bar{X} < .77$ (c) $1.952 < \bar{X} < 2.048$

For what type of alternative hypothesis, if any, would each of these be appropriate?

3. We might assume that weights of 4-month-old pigs are normally distributed with standard deviation $\sigma = 10$ pounds. A standard ration is known to produce an average weight of 120 pounds. If we desire to check whether a new ration is equally good, we might take a sample of 25 pigs, feed them the new ration, check their weights at age 4 months, and test $H_0: \mu = 120$ versus $H_1: \mu < 120$ using the critical region $\bar{X} < 116.08$. Compute the probability of type I error for this test and derive the power function. What is the power of the test if $\mu = 115$ pounds? (That is, what is $Q(115)$?)

4. Suppose that the number of ounces that a bottling machine puts in a coke bottle is a normal random variable with mean μ and variance σ^2. If σ^2 is too large, many bottles will be filled to overflowing; for the setting on the machine controlling the mean, a σ^2 not exceeding $\frac{1}{4}$ is tolerable. Suppose we take a sample of 20 bottles filled by the machine and want to test $H_0: \sigma^2 = \frac{1}{4}$ versus $H_1: \sigma^2 > \frac{1}{4}$. We decide to reject H_0 if

$$\sum_{i=1}^{20} (X_i - \bar{X})^2 > 7.536.$$

What is the probability of type I error for this test?

5. Assume that the time to failure (in hours) of a certain type of vacuum tube is an exponential random variable with parameter λ. We put 10 of these tubes on test and want to test the hypothesis $H_0: \lambda = .001$ versus $H_1: \lambda < .001$. We decide to wait only until the first tube fails and will reject H_0 if the time to this first failure is no larger than 5.13 hours. Show that the probability of type I error for this test is .05. (Thus we are concerned with the minimum sample value of a sample of size 10 of an exponential random variable; the reader might like to review Example 6.2.9 to see the density function for this minimum value.)

6. Assume that X is a uniform random variable on the interval from 0 to θ. We observe one value of X and want to test $H_0: \theta = 1$ versus $H_1: \theta > 1$; we decide to reject H_0 if the sample value exceeds .99. Compute the probability of type I error for this test and draw the power function.

7. Plot the power function, $Q(\lambda)$, for the test described in Exercise 5.

8. Assume, as in Exercise 6, that X is uniform on the interval 0 to θ and that we want to test $H_0: \theta = 1$ versus $H_1: \theta > 1$. A sample of $n = 10$ values of X are observed and H_0 is rejected if $X_{(10)}$ (the maximum sample value) exceeds

$(.99)^{1/10} \doteq .999$. Compute the probability of type I error for this test and compare its power function with that derived in Exercise 6.

9. Again assume the situation described in Exercise 8 (X is uniform, same H_0 and H_1). A sample of $n = 10$ samples values is observed and now H_0 will be rejected only if the minimum sample value, $X_{(1)}$, is larger than $1 - (.01)^{1/10} \doteq .369$. Compute the probability of type I error for this test and compare its power function with those derived in Exercises 6 and 8.

10. In general, the power of a "reasonable" test will increase as n, the sample size, increases. To illustrate this fact, suppose we have a random sample of a normal random variable X whose standard deviation is known to be $\sigma = 1$. Then the sample mean $\bar{X}$ is normal with mean μ (the same as the mean of X) and variance $1/n$. Suppose we want to test $H_0: \mu = 0$ versus $H_1: \mu > 0$. Let c be some fixed positive constant and assume we will reject H_0 if $\bar{x}$ (the observed value of $\bar{X}$) exceeds c.
    (a) Let μ_1 be some fixed alternative value of μ (thus $\mu_1 > 0$). Compute $Q(\mu_1)$ as a function of n.
    (b) Show that $Q(\mu_1)$ is an increasing function of n, no matter which fixed positive value we take for μ_1.

11. For the test described in Exercise 8, again show that the power function is an increasing function of n, for any value specified by H_1 (that is, for any $\theta > 1$).

8.2. Simple Hypotheses

It will be recalled from Section 8.1 that a simple hypothesis is one that completely specifies the density or probability function for a random variable. Suppose that X is a random variable with density f_X indexed by a single parameter θ. Then, if we wanted to test $H_0: \theta = 1$ versus $H_1: \theta = 5$, we would say that we were testing a simple hypothesis versus a simple alternative, since each of the two hypotheses completely specifies the density f_X. Note that if we are testing a simple H_0 versus a simple H_1, then both the probability of a type I error and the probability of a type II error are well defined; since each hypothesis specifies a single value for the unknown parameter, we can quote unique probabilities for both type I and type II errors. These two probabilities will be denoted by α and β, respectively.

It was also mentioned in Section 8.1 that selecting a test for a hypothesis was equivalent to simply selecting a critical region; if we could find some method of constructing a critical region which was good in some sense, this method would yield what we might call a good test of the hypothesis.

In the special case in which we are testing a simple hypothesis against a simple alternative, we would clearly like to use the critical region that minimizes β, the probability of a type II error, among all regions with a probability of type I error equal to α. The Neyman-Pearson lemma (Theorem 8.2.1) shows how we can always construct such a region.

Theorem 8.2.1. (Neyman-Pearson) Assume that X is a random variable whose distribution function is indexed by the parameter θ. X_1, $X_2, \ldots, X_n$ is a random sample of X; $L(\theta)$ is the likelihood function of the sample. We desire to test $H_0: \theta = \theta_0$ versus $H_1: \theta = \theta_1$ where θ_0 and θ_1 are two specific values for θ. Let R (the critical region) be the subset of S such that

$$\frac{L(\theta_1)}{L(\theta_0)} \geq k.$$

Then $\bar{R}$ will consist of those points such that

$$\frac{L(\theta_1)}{L(\theta_0)} < k$$

where k is chosen so that the probability of type I error is α. Then the region R has the smallest possible probability of type II error among all critical regions of size α (probability of type I error). R is called the most powerful critical region for the given α.

Proof: To indicate the proof of this theorem (in the continuous case), we shall use $\int_A L(\theta_0)$ and $\int_A L(\theta_1)$ as shorthand for the integrals of the sample density function over a subset $A \subset S$, assuming θ_0 and θ_1 as the true values of the parameter θ, respectively. Let R be the region defined above; then

$$\int_R L(\theta_0) = \alpha$$

and we want to show that

$$\int_R L(\theta_1) \geq \int_{R^*} L(\theta_1)$$

where R^* is any other region satisfying

$$\int_{R^*} L(\theta_0) = \alpha.$$

Clearly,

$$R = (R \cap R^*) \cup (R \cap \bar{R}^*)$$
$$R^* = (R^* \cap R) \cup (R^* \cap \bar{R});$$

thus

$$\int_{R^*} L(\theta_1) = \int_{R^* \cap R} L(\theta_1) + \int_{R^* \cap \bar{R}} L(\theta_1)$$

and

$$\int_R L(\theta_1) = \int_{R \cap R^*} L(\theta_1) + \int_{R \cap \bar{R}^*} L(\theta_1).$$

Then

$$\int_{R^*} L(\theta_1) - \int_R L(\theta_1) = \int_{R^* \cap \bar{R}} L(\theta_1) - \int_{R \cap \bar{R}^*} L(\theta_1).$$

Now, by definition, $L(\theta_1) \geq kL(\theta_0)$ for points in R and $L(\theta_1) < kL(\theta_0)$ for points in $\bar{R}$. Thus

$$\int_{R^* \cap \bar{R}} L(\theta_1) < k \int_{R^* \cap \bar{R}} L(\theta_0)$$

and

$$\int_{R \cap \bar{R}^*} L(\theta_1) \geq k \int_{R \cap \bar{R}^*} L(\theta_0);$$

then we have

$$\int_{R^* \cap \bar{R}} L(\theta_1) - \int_{R \cap \bar{R}^*} L(\theta_1) \leq k \left(\int_{R^* \cap \bar{R}} L(\theta_0) - \int_{R \cap \bar{R}^*} L(\theta_0) \right).$$

But

$$\int_{R^* \cap \bar{R}} L(\theta_0) - \int_{R \cap \bar{R}^*} L(\theta_0)$$

$$= \int_{R^* \cap \bar{R}} L(\theta_0) + \int_{R^* \cap R} L(\theta_0) - \int_{R^* \cap R} L(\theta_0) - \int_{R \cap \bar{R}^*} L(\theta_0)$$

$$= \int_{R^*} L(\theta_0) - \int_R L(\theta_0) = \alpha - \alpha = 0.$$

Thus

$$\int_{R^*} L(\theta_1) - \int_R L(\theta_1) \leq k(0) = 0$$

and the result is established. The same sort of reasoning is appropriate in the discrete case, using summation rather than integration.   ◄

This theorem then gives a method which will guarantee, among all possible critical regions with probability of type I error equal to α, the smallest possible value for the probability of type II error. Let us apply this theorem to some specific examples and see the form of this best critical region.

EXAMPLE 8.2.1. Suppose that the time to failure of light bulbs made a certain way is a normal random variable with mean μ and standard deviation $\sigma = 100$. We take a sample of 25 of these bulbs and want to test $H_0: \mu = 1500$ versus $H_1: \mu = 1525$. The likelihood function for the sample then is

$$L(\mu) = \left(\frac{1}{2\pi} \right)^{25/2} \left(\frac{1}{100} \right)^{25} e^{-\Sigma(x_i - \mu)^2 / 2(100)^2} ;$$

the critical region specified by the Neyman-Pearson fundamental lemma is derived as follows:

$$\frac{L(\mu_1)}{L(\mu_0)} = e^{-[\Sigma(x_i-1525)^2 - \Sigma(x_i-1500)^2]/2(100)^2}$$

$$= e^{.0625\bar{x} - 94.53}.$$

Now $e^{.0625\bar{x} - 94.53} \geq k$ is equivalent to

$$\bar{x} \geq 16 \log_e k + 1512.5 = C;$$

that is, if we determine a constant C such that the probability that $\bar{X}$ exceeds C is α when H_0 is true, and then reject H_0 if $\bar{X} > C$, we have the test of H_0 versus H_1 of size α which has the smallest possible value of β. If $\mu = 1500$, then $(\bar{X} - 1500)/20$ is a standard normal random variable and

$$P(\bar{X} > C) = 1 - N_Z\left(\frac{C - 1500}{20}\right).$$

If we want the probability of type I error to be α, then C is chosen so that

$$N_Z\left(\frac{C - 1500}{20}\right) = 1 - \alpha;$$

that is

$$\frac{C - 1500}{20} = z_{1-\alpha}$$

(the $100(1 - \alpha)$ percentile of the standard normal distribution) and

$$C = 1500 + 20z_{1-\alpha}.$$

Thus, if we want $\alpha = .05$, we should reject H_0 if $\bar{X} > 1532.9$; if we want $\alpha = .01$, we should reject H_0 if $\bar{X} > 1546.6$. In either of these two cases for the given α, we have the smallest possible value for the probability of a type II error. If $\mu = 1525$, then $(\bar{X} - 1525)/20$ is a standard normal random variable and

$$P[\bar{X} \leq 1500 + 20z_{1-\alpha}] = N_Z\left(\frac{-25 + 20z_{1-\alpha}}{20}\right) = \beta.$$

Thus for $\alpha = .05$, $z_{1-\alpha} = 1.645$ and

$$\beta = N_Z(-1.25 + 1.645) = N_Z(.395) = .6535.$$

Whereas, if $\alpha = .01$, $z_{1-\alpha} = 2.33$ and

$$\beta = N_Z(-1.25 + 2.33) = N_Z(1.08) = .8599.$$

Figure 8.2.1 illustrates the probabilities of occurrence of these two types of errors.

It is to be stressed that the critical region derived in Example 8.2.1 has the smallest β for a given α. We could simply always reject H_0, regardless of what happens on the sample, in which case we would never commit a type

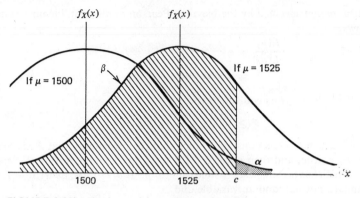

FIGURE 8.2.1   Alpha and beta, Example 8.2.1.

II error and we would have $\beta = 0$ (we would also have $\alpha = 1$ if H_0 is true). The proof of the following theorem is asked for in problem 3 below.

Theorem 8.2.2. Assume that $X_1, X_2, \ldots, X_n$ is a random sample of a normal random variable X with mean μ and known variance σ^2. Then the most powerful (the one with smallest β) critical region of size α for testing $H_0 : \mu = \mu_0$ versus $H_1 : \mu = \mu_1$ (where $\mu_1 > \mu_0$) is specified by

$$\bar{x} \geq \mu_0 + \frac{\sigma}{\sqrt{n}} z_{1-\alpha};$$

if $\mu_1 < \mu_0$, then the most powerful critical region of size α is specified by

$$\bar{x} \leq \mu_0 - \frac{\sigma}{\sqrt{n}} z_{1-\alpha}.$$

Let us now consider an example concerning a test of a hypothesis about the variance of a normal random variable whose mean is known. (It is necessary that we assume that the mean μ is known so that we may have a simple hypothesis and a simple alternative.)

EXAMPLE 8.2.2. A manufacturer of nuts and bolts makes nuts whose interior diameter is, on the average, $\frac{1}{4}$ inch. In order that a nut made by the manufacturer will fit one of his bolts, it is necessary that the interior diameter is neither much bigger than $\frac{1}{4}$ inch nor much smaller than $\frac{1}{4}$ inch. If we assume that the interior diameters of all nuts (of this size) made by this manufacturer are normally distributed with mean $\frac{1}{4}$ inch and unknown standard deviation σ, we might take a random sample of 36 nuts that he has made and test the hypothesis $H_0 : \sigma = .01$ versus $H_1 : \sigma = .02$. Presumably, $\sigma = .02$ would lead to a great number of nuts being too large or too small, whereas $\sigma = .01$ would give essentially all usable nuts.

Let us derive the most powerful test for this case. The likelihood of the sample is

$$L(\sigma^2) = \left(\frac{1}{2\pi\sigma^2}\right)^{18} e^{-\Sigma(x_i-1/4)^2/2\sigma^2}$$

thus

$$\frac{L(\sigma_1^2)}{L(\sigma_0^2)} = \left(\frac{1}{2}\right)^{36} e^{3750\Sigma(x_i-1/4)^2}$$

and

$$(\tfrac{1}{2})^{36} e^{3750\Sigma(x_i-1/4)^2} \geq k$$

is equivalent to

$$\sum\left(x_i - \frac{1}{4}\right)^2 \geq \frac{\log_e k}{3750} - \frac{\log_e (\tfrac{1}{2})^{36}}{3750} = C.$$

Thus, if we can find a value of C such that

$$P[\sum (X_i - \tfrac{1}{4})^2 \geq C] = \alpha,$$

given that H_0 is true, then rejecting H_0 if

$$\sum (X_i - \tfrac{1}{4})^2 \geq C$$

is the most powerful test of size α. We saw in Chapter 6 that if $X_1, X_2, \ldots, X_{36}$ are independent normal random variables, each with mean $\tfrac{1}{4}$ and variance $(.01)^2$, then

$$\frac{\sum (X_i - \tfrac{1}{4})^2}{(.01)^2}$$

is a χ^2 random variable with 36 degrees of freedom. Thus

$$P\left[\frac{\sum (X_i - \tfrac{1}{4})^2}{(.01)^2} > \chi^2_{1-\alpha}\right] = \alpha$$

where $\chi^2_{1-\alpha}$ is the $100(1 - \alpha)$ percentile of the χ^2 distribution with 36 degrees of freedom. Thus, if we take $C = (.01)^2\chi^2_{1-\alpha}$ and reject H_0 whenever $\sum (X_i - \tfrac{1}{4})^2 > C$, we have the most powerful test of H_0 versus H_1. Then, for $\alpha = .05$, we find $\chi^2_{.95} = 50.75$ and we would reject H_0 whenever $\sum (X_i - \tfrac{1}{4})^2 > .005075$; in so doing we have the smallest possible probability of type II error among all tests of size .05.

In each of the two examples studied so far the critical region is entirely specified by the sample value of a statistic; for the more commonly used tests this is generally the case. Such a statistic, whose sample value indicates acceptance or rejection of the null hypothesis, is frequently referred to as a *test statistic*. The test of the hypothesis reduces to merely observing where the sample value of the statistic lies. Theorem 8.2.3 shows the form of the most powerful critical region for tests of simple hypotheses about the standard deviation of a normal random variable with known mean; its proof is asked for in problem 4 below.

Theorem 8.2.3. Assume that $X_1, X_2, \ldots, X_n$ is a random sample of a normal random variable X with known mean μ and unknown variance σ^2. Then the most powerful critical region of size α for testing $H_0: \sigma = \sigma_0$ versus $H_1: \sigma = \sigma_1 (\sigma_1 > \sigma_0)$ is specified by

$$\sum (x_i - \mu)^2 > \sigma_0^2 \chi_{1-\alpha}^2$$

where $\chi_{1-\alpha}^2$ is the $100(1 - \alpha)$ percentile of the χ^2 distribution with n degrees of freedom. If $\sigma_1 < \sigma_0$, the most powerful critical region of size α is specified by

$$\sum (x_i - \mu)^2 < \sigma_0^2 \chi_{\alpha}^2.$$

Let us now consider testing a simple hypothesis versus a simple alternative for the parameter of a discrete variable.

EXAMPLE 8.2.3. Assume that 10 years ago traffic fatality data for accidents occurring in San Francisco were examined and it was found that the number of traffic fatalities within the city limits, per day, was apparently a Poisson random variable with parameter $\lambda = .1$. Due to the increased congestion of the present day, it is suspected that the rate is now .15 per day rather than .1. To investigate this point, assume that we have available the records for the number of traffic fatalities for n recent days and want to test $H_0: \lambda = .1$ versus $H_1: \lambda = .15$. We shall use the Neyman-Pearson fundamental lemma to derive the most powerful test. The likelihood function for the sample is

$$L(\lambda) = \frac{\lambda^{\sum x_i}}{\Pi x_i!} e^{-n\lambda};$$

thus

$$\frac{L(\lambda_1)}{L(\lambda_0)} = \frac{L(.15)}{L(.1)} = (1.5)^{\sum x_i} e^{-.05n}$$

and

$$\frac{L(\lambda_1)}{L(\lambda_0)} \geq k$$

is equivalent to

$$\sum x_i \geq \frac{\log_e k + .05n}{\log_e (1.5)} = C.$$

Thus, to use the most powerful critical region of size α, we would have to determine C such that

$$P(\sum X_i > C) = \alpha,$$

given $\lambda = .1$, and then would reject H_0 if we found that $\sum x_i > C$. We found in Chapter 5 that $\sum X_i$ is a Poisson random variable with parameter $n\lambda$ if $X_1, X_2, \ldots, X_n$ are independent Poisson random variables each with parameter λ. Thus, if we choose C equal to the $100(1 - \alpha)$ percentile of the Poisson distribution with parameter $n/10$, we have the most powerful test. Since the Poisson distribution is discrete, we do not have the complete freedom for choices of α that the continuous cases afford.   Note in particular that if $n = 20$, then $n/10 = 2$, and from the Poisson tables we find that

$$P(\sum X_i > 4) = .0527, \qquad P(\sum X_i > 5) = .0166;$$

thus there is no constant C such that we have α equal to exactly .05 (or .04 or .03 or .02).   This is a minor point, but one which the reader should be aware of.   As n gets larger, of course, there are more available values for α.

It may be evident, especially in the case of testing a simple hypothesis versus a simple alternative, that the choice of which statement is H_0 and which is H_1 is arbitrary.   However, the following general sort of reasoning is frequently applied when choosing H_0.   The Neyman-Pearson fundamental lemma gives a method for finding the most powerful test for arbitrary α, the probability of type I error.   Thus, depending on how much of a gamble we would like to take, we can make the probability of rejecting H_0 when it is true equal to .1 or .05 or .0001.   This causes people to state as H_0 something that they believe the sample will indicate to be false;   then in rejecting H_0 on the basis of the sample, they can make a concrete probability statement about whether they are wrong in so rejecting it and the value of that probability is within their control (since it is the probability of type I error).   If we were to state as H_0 something that we believe the sample would indicate to be true, then in accepting H_0 on the basis of the sample we could only make a type II error and its probability would not be within our control (for a given α);   most people feel that this is a weak position to occupy.

Let us try to make the preceding discussion more concrete on the basis of Example 8.2.3.   There it was suspected that the rate of occurrence of traffic fatalities was now .15 per day rather than .1; thus it was felt that the actual sample data would support $\lambda = .15$ as a conclusion and this was stated as H_1.   We can then select $\alpha = .01$ if we like and, if the sample results bear out our belief that $\lambda = .15$ (and we reject H_0), we know that there is only 1 chance in 100 that we have reached a wrong conclusion.   If, on the other hand, we interchange the two hypotheses with $\alpha = .01$ and again assume that the sample results are consistent with $\lambda = .15$ (now we would accept H_0), the probability that we are wrong in accepting H_0 is β, a quantity quite out of our control in the usual method of constructing tests.   We could, of course, manipulate α in such a way to set β at a level that we like, but it is much more straightforward simply to state our hypotheses (as H_0 and H_1) in a way that the error we want to control is α.

When a rather complete tabulation of the distribution function of the test statistic is available, it is possible to find a sample size n that will set both α and β at desired levels.   In such a case it is quite immaterial which of the two hypotheses is called H_0.   The following example illustrates this technique.

EXAMPLE 8.2.4.   The reader will recall Example 8.2.1 in which we assumed that the time to failure, X, of light bulbs made a certain way was a normal random variable with $\sigma = 100$. We wanted to test $H_0: \mu = 1500$ versus $H_1: \mu = 1525$. We found in that example that the most powerful critical region was of the form $\bar{x} > C$. Suppose now that we want to find n and C such that $\alpha = .01$ and $\beta = .05$. Then

$$1 - \alpha = .99 = N_Z\left(\frac{C - 1500}{100/\sqrt{n}}\right)$$

which implies that

$$\frac{C - 1500}{100/\sqrt{n}} = 2.33,$$

and

$$\beta = .05 = N_Z\left(\frac{C - 1525}{100/\sqrt{n}}\right)$$

which implies that

$$\frac{C - 1525}{100/\sqrt{n}} = -1.64$$

Thus we have two equations in two unknowns (C and n) which we can solve simultaneously. Their solution is $C = 1514.67$, $n = 252.8$; since we cannot take a fractional sample size we should take a sample of 253 bulbs and then reject H_0 if $\bar{X} > 1514.67$. In so doing our actual probability of type I error is slightly less than .01 and our actual probability of type II error is slightly less than .05.

The following example illustrates the sort of statement that can be made about the power of the Neyman-Pearson critical region when testing a simple H_0 against a one-sided alternative H_1.

EXAMPLE 8.2.5.   Let us again return to the light bulb example of 8.2.1. Suppose that we want to test the hypothesis $H_0: \mu = 1500$ versus the composite one-sided alternative $H_1: \mu > 1500$ (the alternative is called *one-sided* since it simply states that μ takes on one of the values in the interval to the right of 1500, those values on one side of 1500). We take a sample of $n = 100$ bulbs and shall reject H_0 if $\bar{X} > 1516.4$ (thus $\alpha = .05$); we are using the Neyman-Pearson critical region, the one which is most powerful against the simple alternative that $\mu = \mu_1$ where μ_1 is any value greater than 1500. Thus, if we were to plot the power function for this test (see Figure 8.2.2), we know that, among all tests with $\alpha = .05$, the value of $Q(\mu_1)$ for this test is the largest possible. That is, if we were to consider any other critical region for $H_0: \mu = 1500$ versus $H_1: \mu > 1500$, with $\alpha = .05$ and n fixed, its power function cannot be above the power function for the test described, for *any* $\mu > 1500$. Such a test is called *uniformly most powerful* and is clearly a desirable test to use. For many common testing problems uniformly most powerful tests do exist, while for others they can be shown not to exist.

The Neyman-Pearson critical region gives the smallest possible value for β, among all tests with the same value for α. Another desirable feature of

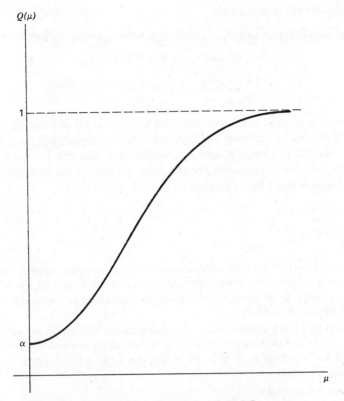

FIGURE 8.2.2   Power function, Example 8.2.5.

such tests is that they will generally be a function of the sufficient statistic for the parameter, if one exists. This result is given in the following theorem.

Theorem 8.2.4.   Assume $X_1, X_2, \ldots, X_n$ is a random sample of X and that the distribution function for X depends on a single unknown parameter θ; assume $Y = g(X_1, X_2, \ldots, X_n)$ is a sufficient statistic for θ. Then the Neyman-Pearson critical region will be defined as a function of Y.

Proof: The Fisher-Neyman factorization criterion (Theorem 7.2.3) says that the likelihood function for the sample can be written

$$L(\theta) = G(g(x_1, x_2, \ldots, x_n), \theta)H(x_1, x_2, \ldots, x_n),$$

where H does not involve θ, if $y = g(x_1, x_2, \ldots, x_n)$ is a sufficient statistic for θ. The Neyman-Pearson critical region is given by those sample values such that

$$\frac{L(\theta_1)}{L(\theta_0)} \geq k;$$

but if Y is a sufficient statistic

$$\frac{L(\theta_1)}{L(\theta_0)} = \frac{G(g(x_1, x_2, \ldots, x_n), \theta_1)H(x_1, x_2, \ldots, x_n)}{G(g(x_1, x_2, \ldots, x_n), \theta_0)H(x_1, x_2, \ldots, x_n)}$$

$$= \frac{G(g(x_1, x_2, \ldots, x_n), \theta_1)}{G(g(x_1, x_2, \ldots, x_n), \theta_0)}$$

since the two H factors are identical and will cancel off in forming the ratio. Thus the inequality defining the Neyman-Pearson region will be defined by g (the sufficient statistic), θ_0 and θ_1; we see then that the Neyman-Pearson region uses all the information contained in the sample about θ (in the sense that the region must be a function of the sufficient statistic).   ◀

EXERCISE 8.2

1. Suppose that X is a Bernoulli random variable with parameter p. We take a random sample of 4 observations of X and want to test $H_0: p = \frac{1}{4}$ versus $H_1: p = \frac{3}{4}$. If we reject H_0 only if we get 4 successes in the sample, compute the values of α and β.

2. Given that X is a uniform random variable on the interval $(0, \theta)$, we might test $H_0: \theta = 1$ versus the alternative $H_1: \theta = 2$ by taking a sample of 2 observations of X and rejecting H_0 if $\bar{X} > .99$. Compute α and β for this test.

3. Prove Theorem 8.2.2.

4. Prove Theorem 8.2.3.

5. Compute the probability of type II error for the test derived in Example 8.2.2.

6. If X is a normal random variable with mean 0, derive the uniformly most powerful test of $H_0: \sigma^2 = 2$ versus $H_1: \sigma^2 > 2$ of size α, based on a random sample of n observations of X.

7. Derive the form of the Neyman-Pearson critical region for testing $H_0: \lambda = \lambda_0$ versus $H_1: \lambda = \lambda_1(\lambda_1 > \lambda_0)$, based on a random sample of n observations of an exponential random variable with parameter λ.

8. Assume that the probability of getting a head on a single flip of a given coin is p and derive the most powerful critical region for testing $H_0: p = \frac{1}{2}$ versus $H_1: p = \frac{3}{4}$, based on a sample of n flips of the coin.

9. Assume we have a random sample of a continuous random variable X and want the most powerful test of the hypothesis that X is a standard normal variable versus the alternative that X is a normal random variable with $\mu = 1$, $\sigma^2 = 1$. What is the critical region for this test?

10. Assume we have a random sample of a continuous random variable X and want to test the simple hypothesis that the density for X is

$$f_X(x) = 2x, \qquad 0 < x < 1$$

versus the simple alternative that the density for X is

$$f_X(x) = 1, \qquad 0 < x < 1.$$

Find the most powerful test for this hypothesis.

11. Assume that the time to failure of a vacuum tube is an exponential random variable with parameter λ and let $R(t)$ be the reliability for this type of tube for t hours (see Example 7.3.5). Find the Neyman-Pearson critical region for testing $H_0: R(1000) = .99$ versus the alternative $H_1: R(1000) = .999$.

12. The number of sales made by a used car salesman, per day, is a Poisson random variable with parameter λ. Given a random sample of the number of sales he made on n days, what is the most powerful test of the hypothesis

$$H_0: p = .1 \text{ versus } H_1: p = .25$$

where p is the probability he makes at least one sale (per day)?

8.3. Composite Hypotheses

In Section 8.2 we were especially concerned with tests of simple hypotheses versus simple alternatives; in such situations the Neyman-Pearson fundamental lemma can be used to find the most powerful test of size α. If H_0 or H_1 or both are composite hypotheses, then unfortunately no such theorem exists. However, there is a general procedure that can be followed to construct a critical region. It is similar to the procedure used in the Neyman-Pearson lemma and generally leads to a good test.

Let us first make note of the fact that if H_0 is composite, then we do not have a unique value for the probability of type I error. That is, the probability of rejecting H_0 when it is true is generally dependent on which of the several possible values of the parameter stated in H_0 is assumed to be the true one. In such situations it is usual to let α denote the largest value of the probability of type I error for all values of the parameter specified by H_0. Then, for all possible tests with the same α, we would prefer the one with the lowest OC curve for all values specified by H_1, if such a test exists.

The *likelihood ratio test criterion* is a method which can be employed to derive a critical region for testing H_0 versus H_1, even if one or both are composite; it generally leads to a good test in the sense of giving small probabilities of type II error for a given α. It is defined as follows.

DEFINITION 8.3.1. Suppose that $X_1, X_2, \ldots, X_n$ is a random sample of a random variable X with a probability law indexed by $\theta_1, \theta_2, \ldots, \theta_r$. To test H_0 versus H_1, where either one or both may be composite hypotheses, let $L(\hat{\omega})$ be the maximum value of the likelihood function of the sample where the parameters $\theta_1, \theta_2, \ldots, \theta_r$ are restricted to values specified by H_0, and let $L(\hat{\Omega})$ be the maximum value of the likelihood function of the sample where the parameters may take on any value specified by the union of H_0 and H_1. Then the critical region R consists of those sample outcomes such that

$$\lambda = \frac{L(\hat{\omega})}{L(\hat{\Omega})} < A,$$

where A is chosen so that $P(\text{type I error}) = \alpha$.

Notice that the likelihood ratio defined above, $L(\hat{\omega})/L(\hat{\Omega})$, will never be greater than 1 since it is the ratio of two maximum values of the likelihood function for the same sample and the denominator allows a range for the parameters which includes the range for the numerator. Thus the constant A will always be less than 1 and can be determined from the probability law of the likelihood ratio. In many common cases the probability law of the ratio is quite straightforward, while in others the exact distribution is quite cumbersome to compute. Luckily, the approximate probability distribution of the ratio is easily determined, for large samples, by the result given in Theorem 8.3.3.

The following example derives the likelihood ratio test that the mean of a normal random variable is equal to a specific value.

EXAMPLE 8.3.1.   Many light bulbs currently sold have a statement like "Average Life—750 Hours" printed on the package containing them. It would seem reasonable (for pre-tested bulbs) that the lifetime of bulbs made by a given manufacturer should be normal in form; the statement on the package then is referring to the mean value of that normal distribution. Suppose n bulbs are purchased and turned on until they burn out; their lifetimes could be assumed to be a random sample of size n from this normal distribution. Let us derive the likelihood ratio test of the hypothesis $H_0: \mu = 750$ versus the alternative $H_1: \mu \neq 750$. The likelihood function for the sample is

$$L(\mu, \sigma^2) = \frac{e^{-\Sigma(x_i - \mu)^2/2\sigma^2}}{(2\pi\sigma^2)^{n/2}}$$

To determine $L(\hat{\omega})$ we must assume H_0 is true and find the maximum value for L, that is, we want to maximize $L(750, \sigma^2)$ with respect to σ^2. Recalling the maximization of L from Section 7.1 it is easily seen that the maximizing value is

$$\hat{\sigma}^2 = \sum (x_i - 750)^2/n$$

and thus

$$L(\hat{\omega}) = e^{-n/2}\left(\frac{n}{2\pi \sum (x_i - 750)^2}\right)^{n/2}$$

To evaluate $L(\hat{\Omega})$ we find that the set of parameter values specified by the union of H_0 and H_1 is the full parameter space; thus the values that will maximize L are the maximum likelihood estimators for μ and σ^2:

$$\hat{\mu} = \bar{x}, \qquad \hat{\sigma}^2 = \sum (x_i - \bar{x})^2/n.$$

We find, then, that

$$L(\hat{\Omega}) = e^{-n/2}\left(\frac{n}{2\pi \sum (x_i - \bar{x})^2}\right)^{n/2}$$

and the likelihood ratio then is

$$\lambda = \frac{L(\hat{\omega})}{L(\hat{\Omega})} = \left(\frac{\sum (x_i - \bar{x})^2}{\sum (x_i - 750)^2}\right)^{n/2}$$

after cancelling common factors. It is easy to verify the identity

$$\begin{aligned}\sum (x_i - 750)^2 &= \sum [(x_i - \bar{x}) + (\bar{x} - 750)]^2 \\ &= \sum [(x_i - \bar{x})^2 + 2(x_i - \bar{x})(\bar{x} - 750) + (\bar{x} - 750)^2] \\ &= \sum (x_i - \bar{x})^2 + n(\bar{x} - 750)^2.\end{aligned}$$

If we substitute this in the denominator of the ratio λ, and divide both numerator and denominator by $\sum (x_i - \bar{x})^2$ we have

$$\lambda^{2/n} = 1 \bigg/ \left(1 + \frac{n(\bar{x} - 750)^2}{\sum (x_i - \bar{x})^2}\right).$$

Definition 8.3.1 states that we should reject H_0 if λ is small (A is chosen to make our probability of type I error equal to α). Notice that if λ (or $\lambda^{2/n}$) is small then $n(\bar{x} - 750)^2/\sum (x_i - \bar{x})^2$ is large; that is, the rejection region is defined as those sample values such that

$$\frac{n(\bar{x} - 750)^2}{\sum (x_i - \bar{x})^2} > k$$

or, equivalently, those such that

$$\frac{|\bar{x} - 750| \sqrt{n}}{\sqrt{\sum (x_i - \bar{x})^2}} > \sqrt{k}.$$

We recall, from Theorem 7.3.3, that

$$\frac{(\bar{x} - 750)\sqrt{n(n - 1)}}{\sqrt{\sum (x_i - \bar{x})^2}}$$

is the observed value of a t random variable if in fact we have a random sample of n observations of a normal random variable with $\mu = 750$. Thus, the likelihood ratio test criterion critical region of $H_0: \mu = 750$ versus $H_1: \mu \neq 750$ is equivalent to

$$|t| > \sqrt{k(n - 1)};$$

thus, if we are to have a probability of type I error equal to α we should reject H_0

only if

$$\frac{|\bar{x} - 750| \sqrt{n(n-1)}}{\sqrt{\sum (x_i - \bar{x})^2}} > t_{1-\alpha/2}$$

where $t_{1-\alpha/2}$ is the $100\left(1 - \dfrac{\alpha}{2}\right)$th percentile of the t distribution with $n - 1$ degrees of freedom.

The preceding example is a special case of the following theorem, which you are asked to prove in Exercise 3 below.

Theorem 8.3.1. $X_1, X_2, \ldots, X_n$ is a random sample of a normal random variable with unknown mean μ and unknown variance σ^2. Then the likelihood ratio test criterion critical region R for a test of size α of H_0 versus H_1 is specified as follows for the stated H_0 and H_1.

Test	H_0	H_1	R
1.	$\mu \leq \mu_0$	$\mu > \mu_0$	$\bar{x} > \mu_0 + \dfrac{S}{\sqrt{n}} t_{1-\alpha}$
2.	$\mu \geq \mu_0$	$\mu < \mu_0$	$\bar{x} < \mu_0 - \dfrac{S}{\sqrt{n}} t_{1-\alpha}$
3.	$\mu = \mu_0$	$\mu \neq \mu_0$	$\left|\dfrac{\bar{x} - \mu_0}{S/\sqrt{n}}\right| > t_{1-\alpha/2}$

The following example derives the likelihood ratio test for the hypothesis that the variance of a normal random variable has a specified value.

EXAMPLE 8.3.2. The manufacturer of a precision scale claims that the standard deviation of measurements made by his machine will not exceed .02 mg. Assuming that repeated measurements made with this machine are normally distributed, let us derive the likelihood ratio test of H_0: $\sigma \leq .02$ versus the alternative H_1: $\sigma > .02$. The likelihood function of the sample is, again,

$$L(\mu, \sigma^2) = \left(\frac{1}{2\pi\sigma^2}\right)^{n/2} e^{-\Sigma(x_i - \mu)^2/2\sigma^2}.$$

To find $L(\hat{\omega})$ we must maximize L under the assumption that $\sigma \leq .02$. The maximizing value for μ is $\bar{x}$ and, if

$$\hat{\sigma}^2 = \sum (x_i - \bar{x})^2/n \leq (.02)^2$$

then $\hat{\sigma}$ is the maximizing value for σ; if

$$\sum (x_i - \bar{x})^2/n > (.02)^2$$

then the maximizing value for σ is .02 (under the restriction $\sigma \leq .02$). Thus

$$L(\hat{\omega}) = e^{-n/2}\left(\frac{n}{2\pi\sum (x_i - \bar{x})^2}\right)^{n/2} \qquad \text{if } \sum(x_i - \bar{x})^2/n \leq (.02)^2$$

$$= \left(\frac{1}{2\pi(.02)^2}\right)^n e^{-\Sigma(x_i-\bar{x})^2/2(.02)^2} \qquad \text{if } \sum (x_i - \bar{x})^2/n > (.02)^2.$$

Again, the union of values specified by H_0 and H_1 is the full parameter space so the values which maximize L are

$$\hat{\mu} = \bar{x}, \qquad \hat{\sigma}^2 = \sum (x_i - \bar{x})^2/n$$

and

$$L(\hat{\Omega}) = \left(\frac{n}{2\pi\sum (x_i - \bar{x})^2}\right)^{n/2} e^{-n/2}.$$

Then we have

$$\lambda = \frac{L(\hat{\omega})}{L(\Omega)} = 1 \qquad \text{if } \sum (x_i - \bar{x})^2/n \leq (.02)^2$$

$$= \left(\frac{\sum (x_i - \bar{x})^2}{n(.02)^2}\right)^{n/2} e^{n/2-\Sigma(x_i-\bar{x})^2/2(.02)^2} \qquad \text{if } \sum (x_i - \bar{x})^2/n > (.02)^2.$$

Figure 8.3.1 gives the graph of λ versus $\sum (x_i - \bar{x})^2/n(.02)^2$. Notice that $\lambda < A$ is equivalent to $\sum (x_i - \bar{x})^2/n(.02)^2 > c$; if in fact H_0 is true, then $\sum (X_i - \bar{X})^2/(.02)^2$

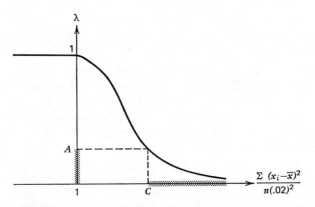

FIGURE 8.3.1   Likelihood ratio test criterion, Example 8.3.2.

is a χ^2 random variable with $n - 1$ degrees of freedom and

$$P(\sum (X_i - \bar{X})^2/n(.02)^2 > C) = P(\chi^2_{n-1} > Cn).$$

Thus, to have a probability of type I error equal to α we should choose $C = \chi^2_{1-\alpha}/n$, where $\chi^2_{1-\alpha}$ is the $100(1 - \alpha)$ percentile of the χ^2 distribution with $n - 1$ degrees of freedom. If, for example, we used the machine $n = 10$ times to weigh the same object and we wanted to test $H_0: \sigma \leq .02$ versus $H_1: \sigma > .02$ with $\alpha = .1$, then

$\chi^2_{.9} = 14.7$ (with 9 degrees of freedom) and we should reject H_0 if

$$\sum (x_i - \bar{x})^2 > 14.7(.02)^2 = 0.00588,$$

where $x_1, x_2, \ldots, x_{10}$ are the 10 observed measurements.

Theorem 8.3.2 gives the likelihood ratio test criterion critical regions for testing hypotheses about the variance of a normal random variable with unknown mean.

Theorem 8.3.2. $X_1, X_2, \ldots, X_n$ is a random sample of a normal random variable X whose mean μ is unknown. Then the likelihood ratio test criterion critical region R for a test of size α of H_0 versus H_1 is specified as follows for the stated H_0 and H_1.

Test	H_0	H_1	R
1.	$\sigma^2 \leq \sigma_0^2$	$\sigma^2 > \sigma_0^2$	$\sum (x_i - \bar{x})^2 > \sigma_0^2 \chi^2_{1-\alpha}$
2.	$\sigma^2 \geq \sigma_0^2$	$\sigma^2 < \sigma_0^2$	$\sum (x_i - \bar{x})^2 < \sigma_0^2 \chi^2_{\alpha}$
3.	$\sigma^2 = \sigma_0^2$	$\sigma^2 \neq \sigma_0^2$	$\sum (x_i - \bar{x})^2 < \sigma_0^2 \chi^2_{\alpha/2}$
		and	$\sum (x_i - \bar{x})^2 > \sigma_0^2 \chi^2_{1-\alpha/2}$

Exercise 4 below asks for the proof of this theorem.

In each of the examples treated so far, we had no difficulty in finding the A such that the probability of type I error was α. In each case we were able to manipulate the inequality describing the critical region to get an inequality for a statistic whose distribution was known if H_0 were true. This known distribution then can effectively be used to choose A. In many problems, however, the exact distribution for λ may be very difficult to obtain and it may not be possible to manipulate the necessary inequality to derive an equivalent statement for a statistic whose distribution is known, if H_0 is true. In cases such as this, the following theorem, which we shall not prove, is of use in describing the critical region.

Theorem 8.3.3. Suppose that X is a random variable whose distribution function satisfies certain regularity conditions and is indexed by k parameters, $\theta_1, \theta_2, \ldots, \theta_k$, and $X_1, X_2, \ldots, X_n$ is a random sample of X. For

$$H_0 : \theta_1 = \theta_1^*, \ \theta_2 = \theta_2^*, \ \ldots, \ \theta_r = \theta_r^*, \qquad \text{where } r \leq k,$$

versus the alternative $H_1 : H_0$ is false, define

$$\lambda = \frac{L(\hat{\omega})}{L(\hat{\Omega})} = g(x_1, x_2, \ldots, x_n)$$

in the usual way. Then, if we define $\Lambda = g(X_1, X_2, \ldots, X_n)$, $-2 \log_e \Lambda$ is approximately a χ^2 random variable with r degrees of freedom, if H_0 is true, for large values of n. (Thus, if $Y = -2 \log_e \Lambda$,

$$\lim_{n \to \infty} F_Y(t) = F_{\chi^2}(t), \qquad \text{for all } t,$$

where $F_{\chi^2}(t)$ is the distribution function of a χ^2 random variable with r degrees of freedom.)

Thus, we can use the χ^2 table to determine the value of A if we have a "large" sample size n. Since

$$P[\Lambda < A] = P[-2 \log_e \Lambda > -2 \log_e A],$$

we simply choose $-2 \log_e A = \chi^2_{1-\alpha}$ or $A = e^{-\chi^2_{1-\alpha}/2}$ where $\chi^2_{1-\alpha}$ is the 100 $(1 - \alpha)$ percentile of the χ^2 distribution with r degrees of freedom.

EXAMPLE 8.3.3. Suppose that we would like to test the hypothesis that the probability that a particular experienced dart player will hit the bull's-eye of a target, on a single toss, is .9. Thus, if we have him toss a dart at the bull's-eye n times, we will have a random sample of n observations of a Bernoulli random variable with parameter p and would like to test $H_0: p = .9$ versus $H_1: p \neq .9$ with $\alpha = .01$, for example. Then the likelihood function for the sample is

$$L(p) = p^{\Sigma x_i}(1 - p)^{n - \Sigma x_i}$$

and

$$L(\hat{\omega}) = L(.9) = (.9)^{\Sigma x_i}(.1)^{n - \Sigma x_i};$$

we also know that the maximum likelihood estimator of p is $\bar{X}$, so

$$L(\Omega) = L(\bar{x}) = \bar{x}^{\Sigma x_i}(1 - \bar{x})^{n - \Sigma x_i}.$$

Then

$$\lambda = \frac{L(\hat{\omega})}{L(\hat{\Omega})} = \left(\frac{.9}{\bar{x}}\right)^{\Sigma x_i}\left(\frac{.1}{1 - \bar{x}}\right)^{n - \Sigma x_i},$$

and we would reject H_0 only if $-2 \log_e \lambda > \chi^2_{.99} = 6\,635$ (for 1 degree of freedom). If, for example, he threw 200 darts at the bull's-eye, 160 of which hit it, then $\bar{x} = .8$,

$$\lambda = (\tfrac{9}{8})^{160}(\tfrac{1}{2})^{40}, \qquad -2 \log_e \lambda = 17.88,$$

so we would reject H_0. The (approximate) probability that we are wrong in so rejecting is .01.

It is interesting to note the close relationship between a confidence interval for the mean of a normal random variable, as discussed in Section 7.3, and the likelihood ratio acceptance region for testing hypotheses about μ. In particular, if $X_1, X_2, \ldots, X_n$ is a random sample of a normal random variable X with mean μ and variance σ^2, the observed $100(1 - \alpha)\%$ confidence interval for μ consisted of the set

$$T = \{\mu : \bar{x} - t_{1-\alpha/2}s/\sqrt{n} \leq \mu \leq \bar{x} + t_{1-\alpha/2}s/\sqrt{n}\}$$
$$= \{\mu : |\bar{x} - \mu|\sqrt{n}/s \leq t_{1-\alpha/2}\};$$

Notice, then, from Theorem 8.3.1, we would accept $H_0 : \mu = \mu_0$, versus the alternative $H_1 : \mu \neq \mu_0$, if and only if $\mu_0 \in T$, given the observed sample values, and the probability of type I error, α, is one minus the confidence coefficient of $1 - \alpha$. In terms of Example 8.3.1 above, to test the hypothesis that $\mu = 750$, with $\alpha = .05$, say, we can observe our sample values, $x_1, x_2, \ldots,$ x_n, use these to compute a 95% confidence interval for μ and then we should accept the hypothesis if and only if the confidence interval covers 750, the hypothesized value for μ; this procedure is totally equivalent to the likelihood ratio test as derived. It can be shown that the same equivalence exists between one-sided confidence intervals for μ and tests of one-sided alternative hypotheses, such as $H_0 : \mu \leq \mu_0$ versus $H_1 : \mu > \mu_0$.

You are asked in Exercise 5 below to show the equivalence between confidence intervals for σ^2, as discussed in Section 7.3, and the likelihood ratio test of a hypothesis about σ^2 (or σ). This equivalence of confidence intervals and tests can prove useful in many cases. If we are able to derive a $100(1 - \alpha)\%$ confidence interval for a parameter θ, the observed interval may be used as the acceptance region for testing $H_0 : \theta = \theta_0$, where θ_0 is any value in the interval; conversely, if we are able to derive a test of $H_0 : \theta = \theta_0$, for a given value of α, the acceptance region for H_0, given the sample values, may be used as a $100(1 - \alpha)\%$ confidence interval for θ.

EXERCISE 8.3

1. Assume X is a normal random variable with unknown mean μ and known variance σ^2. Derive the likelihood ratio test criterion test for (μ_0 is some specified value)
   (a) $H_0 : \mu = \mu_0$ versus $H_1 : \mu \neq \mu_0$
   (b) $H_0 : \mu \leq \mu_0$ versus $H_1 : \mu > \mu_0$
   (c) $H_0 : \mu \geq \mu_0$ versus $H_1 : \mu < \mu_0$.

2. Assume that the annual rainfall at a certain recording station is a normal random variable with mean μ and standard deviation 2 inches. The rainfall recorded (in inches) in each of 5 years was 18.6, 20.4, 17.3, 15.1 and 22.6. Test the hypothesis that $\mu \geq 21$ versus the alternative $\mu < 21$ with $\alpha = .1$.

3. Prove Theorem 8.3.1.

4. Prove Theorem 8.3.2.

5. A producer of frozen fish is being investigated by the Bureau of Fair Trades. Each package of fish which this producer markets carries the claim that it contains 12 ounces of fish; a complaint has been registered that this claim is not true. The Bureau acquires 100 packages of fish marketed by this company and, letting x_i be the observed weight (in ounces) of the i-th package, $i = 1, 2, \ldots, 100$, they find

$$\sum x_i = 1150, \qquad \sum x_i^2 = 13,249.75.$$

It would seem reasonable to assume that the true weights of packages that they market are normally distributed with mean μ and variance σ^2, neither of which is known. With $\alpha = .01$, would the Bureau accept or reject $H_0: \mu \geq 12$ versus $H_1: \mu < 12$, based on this sample?

6. In deciding whether a certain type of plant would be appropriate for hedges, it is of some importance that individual plants exhibit small variability in the amounts they will grow in a year (at the same age). Specifically, we might assume that the growth made by a plant of a specific type and age (for given climatic conditions) is a normal random variable with mean μ and variance σ^2. Then, to decide whether the plant would be appropriate for hedges, we might like to test $H_0: \sigma^2 \geq \frac{1}{4}$ versus $H_1: \sigma^2 < \frac{1}{4}$ with $\alpha = .05$ (measurements made in feet). Suppose that we record the growth of 5 plants of this type for 1 year and find them to be 1.9, 1.1, 2.7, 1.6 and 2.0 feet. Should we accept H_0?

7. Derive the likelihood ratio test criterion critical region for testing the hypothesis that $\lambda \leq \lambda_0$ versus the alternative that $\lambda > \lambda_0$ where λ is the parameter of an exponential random variable. Assume that you have available a random sample of n observations of X and that you want your probability of type I error to be α.

8. The time to failure of a certain vacuum tube is known to be an exponential random variable with parameter λ when used in a particular type of circuit. A thousand of these tubes are placed in operation and the sum of their times to failure is 109,652 hours. With $\alpha = .05$, would you reject $H_0: \lambda \leq .008$ versus $H_1: \lambda > .008$ on the basis of this sample?

9. The number of accidents in an industrial plant per month is assumed to be a Poisson random variable with parameter λ. Given a sample of the number of accidents per month in a given year (thus $n = 12$), derive the likelihood ratio test criterion for testing $H_0: \lambda = \lambda_0$ versus $H_1: \lambda < \lambda_0$ for a fixed α. Suppose that a given industry finds that they had a total of 15 accidents last year. With $\alpha = .1$, should they reject $H_0: \lambda = 2$ versus $H_1: \lambda < 2$? (Use the result of Theorem 8.3.3.)

10. The number of times that an electric light switch can be turned on and off before it fails is a geometric random variable X with parameter p (thus X is the number of turnons until failure). Given a random sample of 10 switches, and that the sum of the number of turnons to failure (for all 10) was 15,169, would you reject $H_0: p = .00005$ versus $H_1: p > .00005$ with $\alpha = .05$? (Use Theorem 8.3.3.)

11. Assume $X_1, X_2, \ldots, X_n$ is a random sample of a normal random variable X with mean μ_X and variance σ^2; $Y_1, Y_2, \ldots, Y_n$ is an independent random sample of a normal random variable with mean μ_Y and variance σ^2 (note that the variances are the same). Show that the likelihood ratio test of $H_0: \mu_X = \mu_Y$ versus $H_1: \mu_X \neq \mu_Y$ can be defined in terms of the random variable

$$\frac{(\bar{X} - \bar{Y})\sqrt{n}}{S\sqrt{2}}$$

where $2(n - 1)S^2 = \sum (X_i - \bar{X})^2 + \sum (Y_i - \bar{Y})^2$. What is the distribution of this random variable if H_0 is true?

12. The test described in Exercise 11 is frequently used in practice to test the hypothesis that two population means are equal. The following situation is an example of that type. One third-grade class of 20 students is taught reading in a traditional manner for one year; at the end of that period they are given a reading achievement test. The sum and sum of squares of their scores were $\sum x_i = 1950$, $\sum x_i^2 = 192,861$. A second third-grade class, also with 20 students, received reading instruction using a new, different method. At the end of a year they were given the same reading achievement test; the sum and sum of squares of their scores were $\sum y_i = 2020$, $\sum y_i^2 = 205,920$. Assuming these are two independent samples from normal distributions with the same variance, test $H_0: \mu_X = \mu_Y$ versus $H_1: \mu_X \neq \mu_Y$ with $\alpha = .05$.

13. Show that in testing a simple hypothesis (say $H_0: \theta = \theta_0$) versus a simple alternative (say $H_1: \theta = \theta_1$) the likelihood ratio test criterion discussed in this section is equivalent to the Neyman-Pearson best test described in Section 8.2.

14. Describe a reasonable test of $H_0: p_1 = p_2$ versus $H_1: p_1 \neq p_2$, given a random sample of size n_1 of a Bernoulli random variable X with parameter p_1 and a second independent random sample of size n_2 of a Bernoulli random Y with parameter p_2.

15. During the 1972 U.S. presidential campaign a nationwide political polling agency announced that in a random sample of 1500 respondents, 59% would vote for the Republican incumbent, 31% would vote for the Democratic challenger and the remaining 10% were undecided. On the same day, the Democratic challenger announced the results of his own survey of 1000 people: 37% would vote for him (the challenger) and 59% would vote for the Republican incumbent, with the remaining 4% undecided. Assume these are two independent random samples of Bernoulli random variables (where p_i is the proportion of people who will vote for the Democratic challenger $i = 1, 2$), and test $H_0: p_1 = p_2$ versus $H_1: p_1 \neq p_2$ using the test derived in Exercise 14, with $\alpha = .1$.

16. Of 292 young women admitted to Lenox Hospital in New York City, it was found that 13 were unknowingly pregnant; of 290 young women admitted to the Los Angeles County-University of Southern California Hospital over the same period, it was found that 12 were pregnant and did not know it.

Test the hypothesis that the proportion of young women in New York City that are unknowingly pregnant is the same as that in Southern California, versus the alternative that they are not the same, with $\alpha = .05$.

17. Show the equivalence of the likelihood ratio tests of Theorem 8.3.2 and the appropriate confidence intervals for σ^2, discussed in Section 7.3.

8.4. Goodness-of-Fit Tests

We have discussed ways of testing what might be called parametric hypotheses about a random variable; that is, we have assumed that the probability law of the random variable was known and discussed testing hypotheses that stated values for the parameters of the probability law. In many practical problems it is desired to test a hypothesis that specifies the probability law for the random variable being sampled versus the alternative that the probability law is not of the stated type. The likelihood ratio methodology discussed in the last section is then not appropriate, since the likelihood function is not well defined, assuming the alternative is true. If we assume that the probability law is not of the form specified by the hypothesis, then it could be any one from a large class of possible functions and we would find it very difficult to maximize the likelihood function over this class of possible alternatives. The fact that the class of alternatives is so large also makes it difficult to make definitive comparative statements about different possible tests of the hypothesis (in terms of probability of type II error). Thus, we shall follow the usual convention and simply not attempt to evaluate the probability of type II error when our hypothesis specifies the distributional form for the random variable we have sampled.

One method of testing such distributional hypotheses that we shall discuss is the χ^2 goodness-of-fit test. All such tests rest upon the multinomial random variable and the following theorem, which we shall present with no proof. (The multinomial random variable was discussed in Section 5.4.)

Theorem 8.4.1. If $(X_1, X_2, \ldots, X_k)$ is a multinomial random variable with parameters $n, p_1, p_2, \ldots, p_k$, then the distribution function of the random variable

$$U = \sum_{i=1}^{k} \frac{(X_i - np_i)^2}{np_i}$$

approaches the χ^2 distribution function with $k - 1$ degrees of freedom as $n \to \infty$. That is,

$$\lim_{n \to \infty} F_U(t) = F_{\chi^2}(t), \quad \text{for every } t,$$

where $F_{\chi^2}(t)$ is the χ^2 distribution function with $k - 1$ degrees of freedom.

To make the result just quoted seem a little more plausible, we might examine the case of the multinomial random variable with $k = 2$. Then each of our n trials falls in one of two classes. If we call the first class success, then X_1 is simply the number of successes and $X_2 = n - X_1$ is the number of failures in the n trials, where p_1 is the probability of a success on a single trial and $p_2 = 1 - p_1$ is the probability of a failure. Then

$$
\begin{aligned}
U &= \frac{(X_1 - np_1)^2}{np_1} + \frac{(X_2 - np_2)^2}{np_2} \\
&= \frac{(X_1 - np_1)^2}{np_1} + \frac{(n - X_1 - n(1 - p_1))^2}{n(1 - p_1)} \\
&= \frac{(X_1 - np_1)^2(1 - p_1) + (-X_1 + np_1)^2 p_1}{np_1(1 - p_1)} \\
&= \frac{(X_1 - np_1)^2}{np_1(1 - p_1)} = V^2,
\end{aligned}
$$

for example. But

$$
V = \frac{X_1 - np_1}{\sqrt{np_1(1 - p_1)}}
$$

is approximately a standard normal random variable for large n, and thus V^2 is approximately the square of a standard normal random variable for large n. Thus, the distribution of V^2 will be a χ^2 with 1 degree of freedom, for large n, which is just what Theorem 8.4.1 claims. The proof of Theorem 8.4.1 for arbitrary k would proceed in quite an analogous fashion. Essentially the k terms that are summed together can be shown to be the same as a sum of squares of $k - 1$ independent variables, each of which tends to be the square of a standard normal random variable as n increases. Because of Theorem 6.2.4, the sum has the χ^2 distribution with $k - 1$ degrees of freedom.

To use Theorem 8.4.1, then, to test the hypothesis that $(X_1, X_2, \ldots, X_k)$ is a multinomial random variable with specified parameters $p_1, p_2, \ldots, p_k$, we would take our sample, compute

$$
u = \sum_{i=1}^{k} \frac{(x_i - np_i)^2}{np_i},
$$

and reject the hypothesis if $u > \chi^2_{1-\alpha}$, the $100(1 - \alpha)$ percentile of the χ^2 distribution with $k - 1$ degrees of freedom. If the hypothesis is true, then the probability we would reject is α; that is, α is our probability of type I error. Many authors agree that the χ^2 approximation is quite good for the distribution of U so long as $np_i \geq 5$ for $i = 1, 2, \ldots, k$. If necessary, some of the classes may be combined to satisfy this requirement. Some of the following examples illustrate this procedure.

EXAMPLE 8.4.1.   Suppose that we are given a particular die and want to test the hypothesis that it is fair, with $\alpha = .05$, based on a sample of 100 rolls. Then, the hypothesis might be written

$$H_0 : p_i = \tfrac{1}{6}, \qquad i = 1, 2, \ldots, 6,$$

and we could let X_i, $i = 1, 2, \ldots, 6$ be the number of rolls on which we observe face i. Then, suppose we observe $x_1 = 20$, $x_2 = 24$, $x_3 = 15$, $x_4 = 21$, $x_5 = 12$ and $x_6 = 8$; the observed value of u is

$$\sum_{i=1}^{6} \frac{(x_i - 16\tfrac{2}{3})^2}{16\tfrac{2}{3}} = 10.94.$$

Since the 95-th percentile of the χ^2 distribution with 5 degrees of freedom is 11.070, we would accept H_0 and conclude that this die is fair. (Note that if we had chosen $\alpha = .1$, we would have rejected H_0; the probability that we would be wrong in such a conclusion would be .1.)

By making use of the following theorem, the multinomial test just described can be applied to testing the hypothesis that the distribution function F_Y of a random variable Y has any specified form.

Theorem 8.4.2.   Assume that Y has distribution function F_Y (either discrete or continuous) and that $Y_1, Y_2, \ldots, Y_n$ is a random sample of Y. Then, if we define

$$I_1 = \{y : y \leq a_1\}, \qquad\qquad I_2 = \{y : a_1 < y \leq a_2\}, \ldots,$$
$$I_{k-1} = \{y : a_{k-2} < y \leq a_{k-1}\}, \qquad I_k = \{y : a_{k-1} < y\},$$

and let X_i be the number of sample values that fall in I_i, $i = 1, 2, \ldots, k$, $(X_1, X_2, \ldots, X_k)$ is the multinomial random variable with parameters n, $p_1, p_2, \ldots, p_k$, where n is the sample size (for the sample of Y) and

$$p_i = P[Y \text{ falls in } I_i], \qquad i = 1, 2, \ldots, k.$$

Proof: Clearly, the first sample value, Y_1, will fall in I_1 or I_2 or $\cdots$ or I_k; the probability that it falls in I_i is

$$p_i = P[Y_1 \text{ falls in } I_i] = P[Y \text{ falls in } I_i]$$

since $F_{Y_1}(t) = F_Y(t)$. Thus, this first sample of Y can be thought of as being a multinomial trial with k possible outcomes whose probabilities of occurrence are $p_1, p_2, \ldots, p_k$. The second sample value, Y_2, can be thought of as a second, independent, multinomial trial with the same probabilities of occurrence; the same is true for each of the succeeding $n - 2$ sample values.

Since $Y_1, Y_2, \ldots, Y_n$ are independent, the trials they generate are independent and thus $(X_1, X_2, \ldots, X_k)$ is the multinomial random variable with parameters $n, p_1, \ldots, p_k$, where X_i is the total number of Y_j's that fall in I_i, $i = 1, 2, \ldots, k$. ◄

Then, no matter what the form of F_Y, we can test the hypothesis that $Y_1, Y_2, \ldots, Y_n$ each has distribution function F_Y by testing the hypothesis that the multinomial random variable $(X_1, X_2, \ldots, X_k)$ generated by Y_1, $Y_2, \ldots, Y_n$ has the appropriate parameters $p_1, p_2, \ldots, p_k$. If we reject this hypothesis, then F_Y is apparently not appropriate, compared with the sample; if we accept this hypothesis, then F_Y would appear to be consistent with the sample values.

EXAMPLE 8.4.2. The number of misprints per page in a printed book is frequently taken to be a Poisson random variable, since they are presumably independent and, if the whole book were done by the same typesetter, the rate should be constant from the first page to the last. A misprint count was made for 100 pages of a recent novel with the numbers of misprints found as given below.

Number of misprints	Number of pages
0	65
1	25
2	8
3	2
total	100

We will test the hypothesis that Y, the number of misprints per page, is a Poisson random variable with parameter $\lambda = .4$ per page with $\alpha = .1$. Since Y is discrete, the natural intervals to use in defining the multinomial random variable would be ones centered around the values the variable may take on. Thus, we define

$$I_1 = \{y: y \leq \tfrac{1}{2}\}, \qquad I_2 = \{y: \tfrac{1}{2} < y \leq 1\tfrac{1}{2}\},$$
$$I_3 = \{y: 1\tfrac{1}{2} < y \leq 2\tfrac{1}{2}\}, \qquad I_4 = \{y: 2\tfrac{1}{2} < y\},$$

Then we define X_1 to be the number of 0's we observe, X_2 the number of 1's, X_3 the number of 2's and X_4 the number of 3's (or higher values). By referring to the Poisson table, we find that $p_1 = .6703$, $p_2 = .2681$, $p_3 = .0536$ and $p_4 = .0080$. Then we note that $np_4 = .80$ which is not at least 5 so we combine the two classes for 2 and 3 misprints per page. Thus (X_1, X_2, X_3), where X_3 now is the number of times we observe 2 or more misprints, is the multinomial random variable with parameters 100, .6703, .2681 and .0616, if the hypothesis is true. The observed value of the test statistic is

$$u = \frac{(65 - 67.03)^2}{67.03} + \frac{(25 - 26.81)^2}{26.81} + \frac{(10 - 6.16)^2}{6.16} = 2.577.$$

Since the 90-th percentile of the χ^2 distribution with 2 degrees of freedom is 4.605, we accept the hypothesis.

In making goodness-of-fit tests for a discrete random variable, such as in the example above, the intervals $I_1, I_2, \ldots, I_k$ that the data are summarized into are actually determined by the discrete values that the variable can take on. In continuous cases, though, the construction of $I_1, I_2, \ldots, I_k$ is generally much more arbitrary. In such cases, there is general agreement that the lengths of the finite intervals used should be equal and that the number used should be as large as possible consistent with the requirement that $np_i \geq 5$, $i = 1, 2, \ldots, k$.

EXAMPLE 8.4.3.   The times to failure of a thousand electron tubes were as given in the following table.

Time to failure	Number
$t \leq 150$	543
$150 < t \leq 300$	258
$300 < t \leq 450$	120
$450 < t \leq 600$	48
$600 < t \leq 750$	20
$750 < t$	11

The manufacturer claims that lives of tubes of this type are exponentially distributed with $\lambda = .005$. Should we reject his claim, with $\alpha = .01$? Defining I_1 through I_6 in the same way as the above data are summarized, the numbers of failure times in these 6 intervals should then be a multinomial random variable with parameters $1000, p_1, p_2, \ldots, p_6$, where

$$p_1 = \int_0^{150} f_Y(t)\,dt = \int_0^{150} .005e^{-.005t}\,dt = .5277,$$

$$p_2 = \int_{150}^{300} f_Y(t)\,dt = \int_{150}^{300} .005e^{-.005t}\,dt = .2492,$$

etc. The remaining probabilities are $p_3 = .1177$, $p_4 = .0556$, $p_5 = .0263$, $p_6 = .0235$. Then, the observed value of u is

$$\frac{(543 - 527.7)^2}{527.7} + \frac{(258 - 249.2)^2}{249.2} + \cdots + \frac{(11 - 23.5)^2}{23.5} = 9.997;$$

the 99-th percentile of the χ^2 distribution with 5 degrees of freedom is 15.1. Thus we would accept his claim.

In the two goodness-of-fit cases examined above, the original hypothesis completely specified the distribution function of the random variable Y which had been sampled; the specification was complete down to the value of the parameter of the probability law. Such complete specification was necessary, of course, in order that we could compute the values of p_1, $p_2, \ldots, p_k$. In many cases it is desirable to be able to test the hypothesis that the distribution function has a specified form without having to also specify the values of the parameters of the distribution function. Thus, we might simply like to test the hypothesis that a random variable Y follows the Poisson probability law, rather than that Y follows the Poisson probability law with parameter $\lambda = .4$; or we might like to test that Y is exponential, rather than that Y is exponential with parameter $\lambda = .005$. Fortunately, a slight adaptation of the test already discussed is all that is necessary to test these less restrictive hypotheses which do not specify the values of the parameters. The following theorem, stated without proof, gives the necessary methodology.

Theorem 8.4.3. Assume that Y has distribution function F_Y which satisfies certain regularity conditions and which is indexed by r unknown parameters, $\theta_1, \theta_2, \ldots, \theta_r$, and that $Y_1, Y_2, \ldots, Y_n$ is a random sample of Y. Let $\hat{\Theta}_1, \hat{\Theta}_2, \ldots, \hat{\Theta}_r$ be the maximum likelihood estimators for θ_1, $\theta_2, \ldots, \theta_r$, respectively, and define $I_1, I_2, \ldots, I_k$ and $X_1, X_2, \ldots, X_k$ as in Theorem 8.4.2. Then, if we define

$$\hat{P}_i = P[Y \text{ falls in } I_i], \qquad i = 1, 2, \ldots, k$$

where $\hat{\Theta}_1, \hat{\Theta}_2, \ldots, \hat{\Theta}_r$ replace $\theta_1, \theta_2, \ldots, \theta_r$ in F_Y, the distribution of the statistic

$$V = \sum_{i=1}^{k} \frac{(X_i - n\hat{P}_i)^2}{n\hat{P}_i}$$

tends to that of a χ^2 random variable with $k - r - 1$ degrees of freedom as n gets large.

Thus, if we want to test the hypothesis that F_Y has a specified form, saying nothing about the values of the parameters, we simply use the data to estimate the parameters and then proceed as we have in previous examples. That is, we would compute the observed value of V, in this case, and reject H_0 if this observed value exceeds the $100(1 - \alpha)$ percentile of the χ^2 distribution with $k - r - 1$ degrees of freedom. Note that we have "lost" 1 degree of freedom for every parameter that we had to estimate.

EXAMPLE 8.4.4.   A manufacturer of yardsticks has a certain production process which turns out yardsticks. He feels that Y, the "true" length of one of his yardsticks, should be a normal random variable. To test the hypothesis that Y is normal, without specifying μ and σ, he randomly selects 300 yardsticks and measures their actual lengths. These 300 measurements are summarized in the following table.

True length y	Frequency x
$35.7 < y \leq 35.8$	10
$35.8 < y \leq 35.9$	30
$35.9 < y \leq 36.0$	104
$36.0 < y \leq 36.1$	97
$36.1 < y \leq 36.2$	51
$36.2 < y \leq 36.3$	8

He may test the hypothesis that Y, the true length of a yardstick made this way, is a normal random variable with $\alpha = .10$ by using the result given as Theorem 8.4.3. Thus, he would first want to use the sample to compute the observed values of the maximum likelihood estimators for μ and σ^2; then he would use these quantities to compute the observed values of the estimators of the probabilities of individual sample values falling into the six classes used to summarize the data above. We find

$$\bar{y} = 36.008, \qquad \frac{n-1}{n} s^2 = .0114, \qquad \sqrt{\frac{n-1}{n} s^2} = .1067$$

Thus, to compute $\hat{p}_1, \hat{p}_2, \ldots, \hat{p}_6$, we simply compute the probability that a normal random variable with mean 36.008 and standard deviation .1067 would fall in each of the six intervals. Using the standard normal table, we find

$$\hat{p}_1 = .0301, \qquad \hat{p}_2 = .1359, \qquad \hat{p}_3 = .3061,$$
$$\hat{p}_4 = .3246, \qquad \hat{p}_5 = .1615, \qquad \hat{p}_6 = .0418.$$

Thus, if the hypothesis were true the numbers we expect in each of these six classes are 9.03, 40.77, 91.83, 97.38, 48.45 and 12.54, respectively, and we find

$$v = \sum_{i=1}^{6} \frac{(x_i - 300\hat{p}_i)^2}{300\hat{p}_i} = 6.341.$$

The 90-th percentile of the χ^2 distribution with 3 degrees of freedom is 6.25 so the hypothesis is rejected. The probability this conclusion is incorrect is .1, since there is a 10% chance that we would observe this large (or larger) a value of V if the hypothesis were true.

EXAMPLE 8.4.5. Let us test the hypothesis that the number of suicides per month in Manhattan, Y, is a Poisson random variable with $\alpha = .05$, based upon the following observed numbers of suicides per month for a 5-year span.

Number of suicides y	Frequency x
0	33
1	17
2	7
3	3

We must first estimate λ, assuming the hypothesis is true; thus we find $\bar{y} = .667$. Then the estimated probabilities of observing 0, 1, 2 and 3 suicides per month are

$$\hat{p}_1 = .5314, \qquad \hat{p}_2 = .3415, \qquad \hat{p}_3 = .1181, \qquad \hat{p}_4 = .0090$$

and the expected frequencies would be 31.884, 20.490, 7.086 and .54, respectively. Since the expected frequency for $Y = 3$ is less than 5, we combine both the observed frequencies and these expected frequencies for $Y = 3$ with those for $Y = 2$. Then,

$$\hat{p}_1 = .5314, \qquad \hat{p}_2 = .3415, \qquad \hat{p}_3 = .1271$$

and we find

$$v = \sum_{i=1}^{3} \frac{(x_i - 60\hat{p}_i)^2}{60\hat{p}_i} = 1.372.$$

The 95-th percentile of the χ^2 distribution with 1 degree of freedom is 3.84 so we accept the hypothesis that Y is a Poisson random variable.

EXERCISE 8.4

1. A college track man put the shot 100 times in practice in a week. The distances he threw it (measured in feet) are recorded in the following table.

Distance y	Frequency x
$y \leq 61$	12
$61 < y \leq 63$	20
$63 < y \leq 65$	40
$65 < y \leq 67$	25
$y > 67$	3

With $\alpha = .01$, test the hypothesis that the distance, Y, which he can put the shot is a normal random variable with $\mu = 63$ feet, $\sigma = 2$ feet.

2. In each of 100 games a baseball player had 4 times at bat. If the number of hits he got were as follows, test the hypothesis that the number of hits he will get in 4 times at bat is a binomial random variable with $n = 4$ and $p = .3$. (Use $\alpha = .05$.)

Number of hits y	Frequency (Number of games) x
0	18
1	45
2	29
3	6
4	2

3. It would seem reasonable to assume that the final digit of a randomly selected telephone number is equally likely to be 0 or 1 or 2 or $\cdots$ or 9. Select a page at random from your telephone book and count the number of 0's, 1's, 2's, ..., 9's that occur as last digits. Then test the hypothesis that the last digit is equally likely to be 0 or 1 or $\cdots$ or 9, with $\alpha = .1$. (You should get a sample of size $n = 280$ or so since this is roughly the number of phone numbers listed on a single page.)

4. A pair of dice was rolled 500 times. The sums that occurred were as recorded in the following table. Test the hypothesis that the dice were fair, with $\alpha = .05$

Sum	Frequency
2, 3, or 4	74
5 or 6	120
7	83
8 or 9	135
10, 11, or 12	88

5. If cars are arriving at a supermarket consistent with a Poisson process, the length of time between successive arrivals is an exponential random variable. Arrival times for all autos in a 2-hour period were recorded and the times between arrivals (in minutes) were as summarized below.

Time between arrivals t	Frequency
$t \leq 1$	40
$1 < t \leq 2$	29
$2 < t \leq 3$	15
$t > 3$	8
Total	92

Test the hypothesis that this time distribution is consistent with an exponential distribution for time between arrivals, with $\alpha = .1$.

6. A hundred students took the same I.Q. test. The scores they made were as summarized below.

Score x	Frequency
$70 < x \leq 90$	8
$90 < x \leq 110$	38
$110 < x \leq 130$	45
$130 < x \leq 150$	9

We might assume that these 100 scores are a random sample of the scores that would be made by all possible people that could take this exam. Test the hypothesis that the scores made by the (conceptually infinite) population would be normally distributed, with $\alpha = .05$.

7. In a recent 72-hour holiday period in the United States there was a total of 290 fatal auto accidents. The number of fatal accidents, per hour, during this period was as follows.

Number per hour	Number of hours
0 or 1	6
2	11
3	15
4	14
5	12
6	8
7 or more	6

Test the hypothesis that the number of accidents per hour, during such a holiday weekend, is a Poisson random variable.

8. A 1972 government report gave the following distribution for the number of days that hospitalized people spent in the hospital in 1971.

Number of days	Number of cases
1	89
2	152
3	105
4–5	165
6–9	221
10–14	124
15–30	106
31 and over	38

Test the hypothesis that this data was selected from a χ^2 distribution with 4 degrees of freedom, with $\alpha = .10$.

9. Between January 1, 1965, and February 9, 1971, inclusive, a period of 2231 days, a total of 163 earthquakes were recorded somewhere in the world, each of magnitude 4 or more on the Richter scale. Knowing the calendar date and time at which each quake occurred, it is then possible to compute the number of days between successive quakes and the 163 quakes give a total of

162 differences, for days between earthquakes. The following table summarizes the number of days between quakes over this period.

Number of days	Frequency
0–4	50
5–9	31
10–14	26
15–19	17
20–24	10
25–29	8
30–34	6
35–39	6
40 or more	8*

* These 8 values were 40, 43, 44, 49, 58, 60, 81, and 109 days.

Test the hypothesis that these values are a random sample of an exponential random variable, with $\alpha = .01$.

10. The scores made by 120 students on a statistics exam are summarized below.

Score	Frequency
60–65	6
65–70	13
70–75	17
75–80	25
80–85	23
85–90	16
90–95	12
95–100	8

Test the hypothesis that these numbers are a random sample of a normal random variable with $\alpha = .05$.

8.5. Contingency Tables

Many times the elements of a sample may be categorized according to two or more criteria and it then is of interest to know whether the methods of classification are independent. For example, suppose that there are exactly two candidates in a presidential election; everyone that casts a vote may be categorized according to the person he votes for and, quite separately, according to party registration (for example, Republican, Democrat, or independent). Presumably these two modes of classification may be dependent upon each other. Or we might classify each person in a sample from a given racial group according to eye color and also according to hair color and then ask whether these two modes of classification are independent. Or we might examine state health statistics and find the number of boys and the number of girls born in the state in each month of a given year and ask whether these two modes of classification are independent (that is, does the proportion of boys born seem to be the same for all months).

In each of the cases just mentioned, the elements of the sample can be classified according to each of two different types of categories. Thus, suppose we take a sample of size n and have two ways of categorizing each element of the sample. Assume that the first method of classification has r levels and the second method has c levels. Then the numbers of sample elements falling into each possible combination of classes from the two methods can be conveniently summarized in tabular form as shown in Table 8.1.

TABLE 8.1

Level of classification 1	Level of classification 2					Total
	1	2	3	$\cdots$	c	
1	x_{11}	x_{12}	x_{13}	$\cdots$	x_{1c}	$x_{1.}$
2	x_{21}	x_{22}	x_{23}	$\cdots$	x_{2c}	$x_{2.}$
3	x_{31}	x_{32}	x_{33}	$\cdots$	x_{3c}	$x_{3.}$
.	.	.	.	$\cdots$	.	.
.	.	.	.	.	.	.
.	.	.	.	.	.	.
r	x_{r1}	x_{r2}	x_{r3}	$\cdots$	x_{rc}	$x_{r.}$
Total	$x_{.1}$	$x_{.2}$	$x_{.3}$		$x_{.c}$	n

Thus, x_{ij} is the observed number of sample elements falling in the i-th level of classification 1 *and* the j-th level of classification 2, $i = 1, 2, \ldots, r$, $j = 1, 2, \ldots, c$. Note that we are using a dot instead of a subscript to denote summation over that subscript; thus

$$x_{1.} = \sum_{j=1}^{c} x_{1j}, \qquad x_{4.} = \sum_{j=1}^{c} x_{4j}, \qquad x_{.3} = \sum_{i=1}^{r} x_{i3},$$

etc. Since our total sample size is n, then

$$\sum_i \sum_j x_{ij} = \sum_i x_{i.} = \sum_j x_{.j} = n.$$

As was mentioned, a frequent question that is asked concerning such samples is whether or not the methods of classification are independent. We are then asking whether the probability that a sample element falls in the i-th level of category 1 is the same for each level of category 2. If so, the two methods of categorizing are independent and, if not, they are not independent. Let us make the same statement in the context of the presidential election example; we are asking whether the probability a person will vote for candidate A is the same, regardless of whether the voter is registered as a Republican, Democrat, or independent. (If the answer to this is yes, then it would appear that party registration does not have much effect on the outcome of a presidential election.) Thus, the question of independence could also be phrased: Are the probabilities of occurrence of the various levels of classification 1 contingent on the levels of classification 2? As we shall see the hypothesis of no contingency (independence) can be tested using a summarization such as given in Table 8.1. Thus, such a table as 8.1 is commonly called a *contingency table*; it simply summarizes the number of sample elements that fall in the i-th level of category 1 for each of the c levels of category 2, $i = 1, 2, \ldots, r$.

Testing the hypothesis of independence in a contingency table is simply a special case of the result given in Theorem 8.4.3. Suppose that we have an infinite population, each element of which falls in exactly one level of classification 1 and in exactly one level of classification 2, where the two modes of classification have a total of r and c levels, respectively. Let p_{ij} be the probability that a randomly selected element falls in the i-th level of classification 1 and the j-th level of classification 2. Then $\sum_i \sum_j p_{ij} = 1$. If we randomly select a sample of n elements from the population and define

$$X_{ij} = \text{number of sample elements in } i\text{-th level of classification}$$
$$1 \text{ and } j\text{-th level of classification 2}$$

for $i = 1, 2, \ldots, r$ and $j = 1, 2, \ldots, c$, then the rc random variables X_{ij} are multinomial with parameters n, p_{ij}, $i = 1, 2, \ldots, r$, $j = 1, 2, \ldots, c$. Furthermore, if the two methods of classification are independent, then the probability that a randomly selected element falls in class ij is the product of the probability that it falls in class i times the probability it falls in class j. That is, we would have

$$p_{ij} = w_i s_j, \qquad i = 1, 2, \ldots, r, \qquad j = 1, 2, \ldots, c$$

where w_i is the probability that a randomly selected element falls in the i-th level of classification and s_j is the probability it falls in the j-th level of classification 2. Then, assuming independence of the two methods of classification (that is, assuming $p_{ij} = w_i s_j$), the maximum likelihood estimator of w_i is

$$\hat{W}_i = \frac{X_{i.}}{n} = \frac{1}{n} \sum X_{ij}, \qquad i = 1, 2, \ldots, r$$

since $X_{1.}, X_{2.}, \ldots, X_{r.}$ would be multinomial with parameters $n, w_1, w_2, \ldots, w_r$. Similarly, the maximum likelihood estimator of s_j is

$$\hat{S}_j = \frac{X_{.j}}{n}, \qquad j = 1, 2, \ldots, c.$$

Then, by Theorem 8.4.3,

$$V = \sum_i \sum_j \frac{(X_{ij} - n\hat{W}_i\hat{S}_j)^2}{n\hat{W}_i\hat{S}_j}$$

is approximately a χ^2 random variable with $(r-1)(c-1)$ degrees of freedom for large n. (We have $(r-1)(c-1)$ degrees of freedom because initially the data can be classified into $k = rc$ classes. If all the p_{ij}'s were specified, then we would have $k - 1 = rc - 1$ degrees of freedom. To estimate the p_{ij}'s, we actually only need to estimate $w_1, w_2, \ldots, w_{r-1}$ and $s_1, s_2, \ldots, s_{c-1}$; since we lose 1 degree of freedom for each parameter estimated we end up with $rc - 1 - (r-1) - (c-1) = (r-1)(c-1)$ degrees of freedom.) The use of this statistic in testing the hypothesis of independence is identical with that in the examples given in Section 8.4.

EXAMPLE 8.5.1. Assume that the medical records of 5000 people (all now deceased) were available for examination. For each person we know whether or not death was caused by lung cancer and whether or not the individual was a smoker. Table 8.2 might result.

TABLE 8.2

	Died of lung cancer	All other causes	Total
Smokers	348	3152	3500
Nonsmokers	82	1418	1500
Total	430	4570	5000

If these records were a random sample from some population, say that of the United States, then we might test the hypothesis that smoking and death by lung cancer are independent, with $\alpha = .01$, by using the V statistic mentioned above. We can label the smoking versus nonsmoking as classification 1 and the lung cancer versus other causes as classification 2. Then we have

$$\hat{w}_1 = \tfrac{3500}{5000} = .7, \qquad \hat{w}_2 = \tfrac{1500}{5000} = .3,$$

$$\hat{s}_1 = \tfrac{430}{5000} = .086, \qquad \hat{s}_2 = \tfrac{4570}{5000} = .914.$$

The observed value of v is

$$v = \frac{[348 - 5000(.7)(.086)]^2}{5000(.7)(.086)} + \cdots + \frac{[1418 - 5000(.3)(.914)]^2}{5000(.3)(.914)} = 26.764;$$

the 99-th percentile of the χ^2 distribution with 1 degree of freedom is 6.63. Thus we would reject the hypothesis of independence.

At this point a remark may be in order concerning the difference between causality and independence. In Example 8.5.1 the hypothetical data led to rejection of the hypothesis of independence between smoking and lung cancer (although the data were hypothetical, actual records are very similar). The fact that the two appear to be dependent is not equivalent to saying that one causes the other. Both classifications may in fact be related to some third factor that causes these two to act as they do. For example, when tracking two satellites in essentially the same orbit about the earth, we would observe a marked association between their relative positions about the earth at various times, yet it is not fact that one causes the other to behave as it does. The same is true for a test of independence in a contingency table; the test indicates whether or not there appears to be an association between two variables or modes of classification, but this association is not necessarily causal in nature.

EXAMPLE 8.5.2.   The captain of a sport fishing boat, who daily takes tourists out rock fishing, is interested in whether the same type of fish is caught with the same frequency throughout the summer months. The only types of fish caught in this area are ling cod, blue fish, and yellowtails. He records the total catch for each of four months of the three types of fish; see Table 8.3.

TABLE 8.3

| | Number of | | | |
	Ling cod	Blue fish	Yellowtail	Total
June	315	1347	620	2282
July	270	1250	514	2034
Aug.	295	1480	710	2485
Sept.	246	1200	494	1940
Total	1126	5277	2338	8741

Testing the independence of the two categories is equivalent to testing that the relative proportions of the three types of fish caught are the same in each of the four months (the frequencies don't change). To test the hypothesis of independence, with $\alpha = .05$, we compute

$$\hat{w}_1 = \tfrac{2282}{8741} = .2611, \qquad \hat{w}_2 = .2327, \qquad \hat{w}_3 = .2843, \qquad \hat{w}_4 = .2219$$

$$\hat{s}_1 = \tfrac{1126}{8741} = .1288, \qquad \hat{s}_2 = .6037, \qquad \hat{s}_3 = .2675.$$

Then we find

$$v = \sum_i \sum_j \frac{(x_{ij} - 8741\,\hat{w}_i \hat{s}_j)^2}{8741\,\hat{w}_i \hat{s}_j} = 11.960$$

Since the 95-th percentile of the χ^2 distribution with 6 degrees of freedom is 12.6, we would accept the hypothesis of independence. Note that if we had chosen $\alpha = .10$ rather than .05, we would reject the hypothesis.

Contingency tables can be generalized to three or more dimensions and the same procedure, in essence, is applicable. In a 3-dimensional contingency table, each of the n elements of the sample would be categorized according to each of three different modes of classification. There are actually four different independence hypotheses that could be tested; we might test that all three are completely independent or that any single mode of classification is independent of the other two (without specifying that these two are independent of each other).

Let us make the foregoing a little plainer by introducing some exact notation. Suppose that we select a sample of n elements; let X_{ijk} be the number of these in the i-th level of classification 1, the j-th level of classification 2, and the k-th level of classification 3, where $i = 1, 2, \ldots, r, j = 1, 2, \ldots, c$, and $k = 1, 2, \ldots, m$. Let p_{ijk} be the probability that a randomly selected element falls into class ijk. Then the X_{ijk}'s are multinomial with parameters n and p_{ijk}. If all three classifications are independent, then $p_{ijk} = w_i s_j r_k$ for all i, j, k, and we could use sample values to estimate the w_i's, s_j's, and r_k's. Then, as before, we could compute the observed value of V and compare this with the $100(1 - \alpha)$ percentile of the χ^2 distribution with

$$rcm - 1 - (r - 1) - (c - 1) - (m - 1) = rcm - (r + c + m) + 2$$

degrees of freedom; if we exceed this percentile we reject the hypothesis of complete independence, otherwise we accept.

If we want to test the hypothesis that the first two modes of classification are independent of the last one (but not necessarily independent of each other), we would test that $p_{ijk} = w_{ij}s_k$. Thus, we could use the data to estimate the w_{ij}'s and s_k's; Then we could compute the observed value of v and compare this quantity with the $100(1 - \alpha)$ percentile of the χ^2 distribution with $rcm - 1 - (rc - 1) - (m - 1) = (rc - 1)(m - 1)$ degrees of freedom.

EXAMPLE 8.5.3.   As an example of a 3-way contingency table, consider the sample data given in Table 8.4. Summarized there are the results of a sample of

TABLE 8.4

	East		West	
	Male	Female	Male	Female
Democrat	183	217	223	227
Republican	196	154	137	113
Other	12	8	14	16

1500 people, listed according to geographic region (east or west of the Mississippi River), party affiliation (Democrat, Republican or other), and sex.

It would be of interest to political scientists to test that these three variables of classification are independent. If they are independent, then it would appear that party affiliation is distributed the same in the east, as the west, and over the two sexes. If they are not independent, it would indicate an imbalance of registration, which might prove worthy of further study. For the test, as described above, then, we have $r = 2$, $c = 3$, $m = 2$, so the test criterion will (approximately) have a χ^2 distribution with $12 - 7 + 2 = 7$ degrees of freedom. We shall choose $\alpha = .05$ and thus will reject the hypothesis of complete independence if the test statistic exceeds 14.1. Letting w_1, w_2 represent the probabilities for east–west, s_1, s_2, s_3 represent the probabilities for party affiliation and r_1, r_2 represent the probabilities for the male–female split, we find

$$\hat{w}_1 = \tfrac{770}{1500} = .513, \qquad \hat{w}_2 = \tfrac{730}{1500} = .487,$$
$$\hat{s}_1 = \tfrac{850}{1500} = .567, \qquad \hat{s}_2 = \tfrac{600}{1500} = .400, \qquad \hat{s}_3 = \tfrac{50}{1500} = .033$$
$$\hat{r}_1 = \tfrac{765}{1500} = .510, \qquad \hat{r}_2 = \tfrac{735}{1500} = .490.$$

Then the observed value for v is

$$v = \sum_{i,j,k} \frac{(x_{ijk} - 1500\,\hat{w}_i \hat{s}_j \hat{r}_k)^2}{1500\,\hat{w}_i \hat{s}_j \hat{r}_k} = 30.88,$$

so we would reject the hypothesis that the three criteria are independent.

EXERCISE 8.5

1. A sample of 2000 medical records was examined and the following data resulted.

	Died of cancer of intestines	Died of all other causes	Totals
Smokers	22	1178	1200
Nonsmokers	26	774	800
Total	48	1952	2000

Assume that these results were the outcome of a random sample from a certain population and test that the two classifications are independent, with $\alpha = .05$.

2. A random sample of 3000 Valencia oranges was selected from the 1967 crop. Each orange was graded for color (light, medium, or dark) and a determination of sugar content was made (this continuous variable was used to call the orange sweet or not sweet). The results are summarized in the following table.

Color	Sweet	Not sweet	Total
Light	1300	200	1500
Medium	500	500	1000
Dark	200	300	500
Total	2000	1000	3000

Test the hypothesis that sweetness and color are independent, with $\alpha = .01$.

3. Records of 10,000 auto accidents were examined to determine the degree of injury to the driver and whether or not he was using a seat belt. The data are summarized below.

	Seat belt	No seat belt	Total
Minor injuries	2500	1500	4000
Major injuries	450	4550	5000
Death	50	950	1000
Total	3000	7000	10,000

Test the hypothesis that severity of injury to the driver is independent of whether the driver wears a seat belt, with $\alpha = .01$.

4. A random sample of 4000 individuals (all male of the same age) yielded the following data.

| Highest education attained | Annual income | | | Total |
	Less than 5000	5000 to 15,000	More than 15,000	
Grade school	350	35	15	400
High school	100	850	50	1000
College	40	1200	760	2000
Graduate	10	415	175	600
Total	500	2500	1000	4000

Test the hypothesis that annual salary (for males of this age) is independent of education attained, with $\alpha = .1$.

5. Suppose that X_{ij}, $i = 1, 2, \ldots, r$, $j = 1, 2, \ldots, s$, is a random sample of a multinomial random variable with parameters n and $p_{ij} = w_i s_j$. Show that the maximum likelihood estimators of the w_i's and s_j's are

$$\hat{W}_i = \frac{X_{i.}}{n}, \qquad \hat{S}_j = \frac{X_{.j}}{n}.$$

6. Suppose that a sample of n individuals is categorized according to each of k different classifications and that each category has two levels. How many degrees of freedom would we have for testing that all k were independent?

7. Use the data presented in Example 8.5.3 to test the hypothesis that the male–female classification is independent of the geographic–party affiliation combinations.

8. Given that we have random samples from each of two Bernoulli distributions, we can use the contingency table approach discussed in this section to test the hypothesis that the two Bernoulli probabilities, p_1 and p_2 are equal. Specifically, assume we have a random sample of size n_1 from a Bernoulli population with parameter p_1 and an independent random sample of size n_2 from a second Bernoulli population with parameter p_2. Then the observations can be arranged in a 2 × 2 table as below.

Number of

	Successes	Failures	Total
Population 1	X_1	$n_1 - X_1$	n_1
Population 2	X_2	$n_2 - X_2$	n_2

X_1 then is a binomial random variable with parameters n_1, p_1 and X_2 is an independent binomial random variable with parameters n_2, p_2.
(a) Define

$$Z = \frac{X_1/n_1 - X_2/n_2}{\sqrt{\left(\dfrac{X_1 + X_2}{n_1 + n_2}\right)\left(1 - \dfrac{X_1 + X_2}{n_1 + n_2}\right)\left(\dfrac{1}{n_1} + \dfrac{1}{n_2}\right)}}$$

and explain why Z is approximately a standard normal random variable if in fact $p_1 = p_2$.
(b) Show that $Z^2 = V$, the χ^2 statistic used to test independence in a 2 × 2 contingency table.

9. Apply the test of $H_0: p_1 = p_2$, described in Exercise 8, to the data given in Exercise 8.3.16, with $\alpha = .05$.

10. Use the data given in Exercise 8.3.15 to set up a 2 × 3 contingency table and test the independence of the two categories (which is the same as testing that the probabilities for the three categories are identical) with $\alpha = .1$.

11. According to the U.S. government, a typical distribution of ages of 1000 people selected at random from the U.S. population in 1900, 1930, and 1965 would be as presented below.

Age

	Under 14	14 to 19	20 to 24	25 to 44	45 to 64	65 and over	Total
1900	322	120	97	282	138	41	1000
1930	273	113	89	295	175	55	1000
1965	289	106	70	242	200	93	1000

With $\alpha = .01$, test the hypothesis that these three distributions are identical.

9

Bayesian Methods

In Chapters 7 and 8 we discussed the definition and use of some classical methods for estimating parameters and testing hypotheses. In this chapter we shall look at techniques based on Bayes theorem (Theorem 2.6.1). A prime difference between these techniques and the classical ones already discussed is the interpretation of the probabilities used. In essentially everything we have studied so far, the probabilities used were interpreted in a frequency sense; they were referring to an experiment which could be repeated an indefinite number of times and, if the probability of occurrence of an event A was .3, we meant that A would be expected to occur in about 30% of these repetitions. This type of interpretation of probability is called objective or frequentist; the numbers called probabilities are measuring relative frequencies of occurrence in repetitions of the basic experiment.

In common English usage, however, probability is frequently used in another, more subjective sense. For example, we have all heard such statements as "it probably will rain tomorrow" or "the chances are 3 out of 5 the Yankees will win the pennant this year" or "most likely the bank robbery was an inside job." In each of these cases the individual making the statement is using his own experience and knowledge as the basis for the statement and is not referring to some experiment which can or will be repeated an indefinite number of times. These are all examples of uses of subjective probabilities, ones that base their validity strictly on the beliefs of the individuals making them. Thus a subjective probability is measuring a person's "degree of belief" in a proposition, which is not necessarily the same as its long-term frequency of occurrence if, indeed, he is referring to something which could be repeated.

The Bayesian techniques which we shall study in this chapter all make use of subjective probabilities measuring degrees of belief about the value or values of unknown parameters. These subjective probabilities are used to define what is called the prior distribution for the parameter. Thus, when using Bayesian methods we shall act as though an unknown parameter is a random variable and has a known prior distribution (prior to taking a sample); this prior distribution summarizes our subjective degree of belief about the unknown value of the parameter. If we are fairly certain of the parameter's value, we shall choose a prior with a small variance; if we are less certain about its value we shall choose a prior with a larger variance. After the prior is specified the sample values are observed and used to compute what is called the posterior distribution of the parameter. This posterior distribution is made up of both the subjective prior information (our degree of belief) about the parameter and the objective sample information. The posterior distribution then is used to construct an estimator of the unknown parameter or to make interval statements about the unknown parameter.

9.1. Prior and Posterior Distributions

In Chapter 7 we examined both the method of moments and maximum likelihood criteria for using a sample to estimate the value of an unknown parameter θ. We assumed that the probability law of a random variable X, which we could sample, was dependent on the unknown value of θ; then the sample information was used to construct a guess or estimate of the value of θ. Neither of these two methods called for any information for estimating θ other than the sample values that occurred. Moreover, if such extraneous information were available, it would not have been possible to make use of it. In many situations additional information is available about the value of the unknown parameter; if this information can be used to construct a prior distribution for the parameter θ, then the Bayesian methods discussed in this chapter may be used to estimate the unknown value of θ.

> DEFINITION 9.1.1. The *prior distribution* of a parameter θ is a probability function or probability density function expressing our degree of belief about the value of θ, prior to observing a sample of a random variable X whose distribution function is indexed by θ.

It should be stressed that the prior distribution for a parameter θ is making use of additional assumed information above and beyond anything to be observed from a random sample of X. Since this information is expressed as a probability function or density function, we shall let Θ be the symbol for the corresponding random variable and then θ would represent its observed value (which would be the same as the true unknown value). Some examples should clarify this notation.

EXAMPLE 9.1.1.   Suppose that we have a brand new 50-cent piece and are interested in estimating p, the probability of getting a head with this coin on a single flip. We know that p must lie between zero and 1. If we are not willing to assume that any values in this interval are more likely than others, we could then reasonably assume a uniform prior for p on the interval $(0, 1)$. This would be written

$$f_P(p) = 1, \qquad 0 < p < 1$$

$$= 0, \qquad \text{otherwise.}$$

This would, in a sense, correspond to an assumption of total ignorance; we feel that all possible values of p are equally likely. On the other hand, we might feel justified in assuming that p must certainly lie in the interval from .4 to .6, since the coin would appear to be quite symmetrical, and that each of these values is equally likely to occur. Our prior for p then would be

$$f_P(p) = 5, \qquad .4 < p < .6$$

$$= 0, \qquad \text{otherwise.}$$

This is a stronger assumption, because we have ruled out values of p below .4 and above .6. Or, we might again reason that the value of p must certainly lie in the interval from .4 to .6 but that only .4, .5, or .6 are possible values for p with .5 being twice as likely. Then we would be assuming the discrete prior

$$p_P(p) = \tfrac{1}{4}, \qquad \text{at } p = .4 \text{ or } .6$$

$$= \tfrac{1}{2}, \qquad \text{at } p = \tfrac{1}{2}$$

$$= 0, \qquad \text{otherwise.}$$

It should be noted that the priors mentioned in Example 9.1.1 were quite arbitrary and dependent upon the sort of assumptions one is willing to make regarding the unknown value of p; many other possible assumed priors could be mentioned. As we shall see, the final answer arrived at in using a Bayes technique is generally dependent upon the particular prior assumed, so the assumption should not be taken lightly. Furthermore, two different people observing the same sample values might very well arrive at different estimates of the value of the same parameter because they assumed different prior information.

EXAMPLE 9.1.2.   Assume that the length of time a light bulb made a certain way will burn is a normal random variable with mean μ and standard deviation 100 hours. Our prior experience with this process and similar ones might lead us to assume a normal prior for μ; that is, we might assume

$$f_M(\mu) = \frac{1}{\sigma_0 \sqrt{2\pi}} e^{-(\mu - \mu_0)^2 / 2\sigma_0^2}.$$

The parameters μ_0 and σ_0^2 are called the *prior mean* and *prior variance* for μ; the smaller the assumed value of σ_0^2 the more certain we are of the unknown value of μ since the prior density then has more area close to μ_0. On the other hand, the bigger the assumed value for σ_0^2, the less certain we are of the unknown value of μ since there is correspondingly less area in a fixed interval about μ_0.

With the foregoing discussion in mind regarding an assumed prior for an unknown parameter θ, let us switch our attention to the probability density function for a continuous random variable X which is indexed by θ. We previously denoted this density simply by $f_X(x)$; this notation does not express the fact that $f_X(x)$ actually is dependent on the value of θ. While discussing Bayesian techniques in this chapter, we shall use a conditional probability notation for such a density to stress the fact that $f_X(x)$ can only be used to compute probabilities for a given value of θ. Thus, we shall write $f_{X \mid \Theta}(x \mid \theta)$ for the density of X. (If X is discrete, we shall analogously use $p_{X \mid \Theta}(x \mid \theta)$ for its probability function in the context of Bayesian methods.) Thus, if X is an exponential random variable with parameter θ, we shall write

$$f_{X \mid \Theta}(x \mid \theta) = \theta e^{-\theta x}, \qquad x > 0, \qquad \theta > 0$$

$$= 0, \qquad \text{otherwise}$$

to stress the conditional dependence of the density on θ. Or, if X is binomial with parameters n and p, we shall denote its probability function by

$$p_{X \mid P}(x \mid p) = \binom{n}{x} p^x (1 - p)^{n-x}, \qquad x = 0, 1, \ldots, n, \qquad 0 \le p \le 1,$$

$$= 0, \qquad \text{otherwise.}$$

Generally the parameter n is known, so it plays no special role in this conditional notation.

We are in essence acting as though the parameter of the probability law for X is itself a random variable when we assume a prior distribution for such a parameter; thus the density and probability functions which we are used to are actually conditional densities or probability functions. The notation is meant to acknowledge this fact.

Our notation for the density function of the elements of a random sample of n observations of X will also be changed accordingly. Thus if $X_1, X_2, \ldots,$ X_n is a random sample of an exponential random variable with parameter θ, the joint density for the sample will be denoted by

$$f_{\mathbf{X}\,|\,\Theta}(\mathbf{x}\,|\,\theta) = \theta^n e^{-\theta \Sigma x_i}.$$

Given a prior density for θ, $f_\Theta(\theta)$, and the conditional density of the elements of a sample, $f_{\mathbf{X}\,|\,\Theta}(\mathbf{x}\,|\,\theta)$, the joint (unconditional) density for the sample and the parameter, is simply the product of these two functions:

$$f_{\mathbf{X},\Theta}(\mathbf{x}, \theta) = f_{\mathbf{X}\,|\,\Theta}(\mathbf{x}\,|\,\theta)f_\Theta(\theta).$$

(The reader might like to review conditional densities in Chapter 5.) Then the marginal density of the sample values, which is independent of θ, is given by the integral of the joint density over the range of Θ. Thus

$$f_{\mathbf{X}}(\mathbf{x}) = \int_{\text{Range of }\Theta} f_{\mathbf{X},\Theta}(\mathbf{x}, \theta)\, d\theta.$$

This will be referred to as the *marginal of the sample*. The posterior density for the parameter θ is defined as follows.

DEFINITION 9.1.2. The *posterior density* for θ is the conditional density of Θ, given the sample values. Thus

$$f_{\Theta\,|\,\mathbf{X}}(\theta\,|\,\mathbf{x}) = \frac{f_{\mathbf{X},\Theta}(\mathbf{x}, \theta)}{f_{\mathbf{X}}(\mathbf{x})}.$$

This posterior density thus is simply the conditional density of Θ, given the sample values; the prior density expresses our degree of belief of the location of the value of θ prior to sampling, and the posterior density expresses our degree of belief of the location of θ, given the results of the sample. Let us consider some examples of these manipulations.

EXAMPLE 9.1.3. Let us return to the coin of Example 9.1.1 and, based on the result of a single flip, see how the posterior of p changes for the three prior presented there. Thus, if we flip this 50-cent piece one time and let $X = 1$ if we get a head and $X = 0$ if we get a tail, the probability function for the sample is

$$p_{X\,|\,P}(x\,|\,p) = 1 - p, \qquad \text{at } x = 0$$

$$= p, \qquad \text{at } x = 1.$$

If we assume our prior to be

$$f_P(p) = 1, \qquad 0 < p < 1,$$

then the joint density function for X and P is

$$f_{X,P}(x, p) = 1 - p, \qquad \text{for } x = 0, \qquad 0 < p < 1$$

$$= p, \qquad \text{for } x = 1, \qquad 0 < p < 1.$$

(Note that this is actually a mixture, corresponding to the joint function of a discrete and a continuous random variable; we shall use f to denote the joint function in such cases.) The marginal for X then is

$$p_X(x) = \int_0^1 (1 - p)\, dp = \tfrac{1}{2}, \qquad \text{for } x = 0$$

$$= \int_0^1 p\, dp = \tfrac{1}{2}, \qquad \text{for } x = 1.$$

The posterior for p then is

$$f_{P|X}(p \mid x) = 2(1 - p), \qquad \text{for } x = 0, \qquad 0 < p < 1$$

$$= 2p \qquad \text{for } x = 1, \qquad 0 < p < 1.$$

Prior to flipping the coin we felt that the probability that p exceeded $\tfrac{1}{2}$ was $\tfrac{1}{2}$; after we flip the coin and get a head ($x = 1$), this same probability is

$$\int_{1/2}^1 2p\, dp = \tfrac{3}{4}.$$

Or, if we get a tail ($x = 0$), this same probability is

$$\int_{1/2}^1 2(1 - p)\, dp = \tfrac{1}{4}.$$

Thus our posterior probability that $p > \tfrac{1}{2}$ is either $\tfrac{3}{4}$ or $\tfrac{1}{4}$, depending on whether we get a head when the coin is flipped. On the other hand, if we take

$$f_P(p) = 5, \qquad .4 < p < .6$$

as our prior for p, then the marginal for X is

$$p_X(x) = \int_{.4}^{.6} 5(1 - p)\, dp = .5, \qquad \text{for } x = 0$$

$$= \int_{.4}^{.6} 5p\, dp = .5, \qquad \text{for } x = 1.$$

Notice that the marginal for X is the same in this case as it was earlier. The posterior for p now is

$$f_{P|X}(p \mid x) = 10(1 - p), \qquad \text{for } x = 0, \qquad .4 < p < .6$$

$$= 10p, \qquad \text{for } x = 1, \qquad .4 < p < .6.$$

Similarly, if we use the discrete prior

$$p_P(p) = \tfrac{1}{4}, \qquad \text{for } p = .4 \text{ or } .6$$

$$= \tfrac{1}{2}, \qquad \text{for } p = .5,$$

then the posterior for p is discrete as well and we find

$$p_{P \mid X}(p \mid x = 0) = .3, \qquad \text{at } p = .4$$
$$= .5, \qquad \text{at } p = .5$$
$$= .2, \qquad \text{at } p = .6$$

and

$$p_{P \mid X}(p \mid x = 1) = .2, \qquad \text{at } p = .4$$
$$= .5, \qquad \text{at } p = .5$$
$$= .3, \qquad \text{at } p = .6.$$

Let us conclude this section with an example of a normal random variable with a known variance and derive the posterior distribution of the mean μ for a normal prior.

EXAMPLE 9.1.4.   Assume that the weights of a certain species of adult hummingbird are normally distributed with unknown mean μ and known variance $\sigma^2 = \frac{1}{4}$. Furthermore, let us assume that our experience with other types of hummingbirds leads us to conclude that μ is undoubtedly very close to 4 ounces; we feel that this additional information is well represented by a normal prior on μ with mean 4 and standard deviation $\frac{1}{10}$. We randomly select a sample of n hummingbirds of the given species; the density function of the sample then is

$$f_{\mathbf{X} \mid M}(\mathbf{x} \mid \mu) = \left(\frac{2}{\pi}\right)^{n/2} e^{-2\Sigma(x_i - \mu)^2}.$$

We recall the identity

$$\Sigma (x_i - \mu)^2 = \Sigma (x_i - \bar{x})^2 + n(\bar{x} - \mu)^2$$

where $\bar{x} = (1/n) \Sigma x_i$. The joint density of the sample values and M then is

$$f_{\mathbf{X}, M}(\mathbf{x}, \mu) = \left(\frac{2}{\pi}\right)^{n/2} e^{-2[\Sigma(x_i - \bar{x})^2 + n(\bar{x} - \mu)^2]} \frac{10}{\sqrt{2\pi}} e^{-50(\mu - 4)^2}$$

$$= K e^{-2n(\bar{x} - \mu)^2 - 50(\mu - 4)^2}$$

where K involves the constant term and the sample values $x_1, x_2, \ldots, x_n$. By completing the square, we find

$$2n(\bar{x} - \mu)^2 + 50(\mu - 4)^2 = (2n + 50)\left[\mu - \frac{n\bar{x} + 100}{n + 25}\right]^2 + \frac{50n(\bar{x} - 4)^2}{n + 25}$$

and thus

$$f_{\mathbf{X}, M}(\mathbf{x}, \mu) = K e^{-50n(\bar{x} - 4)^2/(n + 25)} e^{-(2n + 50)[\mu - (n\bar{x} + 100)/(n + 25)]^2}.$$

Now, to get the marginal for the sample we need to integrate over μ; everything except the exponential term involving μ factors outside the integral sign and the value of the integral itself is

$$\left[\frac{2\sqrt{n + 25}}{\sqrt{2\pi}}\right]^{-1}$$

(since it is the integral over the whole line of a normal density with mean $(n\bar{x} + 100)/(n + 25)$ and variance $1/(4n + 100)$). Thus, the marginal for the sample is

$$f_X(x) = \left[\frac{2\sqrt{n + 25}}{\sqrt{2\pi}}\right]^{-1} Ke^{-50n(\bar{x}-4)^2/(n+25)}$$

and the posterior for μ is

$$f_{M|X}(\mu \mid x) = \frac{2\sqrt{n + 25}}{\sqrt{2\pi}} e^{-(2n+50)[\mu-(n\bar{x}+100)/(n+25)]^2}.$$

That is, the posterior is again normal in form with mean

$$\frac{n\bar{x} + 100}{n + 25} = \frac{4n}{4n + 100}\bar{x} + \frac{100}{4n + 100}4$$

(a weighted sum of the observed sample mean and the prior mean) and variance $1/(4n + 100)$ (which is the reciprocal of the sum of the reciprocals of the variance of $\bar{X}$ and the prior variance). Thus, if our prior on μ is normal, so is the posterior where the posterior mean is a weighted sum of the observed sample mean and the mean of the prior.

EXERCISE 9.1

1. Assume that the probability that an item coming off an assembly line is defective in some respect is a constant p. Suppose that we select 1 item at random from the assembly line (constituting a sample of 1 observation of a Bernoulli random variable with parameter p) and we assume a uniform prior density for p on the interval $(0, 1)$, What is the posterior for p if the selected item is defective? If it is not defective?

2. Repeat problem 1, assuming that p is uniform on (a, b) where $a \geq 0, b \leq 1$.

3. A new psychological test has been developed to measure "intelligence." Assume that the score an individual will make on the test is a normal random variable with unknown mean μ and known variance of 10. Prior information on μ (from similar tests already in use) would lead us to assume a normal prior density for μ with mean 100 and variance 5. Given a random sample of 20 individual scores on the test, compute the posterior density for μ.

4. If in problem 3 we assume that μ has a uniform prior on the interval $(90, 110)$ (rather than the normal prior used there), show that the posterior for μ is

$$\frac{1}{\sqrt{\pi}}\frac{1}{N_Z\left(\dfrac{110 - \bar{x}}{1/\sqrt{2}}\right) - N_Z\left(\dfrac{90 - \bar{x}}{1/\sqrt{2}}\right)} e^{-(\bar{x}-\mu)^2}, \qquad 90 < \mu < 110,$$

where $\bar{x}$ is the observed mean of the 20 scores.

5. Assume that the length of nylon string which can be extruded by a machine with no break occurring is an exponential random variable X with parameter λ. Also assume an exponential prior density on λ with parameter $\frac{1}{1000}$. Given a sample of 1 observation on X (and we find $x = 1100$ feet), compute the posterior density for λ.

6. The time it will take a certain track star to run a 100-yard dash is a uniform random variable on the interval $(\alpha, 10.5)$ measured in seconds. Assume that α is equally likely to be 9.5 or 9.7 seconds and compute the posterior for α, given a sample of 1 time of 10 seconds.

7. A coin is flipped 1 time. The probability p of getting a head is equally likely to be $\frac{1}{4}$, $\frac{1}{2}$, or $\frac{3}{4}$. Compute the posterior for p, given that we get a tail on the single flip. Compute the posterior for p, given that we get a head on the single flip.

8. Assume that we repeatedly receive shipments of 50 items from the same supplier and that, in the past, 5% of the items received have been defective. What would you use as a prior for D, the unknown number of defectives in a shipment of 50 items just received?

9. In the state of California, 52% of all registered voters are Democrats, 42% are Republicans and the remaining 6% are independent or some other party. How would you use this information to construct a prior distribution for p, the proportion of votes that the Republican nominee for governor will receive?

10. Assume, in Exercise 8, that a random sample of 3 items is selected from the shipment of 50 and none of the 3 is defective. What is the posterior distribution for D, the number of defective items in the shipment?

9.2. Bayes Estimators

As might be expected, estimators based on Bayesian principles will depend on the prior and posterior distribution for an unknown parameter. Before discussing Bayesian estimators let us briefly study two additional density functions which prove useful as prior distributions for parameters. The first of these is the *gamma density function*. We shall say that X is a gamma random variable with parameters m and θ if its density function is

$$f_{X \mid \Theta}(x \mid \theta) = \frac{1}{\Gamma(m + 1)} \, \theta^{m+1} x^m e^{-\theta x}, \quad \text{for } x > 0, \quad m > -1, \quad \theta > 0.$$

(This is also referred to as the gamma density function.) The parameter m is generally assumed known. Let us first verify that the integral of $f_{X \mid \Theta}$, over the whole real line, gives 1 as it should. This is most easily accomplished by referring to the *gamma function*:

$$\Gamma(n) = \int_0^\infty t^{n-1} e^{-t} \, dt;$$

this function was mentioned briefly in Exercise 4.4.15. (Also see the discussion following Example 6.2.6.) Since a definite integral is not a function of the variable of integration, note that the only quantity that can vary is n, which occurs in the exponent for t. Thus, the gamma function is a function of n, as indicated. Integration by parts, letting $u = t^{n-1}$, $dv = e^{-t} dt$, shows that

$$\int_0^\infty t^{n-1} e^{-t} dt = (n-1) \int_0^\infty t^{n-2} e^{-t} dt;$$

$$\Gamma(n) = (n-1)\Gamma(n-1)$$

and, if n is a positive integer, repeated use of this result gives

$$\Gamma(n) = (n-1)(n-2) \cdots 2.1\Gamma(1) = (n-1)!$$

since

$$\Gamma(1) = \int_0^\infty t^{1-1} e^{-t} dt = \int_0^\infty e^{-t} dt = 1.$$

The gamma function, with n equal to a positive integer, will generate the factorials. $\Gamma(n)$ can also be shown to exist for noninteger values, for all $n > 0$. Note that if we replace n by $m + 1$, and set the variable of integration t equal to θx we have

$$\Gamma(m+1) = \int_0^\infty \theta^{m+1} x^m e^{-\theta x} dx,$$

that is

$$1 = \int_0^\infty \frac{\theta^{m+1} x^m}{\Gamma(m+1)} e^{-\theta x} dx,$$

as we wanted to show. Thus $f_{X|\Theta}(x \mid \theta)$ is a legitimate density function. It is also easy to see that the gamma moment generating function is

$$m_X(t) = (1 - t/\theta)^{-(m+1)} \quad \text{for} \quad t < \theta.$$

From the moment generating function we find $\mu_X = (m+1)/\theta$, $\sigma_X^2 = (m+1)/\theta^2$; the maximum value for the gamma density occurs with $x = m/\theta$.

We have in fact already seen two special cases of gamma random variables. If $m = 0$, notice that the gamma density is identical with that of an exponential random variable with parameter θ. Thus the exponential distribution is a special case of a gamma density. If $\theta = \frac{1}{2}$, note that the gamma density is identical with the density of a χ^2 random variable with $2m + 2$ degrees of freedom; thus a χ^2 random variable is also a special case of a gamma random variable. We shall see below several examples in which the gamma density is assumed as a prior for an unknown parameter. Before examining them, however, let us study one additional density which will prove useful as a prior for certain parameters.

The *beta random variable* with parameters a and b has density function

$$f_{X \mid A, B}(x \mid a, b)$$

$$= \frac{\Gamma(a + b)}{\Gamma(a)\Gamma(b)} x^{a-1}(1 - x)^{b-1}, \qquad 0 < x < 1, \qquad a > 0, \qquad b > 0.$$

By making the change of variable $t = x/(1 - x)$ it can be seen that the integral of this density gives 1, as of course it must. By evaluating the integrals involved we find

$$\mu_X = a/(a + b), \qquad \sigma_X^2 = ab/(a + b)^2(a + b + 1).$$

The maximum value of the density occurs at $x = (a - 1)/(a + b - 2)$ (unless $a + b = 2$). If we hold the sum $a + b$ fixed, and consider what happens to this density as a increases, note that the maximum value and the mean both increase and the probability mass is shifted closer to 1 (higher probabilities of X being close to 1): as a decreases (with $a + b$ fixed) the probability mass is shifted to the left and, relatively speaking, closer to zero. Thus, by appropriately changing the parameters a and b, the beta density can take on a large number of different shapes, over the interval $(0, 1)$, including for $a = b = 1$ that of a uniform random variable. The beta density function proves useful as a prior (and posterior) density for unknown parameters, in certain cases.

Both the gamma and beta densities are examples of what are called *conjugate* prior distributions, ones which come from the same general family as the distribution sampled. If we have a random sample of a Bernoulli random variable, and want a prior distribution for p, the conjugate prior is the beta given above; that is, the probability function for the sample is

$p^{\Sigma x_i}(1 - p)^{n - \Sigma x_i}$; if we assume as the prior for p $\dfrac{\Gamma(a + b)}{\Gamma(a)\Gamma(b)} p^{a-1}(1 - p)^{b-1}$

for some specific values of a and b note that the functional forms for the sample observations and for the prior on p are the same. Thus the beta density is called the conjugate prior for p for a random sample of a Bernoulli random variable. In the same way, the gamma density is the conjugate prior for σ^2 when sampling from a normal population. In general, conjugate priors make the computation of the posterior simpler, and the posterior density will again have the same functional form as the prior.

Let us now turn our attention to Bayes estimators. The posterior distribution of a parameter θ measures our degree of belief as to the true unknown value of θ; it combines the assumed prior knowledge of θ and the information contained in the sample about θ through Bayes theorem. Then, to give a single point as our best guess of the value of θ, we might logically choose the mean of the posterior distribution. Recall from our discussion of the mean of a random variable, that this number would locate the center of gravity of the posterior distribution and would thus be a likely candidate if we had to give a single number as our guess of the value of θ. This posterior mean value is called the Bayes estimate of θ, as we shall see in Definition 9.2.1 below. Before giving this definition, however, let us note the fact that if the posterior density for θ is not symmetric, other measures of the middle of the posterior might equally well be used as the point estimate of θ. Using the mode of the posterior, the value that maximizes the posterior density, would seem quite logical; it corresponds to the maximum likelihood estimate for a uniform prior as we shall see. Or, we might prefer to use the median of the posterior as our point estimate of θ, since it would be the number that splits our total degree of belief into two equal pieces (in a probability sense). Nevertheless, the commonly used estimate is the mean.

DEFINITION 9.2.1.   The mean of the posterior distribution of θ is called the *Bayes estimate* for θ. This estimate is denoted by $\theta*$ and the corresponding random variable by $\Theta*$.

Thus, as in Chapter 7, the observed value of the estimator, called the estimate, is denoted by a lower case letter and is a function of the observed sample values. The same function of the sample random variables (rather than their observed values) is denoted by a capital letter and called the estimator.

EXAMPLE 9.2.1.   In Example 9.1.3 we used three different priors for p, the probability of getting a head on flipping a coin, and for each we derived the posterior distribution. Let us now compute the mean value for each posterior, which would be the Bayes estimate, using the corresponding prior. For the posterior

$$f_{P \mid X}(p \mid x) = 2(1 - p), \quad x = 0, \quad 0 < p < 1$$
$$= 2p, \quad x = 1, \quad 0 < p < 1$$

we have

$$p* = \tfrac{1}{3}, \quad \text{if } x = 0$$
$$= \tfrac{2}{3}, \quad \text{if } x = 1,$$

since these are the respective means for the two possible x values we might observe in the sample. For the posterior

$$f_{P \mid X}(p \mid x) = 10(1 - p), \quad x = 0, \quad .4 < x < .6$$
$$= 10p, \quad x = 1, \quad .4 < x < .6$$

we find

$$p* = .4933, \quad \text{if } x = 0$$
$$= .5067, \quad \text{if } x = 1.$$

For the discrete posterior

$$p_{P\mid X}(p\mid x = 0) = .3, \qquad \text{at } p = .4$$
$$= .5, \qquad \text{at } p = .5$$
$$= .2, \qquad \text{at } p = .6$$

and

$$p_{P\mid X}(p\mid x = 1) = .2, \qquad \text{at } p = .4$$
$$= .5, \qquad \text{at } p = .5$$
$$= .3, \qquad \text{at } p = .6,$$

we find

$$p^* = .49, \qquad \text{if } x = 0$$
$$= .51, \qquad \text{if } x = 1.$$

Both the maximum likelihood and method of moments estimates would be 0 if $x = 0$, 1 if $x = 1$, so in this case the Bayes estimates seem to give more sensible results. The reason for this will become apparent as we proceed.

EXAMPLE 9.2.2. In Example 9.1.4 we assumed that weights of humming-birds were normally distributed with unknown mean μ and known variance of $\frac{1}{4}$. We also assumed a normal prior on μ with mean 4 and variance $\frac{1}{100}$. The posterior then turned out to be normal with mean $(n\bar{x} + 100)/(n + 25)$. Thus, for this example, the Bayes estimate is

$$\mu^* = \frac{n\bar{x} + 100}{n + 25}$$

and

$$M^* = \frac{n\bar{X} + 100}{n + 25}.$$

The following example illustrates the use of a conjugate (beta) density for an unknown parameter.

EXAMPLE 9.2.3. A production process turns out items, each of which is or is not defective. Assume that the probability of a defective occurring is p (a constant) and that the defectives occur independently. To estimate p, we might assume a uniform prior for p on $(0, 1)$, take a random sample of n items, and let X be the total number of defectives in the sample. Then X is binomial with parameters n and p and we have

$$p_{X\mid P}(x\mid p) = \binom{n}{x} p^x(1 - p)^{n-x}, \qquad x = 0, 1, \ldots, n, \qquad 0 < p < 1$$

$$f_P(p) = 1, \qquad 0 < p < 1.$$

The joint density of X and P is

$$f_{X,P}(x, p) = \binom{n}{x} p^x(1 - p)^{n-x}, \qquad 0 < p < 1, \qquad x = 0, 1, \ldots, n;$$

the marginal for X is

$$p_X(x) = \int_0^1 \binom{n}{x} p^x (1-p)^{n-x} \, dp = \binom{n}{x} \cdot \frac{x!\,(n-x)!}{(n+1)!} = \frac{1}{n+1}$$

for $x = 0, 1, \ldots, n$. Thus, averaging over the values of p, all of which are assumed to be equally likely, the values of X are equally likely to occur. The posterior for P then is

$$f_{P|X}(p \mid x) = \frac{\Gamma(n+2)}{\Gamma(x+1)\Gamma(n-x+1)} p^x (1-p)^{n-x}, \qquad 0 < p < 1,$$

a beta density with parameters $x+1$ and $n-x+1$. The mean of the posterior is $(x+1)/(n+2)$ and the mode (maximum value) of the posterior is x/n; thus the Bayes estimate of p (given x defectives occur in the sample of n) is

$$p^* = \frac{x+1}{n+2}$$

and the Bayes estimator is

$$P^* = \frac{X+1}{n+2}$$

Both the method of moments and maximum likelihood yield X/n as the estimator of p for this case; the Bayes estimator thus is essentially increasing the sample size by 2 and the number of defectives by 1 and forming the same ratio as maximum likelihood to estimate p. Or, as noted above, the mode of the posterior is x/n (this locates the maximum point on the posterior) and thus, in this example, the Bayes estimate is given by the mean (by definition) and the maximum likelihood estimate by the mode of the posterior density.

Theorem 9.2.1 shows that the maximum likelihood estimate of a parameter θ always is given by the mode of the posterior, if we assume a uniform prior on θ.

Theorem 9.2.1. Suppose that X is a random variable with density $f_{X|\Theta}(x \mid \theta)$ and assume a uniform prior for θ on the interval (a, b). Then, if θ^+ locates the mode of the posterior (given a random sample of n observations of X), θ^+ is the maximizing value of the likelihood function for θ in the interval (a, b).

Proof: The posterior is

$$f_{\Theta|X}(\theta \mid x) = \frac{f_{X|\Theta}(x \mid \theta) f_\Theta(\theta)}{f_X(x)}.$$

But if $f_\Theta(\theta)$ is constant for all θ, then maximizing $f_{\Theta|X}(\theta \mid x)$ with respect to θ is equivalent to maximizing $f_{X|\Theta}(x \mid \theta)$ with respect to θ and the likelihood function is

$$L(\theta) = f_{X|\Theta}(x \mid \theta).$$

Thus the mode of the posterior is the maximum likelihood estimate of θ, given a uniform prior.   ◀

Theorem 9.2.1 then gives some indication of the relationship between the Bayes estimator and the maximum likelihood estimator for a special case. It can be shown that the two will, in general, differ by an amount that is small compared to $1/\sqrt{n}$. Thus, for moderately large sample sizes the two criteria will lead to essentially the same estimator.

The following example assumes a conjugate (gamma) prior for the parameter of an exponential distribution (and as has already been noted, the posterior then is also a gamma density).

EXAMPLE 9.2.4.   Suppose that the time to failure T of an electron tube is an exponential random variable with parameter λ. Thus, the density for a random sample of n times is

$$f_{T|\Lambda}(t \mid \lambda) = \lambda^n e^{-\lambda \Sigma t_i}.$$

Assume that we are given prior information that the average time to failure of tubes of this type is most likely to be at least 100 hours; that is, since the mean of T is $1/\lambda$, we are fairly sure that $1/\lambda > 100$ or $\lambda < .01$. We might then arbitrarily say that we are 80% sure that $\lambda < .01$ and adopt an exponential prior for λ that satisfies this statement. (We choose the exponential type of density mainly because it is convenient; it enables us to represent the given information quite easily.) Thus, we assume as a prior for λ

$$f_\Lambda(\lambda) = 160e^{-160\lambda}, \qquad \lambda > 0.$$

Note that

$$\int_0^{.01} 160e^{-160\lambda} \, d\lambda \doteq .798.$$

The joint density of the sample values and λ then is

$$f_{T,\Lambda}(t, \lambda) = 160\lambda^n e^{-\lambda(\Sigma t_i + 160)};$$

the marginal for the sample values is

$$\begin{aligned} f_T(t) &= \int_0^\infty 160\lambda^n e^{-\lambda(\Sigma t_i + 160)} \, d\lambda \\ &= 160 \frac{\Gamma(n + 1)}{(\Sigma t_i + 160)^{n+1}} \end{aligned}$$

and the posterior for λ is

$$f_{\Lambda|T}(\lambda \mid t) = \frac{1}{\Gamma(n + 1)} [\Sigma t_i + 160]^{n+1} \lambda^n e^{-\lambda(\Sigma t_i + 160)}.$$

Thus the posterior is that of a gamma random variable with parameters $m = n$, $\theta = \Sigma t_i + 160$. The mean of the posterior, which is the Bayes estimate of λ, then is $(n + 1)/(\Sigma t_i + 160)$. The mode of the posterior is easily seen to be $n/(\Sigma t_i + 160)$. We recall that the maximum likelihood estimate of λ is $n/\Sigma t_i$; if, for example, we took a sample of $n = 10$ lifetimes and found $\Sigma t_i = 1600$, which agrees perfectly with the assumed prior mean, the maximum likelihood estimate is $\hat{\lambda} = \frac{1}{160}$ (equal to the prior mean) while the Bayes estimate is $\lambda^* = \frac{1}{176}$. Thus, if our sample agrees perfectly with the assumed prior mean, it is the maximum

TABLE 9.1
Priors and Posteriors, Given a Random Sample of Size n

Line	Density of sample (given value of unknown parameter)	Prior density for unknown parameter	Posterior density for unknown parameter
1	$\left(\dfrac{1}{2\pi\sigma_X^2}\right)^{n/2} e^{-\Sigma(x_i-\mu)^2/2\sigma_X^2}$ (σ_X^2 known)	$\dfrac{1}{\sigma_0\sqrt{2\pi}} e^{-(\mu-\mu_0)^2/2\sigma_0^2}$	$\dfrac{a^{1/2}}{\sqrt{2\pi}}\exp\left\{-\dfrac{a}{2}\left[\mu-\dfrac{1}{a}\left(\dfrac{n\bar{x}}{\sigma_X^2}+\dfrac{\mu_0}{\sigma_0^2}\right)\right]^2\right\}$, $a=\dfrac{n}{\sigma_X^2}+\dfrac{1}{\sigma_0^2}$
2	$\left(\dfrac{1}{2\pi\sigma_X^2}\right)^{n/2} e^{-\Sigma(x_i-\mu)^2/2\sigma_X^2}$ (σ_X^2 known)	$\dfrac{1}{b-a}$	$\sqrt{\dfrac{n}{2\pi\sigma_X^2}}\left[Nz\left(\dfrac{b-\bar{x}}{\sigma_X/\sqrt{n}}\right)-Nz\left(\dfrac{a-\bar{x}}{\sigma_X/\sqrt{n}}\right)\right]^{-1}\times e^{-n(\mu-\bar{x}^2)/2\sigma_X^2}$
3	$\left(\dfrac{1}{2\pi\sigma^2}\right)^{n/2} e^{-\Sigma(x_i-\mu_X)^2/2\sigma^2}$ (μ_X known)	$\dfrac{1}{m!}[m\sigma_0^2]^{m+1}\left(\dfrac{1}{\sigma^2}\right)^{m+2} e^{-m\sigma_0^2/\sigma^2}$	$\dfrac{1}{\left(m+\dfrac{n}{2}\right)!}\,b^{n/2+m+1}\left(\dfrac{1}{\sigma^2}\right)^{n/2+m+2} e^{-b/\sigma^2}$, $b=m\sigma_0^2+\dfrac{1}{2}\Sigma(x_i-\mu_X)^2$
4	$\lambda^n e^{-\lambda\Sigma x_i}$	$\dfrac{1}{m!}\left(\dfrac{m+1}{\lambda_0}\right)^{m+1}\lambda^m e^{-(m+1)\lambda/\lambda_0}$	$\dfrac{\{n\bar{x}+[(m+1)/\lambda_0]\}^{n+m+1}}{\Gamma(n+m+1)}\lambda^{n+m}\times e^{-\lambda[n\bar{x}+(m+1)/\lambda_0]}$
5	$\dfrac{\lambda^{\Sigma x_i}}{\Pi x_i!} e^{-n\lambda}$	$\dfrac{1}{m!}\left(\dfrac{m+1}{\lambda_0}\right)^{m+1}\lambda^m e^{-(m+1)\lambda/\lambda_0}$	$\dfrac{[n+(m+1)/\lambda_0]^{c+1}}{c!}\lambda^c e^{-[n+(m+1)/\lambda_0]\lambda}$, $c=m+\Sigma x_i$
6	$p^{\Sigma x}(1-p)^{n-\Sigma x}$	$\dfrac{\Gamma(a+b)}{\Gamma(a)\Gamma(b)} p^{a-1}(1-p)^{b-1}$	$\dfrac{\Gamma(a+b+n)}{\Gamma(a+\Sigma x_i)\Gamma(b+n-\Sigma x_i)} p^{a+\Sigma x_i-1}\times(1-p)^{b+n-\Sigma x_i-1}$
7	$p^n(1-p)^{\Sigma x_i-n}$	$\dfrac{\Gamma(a+b)}{\Gamma(a)\Gamma(b)} p^{a-1}(1-p)^{b-1}$	$\dfrac{\Gamma(a+b+\Sigma x_i)}{\Gamma(a+n)\Gamma(b+\Sigma x_i-n)} p^{a+n-1}\times(1-p)^{b+\Sigma x_i-n-1}$

likelihood estimate that reproduces the prior estimate, rather than the Bayes, for an exponential random variable with an exponential prior for λ.

Table 9.1 gives some common examples of assumed priors for the parameters of some common probability laws and, given a random sample of n observations of the random variable involved, also gives the resulting posterior distribution for the parameter. By changing the particular values of the prior parameters, the prior can be used to represent a variety of assumed situations.

EXAMPLE 9.2.5.   Assume that the number of times, X, that a light switch can be turned to the on position until it fails is a geometric random variable with parameter p. Let us also assume that our prior experience with switches of this type indicates that p is probably very close to .001; we shall use this information to construct a prior for p, using a beta distribution. A beta random variable with parameters a and b has mean $a/(a + b)$ and variance

$$\frac{ab}{(a + b)^2(a + b + 1)},$$

as we have seen. Then, we might use our prior information about p and assume the prior mean to be .001 and the prior standard deviation to be .0005. These assumptions lead to $a = 3.995$ and $b = 3991$ as values for the parameters of the prior. Then, if we have a sample of n observations on X, we see from line 7 of Table 9.1 that the posterior for p is

$$\frac{\Gamma(3995 + \Sigma x_i)}{\Gamma(n + 3.995)\Gamma(3991 + \Sigma x_i - n)} p^{2.995+n}(1 - p)^{3990+\Sigma x_i - n}.$$

The mean of this density (the Bayes estimate) is $(3.995 + n)/(3995 + \Sigma x_i)$. The maximum likelihood estimate is $n/\Sigma x_i$. Thus, if $n = 25$ and we observe that $\Sigma x_i = 25{,}295$, then

$$p^* = \frac{28.995}{29{,}290} = .000990, \qquad \hat{p} = \frac{25}{25{,}295} = .000988;$$

either of the two methods gives essentially the same answer. This is because we assume sample results which are consistent with the assumed prior information (the mean of the prior is .001 and if $p = .001$ we would expect each x_i close to $1/.001 = 1000$ and the sum of 25 sample values to be close to 25,000). Generally, the less consistent the sample results are with the assumed prior, the bigger is the difference between p^* and $\hat{p}$; p^* always tends to "shade" the sample estimate toward the prior information. If, for example, $n = 25$ and $x_i \Sigma = 1000$, then

$$p^* = \frac{28.995}{4995} = .0058, \qquad \hat{p} = \frac{25}{1000} = .025,$$

and the estimates are quite widely separated. If the sample results do differ markedly from what is expected, as in this latter case, it would appear that the prior information is incorrect; if the sample results are accepted as being correct, the maximum likelihood estimate would be the better one to use.

If, in fact, we are absolutely certain that the value of a parameter θ is θ_0, then the prior for θ would be

$$p_\Theta(\theta) = 1, \quad \text{for } \theta = \theta_0$$
$$= 0, \quad \text{otherwise.}$$

In such a case we would certainly not want to take a sample to estimate θ, since we would already be willing to assume that θ is known. However, an interesting property of Bayes estimators is that the sample values cannot change our minds about such a strong assumption, no matter what happens in the sample. To see this, assume that we have a continuous random variable X whose density is indexed by θ; the density function for a random sample of n observations of X then is

$$f_{X|\Theta}(x \mid \theta)$$

and the joint density of X and Θ is

$$f_{X,\Theta}(x, \theta) = f_{X|\Theta}(x \mid \theta), \quad \text{for } \theta = \theta_0$$
$$= 0, \quad \text{otherwise,}$$

and the marginal for X is

$$f_X(x) = f_{X|\Theta}(x \mid \theta_0)$$

since there is only one value of θ (θ_0) to average over. Then the posterior for θ is

$$f_{\Theta|X}(\theta \mid x) = \frac{f_{X,\Theta}(x, \theta)}{f_X(x)}$$
$$= 1, \quad \text{for } \theta = \theta_0$$
$$= 0, \quad \text{otherwise.}$$

Thus the posterior probability that $\Theta = \theta_0$ is 1 and we are still absolutely certain that $\Theta = \theta_0$, whether the sample results are consistent with this value or not.

EXERCISE 9.2

1. If X is a beta random variable with parameters a and b, verify that its mode is $(a - 1)/(a + b - 2)$, its mean is $a/(a + b)$, and that its variance is

$$\frac{ab}{(a + b)^2(a + b + 1)}.$$

2. Derive the moment-generating function for a gamma random variable with parameters m and θ.

3. Find the mode of a gamma random variable with parameters m and θ.

4. Find the mean and variance of X if

$$f_X(x) = \frac{[m\sigma_0^2]^{m+1}}{m!} \left(\frac{1}{x}\right)^{m+2} e^{-m\sigma_0^2/x}, \qquad \text{for } x > 0.$$

(Compare with the prior for σ^2, line 3, Table 9.1.)

5. Find the mean and variance of X if

$$f_X(x) = \frac{1}{m!} \left(\frac{m+1}{\lambda_0}\right)^{m+1} x^m e^{-((m+1)x)/\lambda_0}, \qquad x > 0.$$

(Compare with the prior for λ, lines 4 and 5, Table 9.1.)

6. Verify the posterior for μ that is given on line 2, Table 9.1, given a uniform prior on (a, b) with σ_X^2 known.

7. Given a sample result such that $a \le \bar{x} \le b$, what is the modal value of the posterior discussed in problem 6 above? If we find $\bar{x} < a$ or $\bar{x} > b$, what is the modal value?

8. Verify the posterior for σ^2 given on line 3, Table 9.1, for the given prior. What is the Bayes estimate for σ^2?

9. Verify the posterior for λ given on line 4, Table 9.1, and compute the Bayes estimate of λ.

10. Verify the posterior for λ given on line 5, Table 9.1, and compute the Bayes estimate of λ.

11. Compute the Bayes estimate for μ, given a uniform prior on (a, b) for μ with σ_X^2 known (line 2, Table 9.1). (Hint:

$$\int (\mu - \bar{x}) e^{-(1/2)(\mu - \bar{x})^2} \, d\mu = -e^{-(1/2)(\mu - \bar{x})^2}.)$$

12. If we are certain that $p = .75$, where p is the parameter of a Bernoulli random variable X, show that no matter how many successes or failures in a row we might observe, the posterior probability that $p = .75$ is 1.

13. Find the mode of the posterior distribution for p in Example 9.2.5.

14. If μ is the unknown mean value of a normal random variable, $\sigma^2 = 1$, what is the conjugate prior for μ?

15. Assume the situation given in line 3, Table 9.1, and express the Bayes' estimator for σ^2 as a function of b. Can you choose b so that $\sigma^* = \hat{\sigma}$?

16. Assume the situation given in line 4, Table 9.1 and find the Bayes' estimate of λ in terms of λ_0, m and n. Can you choose λ_0 and m so that $\lambda^* = \hat{\lambda}$ for all n?

17. Assume the situation given in line 5, Table 9.1 and find the Bayes' estimate of λ in terms of λ_0, m and n. Can you choose λ_0 and m so that $\lambda^* = \hat{\lambda}$ for all n?

18. Assume the situation given in line 6, Table 9.1, and find the Bayes' estimate of p in terms of a, b and n. Can you choose a and b so that $p^* = \hat{p}$ for all n?

19. Assume the situation given in line 7, Table 9.1, and find the Bayes' estimate of p in terms of a, b and n. Can you choose a and b so that $\hat{p} = p^*$ for all n?

9.3. Interval Estimation

Using Bayesian methods it is easy to construct interval estimates of parameters. If we have available the posterior distribution of the parameter, we can construct an interval, generally centered at the posterior mean, which contains $100(1 - \alpha)\%$ of the posterior probability. Such an interval is an example of a $100(1 - \alpha)\%$ Bayes interval for the unknown parameter (to contrast it with the non-Bayesian confidence interval approach we studied earlier). The formal definition follows.

DEFINITION 9.3.1. Let

$$f_{\Theta|\mathbf{X}}(\theta \mid \mathbf{x})$$

be the posterior density for θ. If

$$\int_a^b f_{\Theta|\mathbf{X}}(\theta \mid \mathbf{x}) \, d\theta = 1 - \alpha,$$

then the interval (a, b) will be called a $100(1 - \alpha)\%$ *Bayes interval for* θ. (Either a or b may be infinite for one-sided intervals.)

EXAMPLE 9.3.1. Suppose that the "true" weights of 1-year-old trout are normally distributed with mean μ and standard deviation $\sigma = .2$ (in pounds); it is known that the standard deviation does not change with environment although the mean weight μ does. Men from the State Department of Fish and Game stocked a new, man-made lake with freshly hatched trout a year ago and are now interested in an interval estimate of μ, the average weight for 1-year-old trout in this lake. Based on their experience with this fish in similar circumstances, they assume a normal prior with mean $\mu_0 = 7$ pounds and standard deviation $\sigma_0 = \frac{1}{3}$ pound. They randomly select a sample of 30 fish from the lake and then compute the posterior density for μ and use it to derive their interval estimate for μ. From line 1, Table 9.1, we see that the posterior for μ is normal with mean $(750\bar{x} + 63)/759$ and variance $1/759$ where $\bar{x}$ is the observed average weight of the 30 fish in the sample. Then, a 90% Bayes interval for μ (centered at the posterior mean) would go from

$$\frac{750\bar{x} + 63}{759} - \frac{1.64}{\sqrt{759}}$$

to

$$\frac{750\bar{x} + 63}{759} + \frac{1.64}{\sqrt{759}} \, ;$$

ignoring the prior information about μ, the observed value of a 90% confidence interval for μ would be from

$$\bar{x} - (1.64)\frac{0.2}{\sqrt{30}}$$

to

$$\bar{x} + (1.64)\frac{0.2}{\sqrt{30}} \, .$$

Notice that the Bayes interval is slightly shorter than the confidence interval because the additional assumed prior information is equivalent to a slightly larger sample size if no prior information is assumed.

EXAMPLE 9.3.2. The number of fatal traffic accidents per week in the state of California is a Poisson random variable with parameter λ. Let us assume that there has been an average of 80 fatal accidents per week in the last 4 years. We assume a prior for λ having mean 80 and (arbitrarily) variance equal to 20; if we use the gamma prior given on line 5 of Table 9.1, we then would choose $\lambda_0 = 80$, $m = 319$ to satisfy the assumed prior mean and variance. Assume that we take a sample of 20 weeks and want to compute a 95% Bayes interval for λ, the true rate per week. From line 5, Table 9.1, the posterior density function for λ is

$$\frac{(24)^{320+\Sigma x_i}}{(319 + \Sigma x_i)!} \lambda^{319+\Sigma x_i} e^{-24\lambda};$$

once the observed number of fatal accidents is known, this density can then be used to compute the desired 95% Bayes interval.

If in Example 9.3.2 we wanted a lower 95% Bayes interval for λ, we would want to find a such that

$$\int_a^\infty \frac{(24)^{d+1}}{\Gamma(d+1)} \lambda^d e^{-24\lambda} \, d\lambda = .95,$$

where $d = 319 + \Sigma x_i$. Repeated integration by parts shows this integral to equal

$$e^{-24a} \sum_{j=0}^d \frac{(24a)^j}{j!}$$

so a must satisfy the equation

$$e^{-24a} \sum_{j=0}^d \frac{(24a)^j}{j!} = .95.$$

Given large tables of the Poisson distribution function, this equation can be solved numerically; note it says that the Poisson distribution function should have value .95, at the argument set equal to $d = 319 + \Sigma x_i$; that is, we select that Poisson random variable W, say, whose distribution function equals .95 at $319 + \Sigma x_i$. Then a is gotten by dividing the parameter for W by 24.

As we know, the Poisson distribution function is approximated by the normal, if the Poisson parameter is large. In fact, $F_W(t) \doteq N_Z\left(\frac{t + \frac{1}{2} - \lambda}{\sqrt{\lambda}}\right)$

where W is a Poisson random variable with parameter λ. Notice that the above equation then is, in terms of the normal approximation,

$$N_Z\left(\frac{319\frac{1}{2} + \Sigma x_i - 24a}{\sqrt{24a}}\right) = .95;$$

but this implies we should have

$$\frac{319\frac{1}{2} + \sum x_i - 24a}{\sqrt{24a}} = 1.64.$$

Solving this equation for a gives

$$a = \frac{641.69 + 2\sum x_i - \sqrt{3444.54 + 10.76\sum x_i}}{48}$$

as the lower limit for the 95% Bayes interval. It is interesting to note that the classical approach described in Exercise 7.3.13 leads to the equation

$$\frac{\sum x_i + \frac{1}{2} - 20a}{\sqrt{20a}} = 1.64$$

for this same problem. The solution for a is

$$a = \frac{3.69 + 2\sum x_i - \sqrt{12.60 + 10.76\sum x_i}}{40}$$

For values we would expect to observe for $\sum X_i$, assuming the prior is correct, these two limits are very close. In fact, if we observe $\sum x_i = 1600$, which is the expected value of $\sum X_i$ under the assumption that each X has expected value 80, the above equations give $a = 77.04$ for the Bayes limit and $a = 76.81$ for the classical procedure, a very small difference.

In problems where a parameter has a beta prior (and posterior) distribution, again a normal approximation is useful in evaluating posterior confidence limits. Suppose a parameter p has the beta posterior with parameters a and b, both integers. Repeated integration by parts shows that

$$\int_0^t \frac{\Gamma(a+b)}{\Gamma(a)\Gamma(b)} p^{a-1}(1-p)^{b-1}\,dp = \sum_{i=a}^{a+b-1} \binom{a+b-1}{i} t^i(1-t)^{a+b-1-i};$$

so long as $a + b$ is "large", this sum of binomial probabilities is approximately equal to

$$1 - N_Z\left(\frac{a - \frac{1}{2} - (a+b-1)t}{\sqrt{(a+b-1)t(1-t)}}\right).$$

The following example illustrates these computations.

EXAMPLE 9.3.3.   An assembly line receives the screws it uses in lots of 1000 from its supplier. Typically, about 5% of the screws received have been defective. When a new lot is received, a sample of 20 screws is selected and each is examined to see if it is defective or not. A Bayesian interval for p, the proportion of defectives in the lot is then constructed as follows. It is assumed that P, the true proportion of defectives in the lot is a beta random variable with parameters $a = 24$, $b = 456$; thus

$$\mu_P = \frac{24}{480} = .05, \qquad \sigma_P = \sqrt{\frac{(24)(456)}{(480)^2(481)}} \doteq .01.$$

Then, given a sample of $n = 20$ screws of which $\Sigma \, x_i$, are defective, the posterior distribution for P is again beta with parameters $a = 23 + \Sigma \, x_i$, $b = 475 - \Sigma \, x_i$, from Table 9.1, line 6. Then, the upper limit c_2 for a two-sided 95% Bayesian interval for p is defined by

$$.975 = \int_0^{c_2} f_{P|\mathbf{x}}(p \mid \mathbf{x}) \, dp \doteq 1 - N_Z \left(\frac{22\frac{1}{2} + \Sigma \, x_i - 497c_2}{\sqrt{(497)c_2(1 - c_2)}} \right)$$

and the lower limit is defined by

$$.025 = \int_0^{c_1} f_{P|\mathbf{x}}(p \mid \mathbf{x}) \, dp \doteq 1 - N_Z \left(\frac{22\frac{1}{2} + \Sigma \, x_i - 497c_1}{\sqrt{497c_1(1 - c_1)}} \right).$$

Thus we require

$$\frac{22\frac{1}{2} + \Sigma \, x_i - 497c_2}{\sqrt{497c_2(1 - c_2)}} = -1.96$$

and

$$\frac{22\frac{1}{2} + \Sigma \, x_i - 497c_1}{\sqrt{497c_1(1 - c_1)}} = 1.96.$$

Notice that these equations are very similar to the ones we solved for p in Section 7.4, determining the classical (approximate) confidence interval for p. The solutions are

$$c_1 = .049 + .002 \, \Sigma \, x_i - \sqrt{(.344)(10)^{-3} + (.139)(10)^{-4} \, \Sigma \, x_i - (.308)(10)^{-7} \, (\Sigma \, x_i)^2}$$

$$c_2 = .049 + .002 \, \Sigma \, x_i + \sqrt{(.344)(10)^{-3} + (.139)(10)^{-4} \, \Sigma \, x_i - (.308)(10)^{-7} \, (\Sigma \, x_i)^2}.$$

Thus, if we observe $\Sigma \, x_i = 1$ defective in the sample of 20 these limits are $c_1 = .032$, $c_2 = .070$; if we observe $\Sigma \, x_i = 0$ defectives in the sample these limits are $c_1 = .030$, $c_2 = .068$.

As we have seen, the intervals computed from the posterior distribution for a parameter may be quite similar in form to the classical confidence intervals, for the same confidence coefficient. The interpretation of the two is quite different, however. If we repeatedly took samples of the same size, and computed a $100(1 - \alpha)\%$ confidence interval for each, then in the long run the proportion $1 - \alpha$ of the intervals would cover the true parameter value. The Bayesian intervals have the property that they cover the proportion $1 - \alpha$ of the posterior distribution for the parameter. This does not say anything about their behavior over repeated samples of the same size. If the prior distribution is poorly chosen, the proportion of repeated Bayesian intervals that cover the true parameter value may be considerably less than $1 - \alpha$. The reason that the confidence intervals and Bayesian intervals may be similar is that the posterior distribution does depend on the sample information, as well as the subjective prior information. As the sample size gets larger and larger, the posterior becomes less dependent on the assumed prior (as long as it was sensible in the first place) and the Bayesian intervals will converge to the same values as the confidence intervals. For small values of n they may differ appreciably.

Bayesian procedures may also be adapted for use in hypothesis testing situations. In Example 9.3.3 above the buyer of the shipment would undoubtedly like to test $H_0 : p \leq .05$ versus $H_1 : p > .05$, for each lot. If H_0 is rejected for a particular lot, then the lot would be rejected as containing too many defectives and would be returned to the supplier. Note that we can construct a rule for rejecting H_0 by observing whether or not the hypothesized value (.05 in this case) is covered by the observed confidence interval. If it is included in the confidence interval, H_0 is accepted, and if not, H_0 is rejected. The following example applies this reasoning to the situation described in Example 9.3.3.

EXAMPLE 9.3.4. Let us return to the situation described in Example 9.3.3 and assume that the buyer of the lot of screws would like to test the hypothesis that the proportion of defectives does not exceed .05, versus the alternative that the proportion exceeds .05. He is willing to have a probability of .025 of incorrectly rejecting $H_0 : p \leq .05$ and is willing to assume the prior distribution for P described in Example 9.3.3. The equations given for c_1 and c_2 in Example 9.3.3 bracket the middle 95 % of the posterior distribution for P (2.5 % lies below c_1 and 2.5 % above c_2). Then, if the full interval from c_1 to c_2 lies to the right of .05 (which just says $c_1 > .05$) the interval does not include any of the values for p specified by H_0, and he would like to reject H_0 (note that in using this rule the values specified by H_0 have probability no larger than .025 of occurring). Thus, the buyer would like to reject H_0 if $c_1 > .05$, which is equivalent to $\Sigma x_i > 11.76$ so the buyer would reject H_0 only if he found 12 or more defectives in the sample of 20. A binomial table shows that a binomial random variable with parameters $n = 20$, $p = .05$, will exceed 3 with probability .0159. Thus, the classical approach would reject $H_0 : p \leq .05$ with $\Sigma x_i \geq 4$, with $\alpha = .0159$. The reason for the big difference between the Bayesian and classical results in this problem is that n is relatively small compared to the very exact, tight prior assumed for P in the Bayesian approach. The smaller the prior variance (it was set at .01) the larger n must be to appreciably affect the resulting posterior distribution.

EXERCISE 9.3

1. The time necessary for a barber to give a haircut to a customer is a normal random variable X with $\sigma_X = 2$ minutes. Assuming a normal prior on μ with mean 20 and standard deviation 1 and that the total observed length of time for him to give 25 haircuts was 8 hours, construct a 90 % Bayes interval for μ.

2. Assume that the number of cars per hour that arrive at a service station for gasoline is a Poisson random variable with parameter λ. Also assume an exponential prior for λ with parameter .05 and that a total of 230 cars arrived during a 10-hour period. Evaluate a 99 % 2-sided Bayes interval for λ.

3. In an industrial process for making bolts it is critical that the average length of the bolts should be 2 inches. It is also very critical that the variability of lengths should be small, since bolts that are too short or too long are not usable. Assume that for a given machine setting the lengths of the bolts produced are normally distributed with mean 2 and unknown variance σ^2. We are given prior information that the average value of σ^2 should be about .0001 and the variance of σ^2 about .000001. Find the parameters of the prior for σ^2 of the form given in Table 9.1; assuming that a sample of 100 bolt lengths yields

$$\Sigma\, x_i = 198.25, \qquad \Sigma\, x_i^2 = 393.0008,$$

derive the posterior for σ^2 and evaluate the 95 % 2-sided Bayes interval for σ^2.

4. Let p_k be the proportion of married couples that have k children, $k = 0, 1, 2, \ldots$. If you assume that p_k is uniformly distributed on (a_k, b_k), $(0 \le a_k < b_k \le 1)$, what is the posterior density for p_k, given that in a random sample of n families, x_k had k children? Evaluate a $100(1 - \alpha)\%$ Bayes interval for p_k.

5. The time between births at a large hospital is an exponential random variable with parameter λ. Assume that the average observed time (between births) over the past month has been 45 minutes. If you assume an exponential prior for λ, what is the posterior for λ, given a random sample of n births times? Evaluate a 95 % Bayes interval for λ.

6. In testing hypotheses by using Bayesian methods, it would perhaps seem illogical to test that the unknown parameter is equal to any value other than its prior mean. Suppose then that X is a normal random variable with unknown mean μ and known variance σ_X^2; assume a normal prior for μ with prior mean μ_0 and prior variance σ_0^2. If we take a random sample of n observations of X, show that we shall reject $H_0: \mu = \mu_0$ if either

$$\bar{x} > \mu_0 + z_{1-\alpha/2}\,\frac{\sigma_X^2}{n}\,\sqrt{a}$$

or

$$\bar{x} < \mu_0 - z_{1-\alpha/2}\,\frac{\sigma_X^2}{n}\,\sqrt{a}$$

where $a = n/\sigma_X^2 + 1/\sigma_0^2$.

7. Use repeated integration by parts to show that

$$\int_a^\infty \frac{\theta^{m+1}\,x^m}{\Gamma(m+1)}\,e^{-\theta x}\,dx = \sum_{j=0}^m \frac{(\theta a)^j}{j!}\,e^{-\theta a}$$

for any $a > 0$ and m a positive integer.

8. Use repeated integration by parts to show that

$$\int_0^t \frac{\Gamma(a+b)}{\Gamma(a)\Gamma(b)} \, x^{a-1}(1-x)^{b-1} \, dx = \sum_{i=a}^{a+b-1} \binom{a+b-1}{i} t^i (1-t)^{a+b-1-i}$$

for a and b positive integers, $0 < t < 1$.

9. In Example 9.3.3, assume that the prior for P is uniform on $(0, 1)$ and, for a sample of $n = 20$, compare the 95% posterior limits for p with c_1 and c_2 computed in the example.

10. Use the confidence limits derived in Exercise 9 to test $H_0: p \le .05$ versus $H_1: p > .05$ and compare the number of defectives that defines the critical region with the value derived in Example 9.3.4.

10

Least Squares and Regression Theory

A commonly occurring problem in many fields of endeavor is that of estimating the relationship between two variables. For example, the weight of a baby (Y) is undoubtedly related to its age (x); the salary (Y) which an individual of a given age is paid may very likely be related to the number of years (x) of formal education he has had; the cumulative grade point average (Y) which a college student maintains is probably related to the score (x) he made on an entrance examination; the maintenance cost (Y) per year for an automobile is undoubtedly related to its age (x). The methods of least squares and regression theory address the problem of estimating the relationship between a variable Y and an "independently varying" variable x (or more than one x), given a sample of Y values with their associated values of x. The assumed functional form of the relationship between Y and x can in theory be anything; we shall concentrate our effort on a linear relation but shall indicate procedures for more general models.

10.1. Least Squares Estimation

Let us first of all clarify what we shall mean by a relationship between a random variable Y and a variable x. We shall assume that for any given value of an "independent" variable x there exists a distribution or population of associated y values; this distribution of associated y values, for a given x, is described by a probability density function $f_Y(y \mid x)$ (or, if the population

359

has a discrete number of possible values, by a probability function $p_Y(y \mid x)$). For the time being we shall assume that $f_Y(y \mid x)$ depends on the associated x value only through its mean value. That is, $E(Y \mid x)$ is a function of x but any higher moments of Y do not depend on x. Thus, we shall assume that the mean value of a random variable Y depends upon the associated x value but that the variance of Y does not.

The particular functional form of the dependence of $E[Y \mid x]$ is arbitrary, of course, and varies from one application to another. In this section we shall study the special case where we assume that

$$E[Y \mid x] = a + bx$$

and that the higher moments of Y do not depend on x. The two constants, a and b, are unknown and are to be estimated from a sample of values of Y with their associated x values.

Let us note that we can then define the random variable

$$E = Y - (a + bx)$$

which would have mean zero, no matter what the values of a and b, and which would have variance σ^2 identical with the variance of Y. Then we can write

$$Y = a + bx + E,$$

which is a standard expression in treatments of least squares. E is called the residual error and represents the deviations of the Y values about their mean. Now suppose that we have available a random sample of n pairs (Y_1, x_1), $(Y_2, x_2), \ldots, (Y_n, x_n)$ which we have written as 2-tuples to stress the fact that each sample of Y that we take has an associated value of x. The values of x may or may not be all distinct but, as we shall see, we must have at least two different values of x represented if we are to estimate both a and b. The method of least squares estimation of a and b specifies that we should take as our estimates of a and b those values that minimize

$$\sum_{i=1}^{n} (y_i - a - bx_i)^2$$

where the lower case y_i's are the observed sample Y values and the x_i's are the associated values of the independent variable x. Thus we are minimizing the sum of squares of the residuals when applying the method of least squares.

Figure 10.1.1 gives a graphical presentation of this procedure. We can plot the n sample pairs $(x_1, y_1), (x_2, y_2), \ldots, (x_n, y_n)$ as pictured. Because of the fact that the y_i's are observed values of random variables, these points will not necessarily lie on a straight line, even if the mean value of Y is linearly related to x. The method of least squares then says that we should

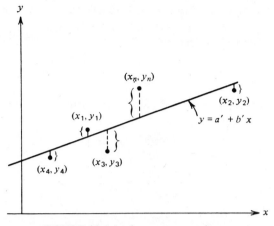

FIGURE 10.1.1   Least squares line.

consider all possible straight lines that we might plot in this plane and choose the one that minimizes the sum of squares of deviations between the observed sample points and the line which we draw. The line which does this minimization is called the least squares line and is denoted by $y = a' + b'x$; a' and b' are called the least squares estimates of a and b. If we choose any other line

$$y = a^* + b^*x,$$

the sum of squares of deviations of the y_i's about this second line is necessarily larger. The line $y = a' + b'x$ is also called the regression of Y on x (see Chapter 5, Section 5.3 where both Y and X were random variables).

Theorem 10.1.1.   Let $(y_1, x_1), (y_2, x_2), \ldots, (y_n, x_n)$ be the observed values of a random variable Y, with their associated x values, where

$$E[Y \mid x] = a + bx.$$

Then the least squares line is given by $y = a' + b'x$ where

$$b' = \frac{\sum (x_i - \bar{x})(y_i - \bar{y})}{\sum (x_i - \bar{x})^2}$$

and

$$a' = \bar{y} - b'\bar{x}.$$

Proof: The quantity to be minimized, with respect to a and b, is

$$Q = \sum_{i=1}^{n} (y_i - a - bx_i)^2.$$

To accomplish this we need to derive the partial derivatives of Q with respect to a and b, set them equal to zero, and solve the resulting equations simultaneously for a' and b'. Then, if

$$\frac{\partial^2 Q}{\partial a^2} > 0$$

and if the determinant of second partial derivatives

$$\begin{vmatrix} \dfrac{\partial^2 Q}{\partial a^2} & \dfrac{\partial^2 Q}{\partial a\,\partial b} \\[2ex] \dfrac{\partial^2 Q}{\partial a\,\partial b} & \dfrac{\partial^2 Q}{\partial b^2} \end{vmatrix}$$

is positive at a', b', we have in fact located the minimum value for Q. Thus,

$$\frac{\partial Q}{\partial a} = -2 \sum_{i=1}^{n} (y_i - a - bx_i)$$

$$\frac{\partial Q}{\partial b} = -2 \sum_{i=1}^{n} x_i(y_i - a - bx_i).$$

The equations to be solved for a' and b' (called the *normal equations*) then are

$$na' + b' \sum x_i = \sum y_i$$
$$a' \sum x_i + b' \sum x_i^2 = \sum x_i y_i.$$

From the first equation we have

$$a' = \bar{y} - b'\bar{x}$$

and substituting this result in the second yields

$$(\bar{y} - b'\bar{x}) \sum x_i + b' \sum x_i^2 = \sum x_i y_i$$

or

$$b'[\sum x_i^2 - \bar{x} \sum x_i] = \sum x_i y_i - \bar{y} \sum x_i.$$

Thus

$$b' = \frac{\sum x_i y_i - \bar{y} \sum x_i}{\sum x_i^2 - \bar{x} \sum x_i} = \frac{\sum (x_i - \bar{x})(y_i - \bar{y})}{\sum (x_i - \bar{x})^2}.$$

The second partials are

$$\frac{\partial^2 Q}{\partial a^2} = 2n, \qquad \frac{\partial^2 Q}{\partial a\,\partial b} = 2 \sum x_i, \qquad \frac{\partial^2 Q}{\partial b^2} = 2 \sum x_i^2.$$

Thus $\partial^2 Q/\partial a^2 = 2n > 0$ and

$$\begin{vmatrix} 2n & 2\sum x_i \\ 2\sum x_i & 2\sum x_i^2 \end{vmatrix} = 4n\sum x_i^2 - 4(\sum x_i)^2$$

$$= 4n\sum (x_i - \bar{x})^2,$$

which is positive if we have at least two different values represented in x_1, $x_2, \ldots, x_n$, for any values of a and b. Thus the line $y = a' + b'x$ does give the minimum value of Q. ◄

We now examine a numerical example.

EXAMPLE 10.1.1.   Suppose that it is medically plausible that the weight of a baby is a linear function of its age, on the average, for ages between 1 month and 6 months and that the higher moments of the weights do not depend on x. We are assuming then that for babies aged x months there is a distribution of weights; for any fixed x the mean of this distribution is given by $a + bx$, $1 \le x \le 6$, where a and b are unknown constants, and there is no further dependence between the weights and the age. The following sample of 10 weights (in pounds) and ages (in months) was observed:

y_i	x_i
14	3
16.5	5
14	2
10	1
16	4.5
20	5.5
22	6
15	3.5
16.5	4
12	2.5

We find that $n = 10$, $\sum x_i = 37$, $\sum x_i^2 = 160$, $\sum y_i = 156$, and $\sum x_i y_i = 625$; thus $\bar{x} = 3.7$, $\bar{y} = 15.6$, $\sum (x_i - \bar{x})^2 = 23.1$, $\sum (x_i - \bar{x})(y_i - \bar{y}) = 47.8$, $b' = \dfrac{47.8}{23.1} = 2.07$. $a' = 15.6 - (2.07)(3.7) = 7.94$, and the least squares line is $y = 7.94 + 2.07x$.

The original data and this line are plotted in Figure 10.1.2. By referring either to the equation or its graph, we would estimate the average weight of 2-month-old babies to be 12.08 pounds; we would estimate the average weight of 5-month-old babies to be 18.29 pounds.

We have seen what the estimates of a and b turn out to be as a function of the observed sample values, using the least squares criterion; as always the

FIGURE 10.1.2   Data for Example 10.1.1.

least squares *estimators* are defined as the same functions of the random variables rather than their observed values. Thus

$$B' = \frac{\sum (x_i - \bar{x})(Y_i - \bar{Y})}{\sum (x_i - \bar{x})^2}$$

$$A' = \bar{Y} - B'\bar{x},$$

since we are assuming that there is a distribution of y values for any given x. We would then like to investigate questions regarding the expected values of these estimators, their variances, and their covariance. The following theorems give these expected values, variances, and covariances.

Theorem 10.1.2. Assume that Y is a random variable with mean $a + bx$, for a given value of x, and let (Y_1, x_1), (Y_2, x_2), . . . , (Y_n, x_n) be a random sample of size n of Y (with their associated x values). Define

$$B' = \frac{\sum (x_i - \bar{x})(Y_i - \bar{Y})}{\sum (x_i - \bar{x})^2}$$

$$A' = \bar{Y} - B'\bar{x};$$

then A' and B' are unbiased estimators of a and b.

Proof:

$$E[B'] = \frac{\sum (x_i - \bar{x})E[Y_i - \bar{Y}]}{\sum (x_i - \bar{x})^2}$$

$$= \frac{\sum (x_i - \bar{x})[b(x_i - \bar{x})]}{\sum (x_i - \bar{x})^2}$$

$$= b,$$

since

$$E[\bar{Y}] = \frac{1}{n} \ [\sum Y_i]$$

$$= \frac{1}{n} \sum (a + bx_i) = a + b\bar{x}$$

and $E[Y_i - \bar{Y}] = [a + bx_i] - [a + b\bar{x}] = b(x_i - \bar{x})$. We also have

$$E[A'] = E[\bar{Y} - B'\bar{x}]$$

$$= [a + b\bar{x}] - b\bar{x} = a$$

and the theorem is proved.   ◀

Theorem 10.1.3. Again, assume that Y is a random variable with mean $a + bx$, for a given value of x, and that the variance of Y is σ^2 for any x (that is, each distribution of Y values, for differing values of x, has the same variance). Let $(Y_1, x_1), (Y_2, x_2), \ldots , (Y_n, x_n)$ be a random sample of Y with their associated values of x. Then

$$V(A') = \frac{\sigma^2 \sum x_i^2}{n \sum (x_i - \bar{x})^2}$$

$$V(B') = \frac{\sigma^2}{\sum (x_i - \bar{x})^2}$$

$$\mathrm{Cov}(A', B') = - \frac{\sigma^2 \bar{x}}{\sum (x_i - \bar{x})^2},$$

where A' and B' are the least squares estimators.

Proof: Let us note first of all that

$$\sum (x_i - \bar{x})(Y_i - \bar{Y}) = \sum (x_i - \bar{x})Y_i - \bar{Y} \sum (x_i - \bar{x})$$

$$= \sum (x_i - \bar{x})Y_i,$$

since $\sum (x_i - \bar{x}) = 0$; then we can write

$$B' = \frac{\sum (x_i - \bar{x})Y_i}{\sum (x_i - \bar{x})^2}.$$

Now we know that if $Y_1, Y_2, \ldots, Y_n$ are independent random variables, each with variance σ^2 and if

$$Z = \sum_{i=1}^{n} c_i Y_i,$$

then

$$V(Z) = \sigma^2 \sum_{i=1}^{n} c_i^2.$$

Thus

$$B' = \sum_{i=1}^{n} c_i Y_i$$

where

$$c_i = \frac{x_i - \bar{x}}{\sum (x_i - \bar{x})^2}$$

and then

$$V(B') = \sigma^2 \sum c_i^2 = \sigma^2 \frac{\sum (x_i - \bar{x})^2}{[\sum (x_i - \bar{x})^2]^2} = \frac{\sigma^2}{\sum (x_i - \bar{x})^2}.$$

(We have written $V(Z)$ as shorthand for the variance of Z for any random variable Z.) Let us now turn to the variance of A'. Since $A' = \bar{Y} - B'\bar{x}$, we have

$$V(A') = V(\bar{Y}) + \bar{x}^2 V(B') - 2\bar{x} \operatorname{Cov}(\bar{Y}, B').$$

Now $V(\bar{Y}) = \sigma^2/n$ and we have just evaluated $V(B')$, so the one remaining quantity we need to get $V(A')$ is the covariance between $\bar{Y}$ and B'. You are asked below to show that if $Y_1, Y_2, \ldots, Y_n$ are independent random variables, each with variance σ^2, and $Z_1 = \sum c_i Y_i$, $Z_2 = \sum d_i Y_i$, then $\operatorname{Cov}(Z_1, Z_2) = \sigma^2 \sum c_i d_i$. Thus

$$\operatorname{Cov}(\bar{Y}, B') = \sigma^2 \sum \frac{1}{n} \frac{(x_i - \bar{x})}{\sum (x_i - \bar{x})^2} = 0$$

and we have

$$V(A') = V(\bar{Y}) + \bar{x}^2 V(B')$$

$$= \sigma^2 \left[\frac{1}{n} + \frac{\bar{x}^2}{\sum (x_i - \bar{x})^2} \right] = \frac{\sigma^2}{n} \frac{\sum x_i^2}{\sum (x_i - \bar{x})^2}.$$

$$\operatorname{Cov}(A', B') = \operatorname{Cov}(\bar{Y} - B'\bar{x}, B')$$

$$= \operatorname{Cov}(\bar{Y}, B') - \bar{x} \operatorname{Cov}(B', B')$$

$$= 0 - \bar{x} V(B')$$

$$= -\sigma^2 \frac{\bar{x}}{\sum (x_i - \bar{x})^2},$$

and the theorem is proved.   ◄

Notice then that the variances and the covariance of our estimators of a and b depend upon the values of x. If we are free to choose the x values, we can select them in optimal ways in terms of accuracy or variability of our estimators of a and b. Since

$$V(B') = \frac{\sigma^2}{\sum (x_i - \bar{x})^2},$$

it is plain that we shall get the most precise (least variable) estimate of the slope b if we make $\sum (x_i - \bar{x})^2$ as large as possible. In Example 10.1.1, we assumed that the average weight of babies of age x months was $a + bx$ for x between 1 and 6. To get the most precise estimate of b, we should choose half the babies in the sample to be 1 month old and the other half to be 6 months old because these values for x would maximize $\sum (x_i - \bar{x})^2$. It is intuitively plausible, regardless of the result about $V(B')$, that this distribution of x's should give the best estimate of the slope since, in taking half the values at each of the two end points of the interval, we are getting the most precise estimates we can of the average weights at both 1 month and 6 months. Then, because we assume the mean of Y to be $a + bx$ over this interval, the most precise estimates at the end points are connected by a straight line to give the most precise estimate of the slope. In general, if we can choose the x values, we shall get the smallest variance when estimating b by maximizing $\sum (x_i - \bar{x})^2$. Along the same lines, we should note that $V(A')$ is minimized if $\bar{x}$ is zero; at the same time, A' and B' are uncorrelated if $\bar{x} = 0$.

The method of least squares does not contain any rationale for estimating σ^2, the variance of Y. However, the following theorem presents an unbiased estimator for σ^2 whose observed value can be computed from a sample.

Theorem 10.1.4.   Assume that $E[Y \mid x] = a + bx$ and that $V(Y \mid x) = \sigma^2$. Then

$$(S')^2 = \frac{1}{n-2} \sum_{i=1}^{n} (Y_i - A' - B'x_i)^2$$

$$= \frac{1}{n-2} [\sum (Y_i - \bar{Y})^2 - (B')^2 \sum (x_i - \bar{x})^2]$$

is an unbiased estimator of σ^2.

Proof: Let us first verify the identity given above. Clearly

$$Y_i - A' - B'x_i = Y_i - (\bar{Y} - B'\bar{x}) - B'x_i$$

$$= Y_i - \bar{Y} - B'(x_i - \bar{x});$$

thus,

$$\sum (Y_i - A' - B'x_i)^2$$
$$= \sum (Y_i - \bar{Y} - B'(x_i - \bar{x}))^2$$
$$= \sum (Y_i - \bar{Y})^2 - 2B' \sum (x_i - \bar{x})(Y_i - \bar{Y}) + (B')^2 \sum (x_i - \bar{x})^2$$
$$= \sum (Y_i - \bar{Y})^2 - (B')^2 \sum (x_i - \bar{x})^2$$

since

$$\sum (x_i - \bar{x})(Y_i - \bar{Y}) = B' \sum (x_i - \bar{x})^2.$$

Clearly

$$Y_1 - \bar{Y} = \frac{n-1}{n} Y_1 - \frac{1}{n} \sum_2^n Y_i$$

and thus

$$V(Y_1 - \bar{Y}) = \frac{(n-1)^2}{n^2} \sigma^2 + \frac{1}{n^2} (n-1)\sigma^2$$

$$= \frac{n-1}{n} \sigma^2.$$

Similarly

$$V(Y_i - \bar{Y}) = \frac{n-1}{n} \sigma^2, \qquad i = 2, 3, \ldots, n.$$

Now $E[Y_i - \bar{Y}] = b(x_i - \bar{x})$ and thus

$$E[\sum (Y_i - \bar{Y})^2] = \sum \left(\frac{n-1}{n} \sigma^2 + b^2(x_i - \bar{x})^2 \right)$$

$$= (n-1)\sigma^2 + b^2 \sum (x_i - \bar{x})^2$$

since the expected value of the square of a variable is its variance plus the square of its mean. Similarly

$$\sum (x_i - \bar{x})^2 E[(B')^2] = \sum (x_i - \bar{x})^2 \left[\frac{\sigma^2}{\sum (x_i - \bar{x})^2} + b^2 \right]$$

$$= \sigma^2 + b^2 \sum (x_i - \bar{x})^2;$$

thus, combining these two results, we have

$$E[(S')^2] = \frac{1}{n-2} E[\sum (Y_i - \bar{Y})^2 - (B')^2 \sum (x_i - \bar{x})^2]$$

$$= \frac{1}{n-2} [(n-1)\sigma^2 + b^2 \sum (x_i - \bar{x})^2 - \sigma^2 - b^2 \sum (x_i - \bar{x})^2]$$

$$= \sigma^2,$$

and $(S')^2$ is an unbiased estimator of σ^2 as was claimed. ◀

It will be recalled that if we had a random sample $Y_1, Y_2, \ldots, Y_n$ of a random variable Y, whose mean was a constant μ and not related to some additional variable x, then we constructed an unbiased estimator of σ^2 by taking the sum of squares of the Y_i's about their mean $\overline{Y}$ (which estimates μ) and dividing by $n - 1$. Now, in the least squares model where the mean value of Y is dependent on another variable x, we estimate σ^2 by taking the sum of squares of the deviations of the Y_i's about the estimated line, $A' + B'x$, which for any x gives the estimator of the mean of Y, and dividing by $n - 2$. In the former case the divisor is $n - 1$ since only one statistic $(\overline{Y})$ is needed to estimate the mean of Y; in the latter case the divisor is $n - 2$ since two statistics $(A'$ and $B')$ are needed to estimate the mean of Y for a given x.

EXAMPLE 10.1.2. Assume that the cumulative grade point average a student achieves at the end of the freshman year is a linear function of the score, x, that he made on an entrance exam. That is, we are assuming that if we look at all students scoring x on the entrance exam, we have a distribution of cumulative grade point averages achieved at the end of the freshman year; furthermore the means of these distributions, for varying values of x, lie on the straight line $a + bx$. As before, we also assume that the higher moments are all independent of x. The entrance exam scores can range from 60 to 100, say, and the grade point average is computed on the basis of 4 points for an A, 3 points for a B, 2 for a C, 1 for a D, and 0 for an F.

Suppose that we want to estimate a and b and especially want a precise estimate of b; to do so we would want our sample x values separated as widely as possible. When examining the entrance exam records, we find that there are relatively few scores at 60 and 100 but that for $x = 65$ and $x = 95$ there are a lot of students to choose from. Accordingly, we select at random 4 students who scored 65 and 4 students who scored 95 and find their grade point averages to be 2.25, 2.17, 2.67, 2.42 for $x = 65$ and 3.63, 2.95, 3.02, 3.49 for $x = 95$. (In actual practice, of course, we would want to take a larger sample than this to get good estimates of a and b, but for illustrative purposes n is kept small.) Then we find that $\bar{x} = 80$, $\bar{y} = 2.825$, $\sum (x_i - \bar{x})^2 = 1800$, $\sum (y_i - \bar{y})^2 = 2.0916$, $\sum (x_i - \bar{x})(y_i - \bar{y}) = 53.7$, $b' = \frac{53.7}{1800} = .030$, and $a' = 2.825 - .030(80) = .425$. Thus $y = .425 + .030x$ is the least squares line.

Then we estimate that if a student scores $x = 75$ on the entrance exam, he will on the average have a $.425 + (.03)(75) = 2.675$ grade point at the end of his freshman year; a student scoring $x = 90$ on the entrance exam should on the average have a $.425 + (.03)(90) = 3.125$ grade point at the end of his freshman year. Our estimate of σ^2 is

$$(s')^2 = \tfrac{1}{6}[2.0916 - (.03)^2(1800)] = .079$$

and our estimate of σ is

$$s' = \sqrt{.079} = .280.$$

The estimated variance of B' is $(.079)(\frac{1}{1800}) = .0000439$; the estimated variance of A' is $(.079)\left(\frac{1}{6} + \frac{(80)^2}{1800}\right) = .294$; the estimated covariance between A' and B' is $(.079)\left(\frac{-80}{1800}\right) = -.00351$.

When selecting only two different x values, as in this example, we are relying very heavily on the assumed linearity of $E[Y \mid x]$. Since we did not observe x at intermediate points between 65 and 95, we would not be able to detect from the sample whether this linear assumption appeared correct.

The Gauss-Markov theorem, which we shall now state in a restricted context, is the major reason why the method of least squares is so frequently employed when estimating the relationship between the mean of a random variable Y and an independent variable x. We shall state the theorem only in the context of a linear relationship between $E[Y \mid x]$ and x, but it can be extended to more general cases.

Theorem 10.1.5. Assume that $E[Y \mid x] = a + bx$, for given values of x, and that the variance of Y is σ^2 for all x. Then if (Y_1, x_1), (Y_2, x_2), ..., (Y_n, x_n) is a random sample of Y with their associated x values, the least squares estimators A' and B' are the best linear unbiased estimators of a and b.

The remarkable thing about this theorem is the fact that the actual distribution of the Y values does not have to be known for us to compute the best linear unbiased estimators of a and b. It will be recalled that "best linear unbiased" means that among all linear functions of the Y_i's that are unbiased estimators, none has a smaller variance. However, it is important to realize that we assume the variance of Y to be a constant for all values of x in order that A' and B' have this property; if this is not satisfied, then the best linear unbiased estimators are altered somewhat from the above values (see Exercise 6 below).

EXERCISE 10.1

1. Eight 1-foot-tall Monterey pine trees were planted in similar controlled environments but were subjected to differing amounts of irrigation to simulate the effect of differing amounts of rainfall. At the end of 1 year, their achieved heights were measured. The following table presents their measured heights (in inches) at the end of a year (y_i) and the amount of rainfall (in inches) simulated for each (x_i). Assume that Y, the height of a tree at the end of a year (starting from a 1-foot height), is a random variable with mean $a + bx$, where x is the rainfall, and with constant variance σ^2 for all x. Compute the

best linear unbiased estimates of a and b and compute an unbiased estimate of σ^2 and of $V(A')$ and $V(B')$.

y_i	x_i
19	10
22	14
25	18
31	22
33	26
39	30
44	34
45	38

2. Show that

$$\Sigma\, x_i y_i - \bar{x}\, \Sigma\, y_i = \Sigma\, x_i y_i - \bar{y}\, \Sigma\, x_i$$
$$= \Sigma\, (x_i - \bar{x})(y_i - \bar{y}).$$

3. A random sample of 10 men aged 30 yielded the following information on their current annual salary (y_i) (in thousands of dollars) and the number of years of formal schooling they had (x_i). Assume that the mean of the distribution of salaries is a linear function of the number of years of schooling and

y_i	x_i
7.0	10
6.2	12
8.1	13
7.5	14
6.5	16
10.5	16
8.0	16
13.2	16
12.8	18
16.5	20

that the variance of Y is independent of x; compute the best linear unbiased estimates of a and b. Also estimate σ^2 and $V(A')$, $V(B')$.

4. Assume that Y is a random variable with $E[Y \mid x] = bx$ for any x, and that the variance of Y is σ^2, independent of x. Assuming (Y_1, x_1), (Y_2, x_2), . . . , (Y_n, x_n) is a random sample of Y, with the associated x values, derive the least squares estimator of b. Define an unbiased estimator of σ^2 and show that it is unbiased.

5. Show that the least squares estimator of b derived in problem 4 is the best linear unbiased estimator of b, in the context of the assumptions given there. (Hint:

$$B' = \frac{\Sigma x_i Y_i}{\Sigma x_i^2} = \Sigma c_i Y_i, \qquad \text{where } c_i = \frac{x_i}{\Sigma x_i^2}.$$

If B is any other linear unbiased estimator of b, then

$$B = \Sigma d_i Y_i = \Sigma (c_i - f_i) Y_i, \qquad \text{with } \Sigma f_i x_i = 0.$$

Compute $V(B)$.)

6. Suppose that $E[Y \mid x] = bx$ and that $(Y_1, x_1), (Y_2, x_2), \ldots, (Y_n, x_n)$ is a random sample of Y with their associated x values. Assume that $V(Y_i) = \sigma_i^2$; that is, the variances of the Y values are now conceivably dependent upon the x values. Show that

$$B' = \frac{\Sigma x_i Y_i / \sigma_i^2}{\Sigma x_i^2 / \sigma_i^2}$$

is the best linear unbiased estimator of b. (Hint: Define

$$Z_i = \frac{Y_i}{\sigma_i}, \qquad w_i = \frac{x_i}{\sigma_i}$$

and apply the result of problem 5.)

7. Suppose that everything is as given in problem 6, with $x_i > 0$, $\sigma_i^2 = \sigma^2 x_i$. Evaluate the best linear unbiased estimator of b.

8. Again, suppose that everything is as given in problem 6 and now assume that $\sigma_i^2 = \sigma^2 x_i^2$; evaluate the best linear unbiased estimator of b.

9. Show that the least squares line $y = a' + b'x$ always goes through the point whose coordinates are $(\bar{x}, \bar{y})$.

10. Assume that $E[Y \mid x] = a + b/x$, for $x > 0$, and derive the least squares estimates of a and b.

11. Assume Y is a random variable, and for a given value of x, $E[Y \mid x] = a + bx$. Then we can write $Y = a + bx + E$, as above, and given a random sample (Y_i, x_i), $i = 1, 2, \ldots, n$, the least squares estimators are the values that minimize $\Sigma_{i=1}^{n} E_i^2$; these estimators are unique and have desirable properties. A competing rationale for least squares would be to use the values which minimize the absolute deviations, $\Sigma |E_i|$, rather than the sum of their squares. These values are different than A' and B', in general, and for certain observed (y_i, x_i) may not be unique.

(a) Assume the observed values, (y_i, x_i), lie on a straight line and show that the least squares estimators are identical with the estimators which minimize the sum of the absolute deviations.

(b) Assume $n = 3$ and the observed sample values are $(0, -1)$, $(1, 0)$, $(0, 1)$ (the first coordinate is the y value, the second is the x value). Show that the two lines

(i) $y = 1/3 + x/3$

(ii) $y = 1/3 - x/3$

have the same value for $\sum |E_i|$. Can you find a line with a smaller value for $\sum |E_i|$?

(c) Compute the least squares line for these data points.

12. Assume $Y_1, Y_2, \ldots, Y_n$ are independent random variables, each with variance σ^2, and define $Z_1 = \sum c_i Y_i$, $Z_2 = \sum d_i Y_i$. Compute $V(Z_1)$, $V(Z_2)$ and cov (Z_1, Z_2).

13. If Y is a random variable with variance σ^2, show that cov $(Y, Y) = \sigma^2$.

14. Assume $E[Y \mid x] = a + bx$, $V(Y \mid x) = \sigma^2$ and that a sample of n observations (Y_i, x_i), $i = 1, 2, \ldots, n$, has been used to find the least squares estimator

$$Y' = A' + B'x.$$

Y' then can be used to estimate the mean value for the Y distribution for any x between min (x_i) and max (x_i). Compute the variance of Y' as an estimator of the mean Y value for $x = x^*$ (some specific value for x).

15. Suppose we assume $E[Y \mid x] = a + bx$, $V[Y \mid x] = \sigma^2$ for all x and that we are free to choose the x values in the sample. We decide to take samples of n_1 Y values with $x = x_1$, n_2 Y values with $x = x_2$ and n_3 Y values with $x = x_3$. Then we use a double subscript to identify the Y values and Y_{ij} is the j-th observation with x_i, $i = 1, 2, 3, j = 1, 2, \ldots, n_i$. Define $\bar{Y}_i = \sum_{j=1}^{n_i} Y_{ij}/n_i$ and $\bar{Y} = \sum_i \sum_j Y_{ij}/n$, where $n = \sum n_i$. Show that the least squares estimator for b is

$$b' = \frac{\sum_{i=1}^{3} n_i(x_i - \bar{x})(\bar{Y}_i - \bar{Y})}{\sum n_i(x_i - \bar{x})^2}.$$

16. Assume the same situation as described in 15 and that $n_1 = n_2 = n_3$. Show that the least squares line for the full set of data is identical with the least squares line determined by $(\bar{Y}_1, x_1)$, $(\bar{Y}_2, x_2)$, $(\bar{Y}_3, x_3)$. Is this still true if the three n_i's are not equal?

17. Again assume the situation described in Exercise 15 and assume we want to estimate the mean of the Y population for $x = x_2$. If the three n_i's are not equal then we have three estimators available, namely $\bar{Y}_2$, $Y_1' = A_1' + B_1'x_2$ and $Y_2' = A_2' + B_2'x_2$, where Y_1' is the least squares line as described in Exercise 15 and Y_2' is the least squares line determined by the means $(\bar{Y}_1, x_1)$, $(\bar{Y}_2, x_2)$, $(\bar{Y}_3, x_3)$ as described in 16.

(a) Show that all three estimators are unbiased.

(b) Which of the three estimators has the smallest variance?

18. If we assume $E[Y \mid x] = a + bx$, $V[Y \mid x] = \sigma^2$, and we have a sample of size n of (Y_i, x_i), note that we could work with the new variables $Z_i = Y_i - \bar{Y}$, $w_i = x_i - \bar{x}$.

   (a) Show that $E[Z_i \mid w_i] = bw_i$ and that $V[Z_i \mid w_i] = \dfrac{n-1}{n} \sigma^2$, a constant for all w_i.

   (b) What is the least squares estimator for b, in terms of the (Z_i, w_i)'s and in terms of the (Y_i, x_i)'s? This correspondence, together with the result given in Exercise 5 proves Theorem 10.1.5 for the least squares estimator for b.

19. Assume the situation described in Exercise 18, but now let $Z_i = Y_i - B'x_i$, where B' is the usual least squares estimator for b. Show that $E[Z_i \mid x_i] = a$ and $V[Z_i \mid x_i] = \sigma^2 g$ where $g = \sum x_i^2 / n \sum (x_i - \bar{x})^2$, a constant for all x_i. This result, together with Theorem 7.2.1 proves Theorem 10.1.5 for the least squares estimator of a.

20. n different people counted the traffic on the same street, at the same time of day, on separate independent occasions. Unfortunately, they did not all observe the traffic for the same length of time. Person i observed it for x_i minutes, $i = 1, 2, \ldots, n$. Assume that the traffic on this road passes the observation point like events in a Poisson process, with a rate of λ cars per minute. Thus, letting Y_i be the number of cars to pass the observation point when person i was there, Y_i is a Poisson random variable with parameter λx_i and $E[Y_i] = V[Y_i] = \lambda x_i$, $i = 1, 2, \ldots, n$. What is the best linear unbiased estimator for λ? (Hint: Reread Exercise 7.)

21. For the situation described in Exercise 20 a reasonable heuristic estimator for λ could be constructed by reasoning as follows:

$$E\left[\frac{Y_i}{\sqrt{x_i}}\right] = \lambda\sqrt{x_i}, \qquad V\left[\frac{Y_i}{\sqrt{x_i}}\right] = \lambda;$$

then $\sum \dfrac{Y_i}{\sqrt{x_i}} \Big/ \sum \sqrt{x_i}$ is an unbiased estimator for λ. Show that this estimator has a larger variance than the one given above in Exercise 20.

10.2. Interval Estimation and Tests of Hypotheses

We ended the last section with a theorem giving an optimum property of the least squares estimates, no matter what the form of the underlying distribution for Y. In this section we shall make the assumption that Y is normally distributed, with mean $a + bx$ and constant variance σ^2. Under this distributional assumption, we shall find it easy to construct confidence intervals for a and b and for $a + bx$, the mean value of Y, and to test hypotheses about these quantities. We shall first investigate the maximum likelihood estimators of a, b, and σ^2 for a normal assumption on Y.

Theorem 10.2.1. Y is a normal random variable with mean $a + bx$, for a specified value of x, and with variance σ^2 for any x. The maximum likelihood estimators of a, b, and σ^2 are

$$\hat{A} = \bar{Y} - \hat{B}\bar{x}$$

$$\hat{B} = \frac{\sum (x_i - \bar{x})(Y_i - \bar{Y})}{\sum (x_i - \bar{x})^2}$$

$$\hat{\Sigma}^2 = \frac{1}{n}\sum (Y_i - \hat{A} - \hat{B}x_i)^2,$$

given a random sample $(Y_1, x_1), (Y_2, x_2), \ldots, (Y_n, x_n)$ of Y with their associated x values.

Proof: The likelihood function for the sample is

$$L(a, b, \sigma^2) = \left(\frac{1}{2\pi\sigma^2}\right)^{n/2} e^{(-1/2\sigma^2)\Sigma(y_i - a - bx_i)^2}.$$

Then

$$\log_e L = -\frac{n}{2}\log_e 2\pi\sigma^2 - \frac{1}{2\sigma^2}\sum (y_i - a - bx_i)^2$$

and

$$\frac{\partial \log_e L}{\partial a} = \frac{1}{\sigma^2}\sum (y_i - a - bx_i)$$

$$\frac{\partial \log_e L}{\partial b} = \frac{1}{\sigma^2}\sum x_i(y_i - a - bx_i)$$

$$\frac{\partial \log_e L}{\partial \sigma^2} = -\frac{n}{2\sigma^2} + \frac{1}{2\sigma^4}\sum (y_i - a - bx_i)^2.$$

We note that if we set these three equations equal to zero and multiply through by σ^2, the first two equations are identical with the least squares normal equations that we solved in Section 10.1. Thus, their solution is given by

$$\hat{a} = \bar{y} - \hat{b}\bar{x}$$

$$\hat{b} = \frac{\sum (x_i - \bar{x})(y_i - \bar{y})}{\sum (x_i - \bar{x})^2} ;$$

once the first two equations are solved for $\hat{a}$ and $\hat{b}$, their solutions can be put into the third equation, yielding

$$\hat{\sigma}^2 = \frac{1}{n} \sum (y_i - \hat{a} - \hat{b}x_i)^2.$$

Thus we have the desired result upon replacing the sample values $y_1, y_2, \ldots,$ y_n with the random variables $Y_1, Y_2, \ldots, Y_n$. ◄

We note immediately that the maximum likelihood estimators of a and b are identical with the least squares estimators of these same parameters. Thus we can say immediately that $\hat{A}$ and $\hat{B}$ are the best linear unbiased estimators of a and b, respectively. We shall use $\hat{A}$ and $\hat{B}$, rather than A' and B', to denote these estimators from now on. We also know that

$$E[n\hat{\Sigma}^2] = (n - 2)\sigma^2$$

so that $\hat{\Sigma}^2$ is a biased estimator of σ^2. From now on we shall use only the unbiased estimator of σ^2 (gotten by dividing by $n - 2$ rather than n) and shall denote it by S^2 rather than $(S')^2$.

The estimator $\hat{B}$ of b is a linear combination of normal random variables, under the assumption of normality for Y, and thus is itself a normal random variable; its mean is b and its variance is

$$\frac{\sigma^2}{\sum (x_i - \bar{x})^2}.$$

Similarly, since $\hat{A}$ is a linear combination of normally distributed random variables, it is itself normally distributed with mean a and variance

$$\frac{\sum x_i^2}{n \sum (x_i - \bar{x})^2} \sigma^2.$$

It can also be shown that $[(n - 2)/\sigma^2] S^2$ is a χ^2 random variable with $n - 2$ degrees of freedom, under our normal assumption, and that S^2 is distributed independently of $\hat{A}$ and $\hat{B}$. These facts will enable us to put confidence intervals about a and b and to test hypotheses about a and b. The following theorem gives the basic result for these techniques.

Theorem 10.2.2. Let Y be a normal random variable with mean $a + bx$, for each fixed value of x, and with constant variance σ^2. Let $\hat{A}$ and $\hat{B}$ be the maximum likelihood (least squares) estimators of a and b and let S^2 be the unbiased estimator of σ^2 already defined. Then

$$\frac{\hat{A} - a}{S \sqrt{\dfrac{\sum x_i^2}{n \sum (x_i - \bar{x})^2}}} \quad \text{and} \quad \frac{\hat{B} - b}{S \sqrt{\dfrac{1}{\sum (x_i - \bar{x})^2}}}$$

each have the t-distribution with $n - 2$ degrees of freedom.

Proof: We are given that

$$\frac{\hat{A} - a}{\sigma\sqrt{\dfrac{\sum x_i^2}{n \sum (x_i - \bar{x})^2}}} \quad \text{and} \quad \frac{\hat{B} - b}{\sigma\sqrt{\dfrac{1}{\sum (x_i - \bar{x})^2}}}$$

are each standard normal random variables; $[(n - 2)/\sigma^2]\, S^2$ is a χ^2 random variable with $n - 2$ degrees of freedom and is independent of both of the standard normal random variables just mentioned. Thus,

$$\frac{\hat{A} - a}{\sigma\sqrt{\dfrac{\sum x_i^2}{n \sum (x_i - \bar{x})^2}}} \Bigg/ \sqrt{\frac{(n - 2)S^2}{\sigma^2(n - 2)}} = \frac{\hat{A} - a}{S\sqrt{\dfrac{\sum x_i^2}{n \sum (x_i - \bar{x})^2}}}$$

and

$$\frac{\hat{B} - b}{\sigma\sqrt{\dfrac{1}{\sum (x_i - \bar{x})^2}}} \Bigg/ \sqrt{\frac{(n - 2)S^2}{\sigma^2(n-2)}} = \frac{\hat{B} - b}{S\sqrt{\dfrac{1}{\sum (x_i - \bar{x})^2}}}$$

each have the t-distribution with $n - 2$ degrees of freedom. (See Theorem 7.4.3.) ◄

Constructing a confidence interval for a or b, or testing a hypothesis about either, then will proceed in exactly the same way as we saw for the mean of a normal random variable in Chapters 7 and 8. Specifically, the statistics

$$L = \hat{B} - t_{1-\alpha/2} S/\sqrt{\sum (x_i - \bar{x})^2}$$
$$U = \hat{B} + t_{1-\alpha/2} S/\sqrt{\sum (x_i - \bar{x})^2}$$

where $t_{1-\alpha/2}$ is the $100\,(1 - \alpha/2)$ percentile of the t-distribution with $n - 2$ degrees of freedom, form a $100(1 - \alpha)\%$ confidence interval for b. A $100(1 - \alpha)\%$ confidence interval for a is defined similarly.

We can also construct $100(1 - \alpha)\%$ confidence bands about the regression line $\hat{Y}_x = \hat{A} + \hat{B}x$; that is, for any given constant value of x, $\hat{Y}_x$ is normally distributed with mean $a + bx$ and variance

$$\begin{aligned}
V(\hat{Y}_x) &= V(\hat{A} + \hat{B}x) \\
&= V(\hat{A}) + x^2 V(\hat{B}) + 2x \operatorname{Cov}(\hat{A}, \hat{B}) \\
&= \frac{\sigma^2}{\sum (x_i - \bar{x})^2}\left[\frac{1}{n}\sum x_i^2 + x^2 - 2x\bar{x}\right] \\
&= \frac{\sigma^2}{\sum (x_i - \bar{x})^2}\left[\frac{1}{n}\sum (x_i - \bar{x})^2 + (x - \bar{x})^2\right] \\
&= \sigma^2\left[\frac{1}{n} + \frac{(x - \bar{x})^2}{\sum (x_i - \bar{x})^2}\right].
\end{aligned}$$

Thus,

$$\frac{\hat{Y}_x - (a + bx)}{S\sqrt{\dfrac{1}{n} + \dfrac{(x - \bar{x})^2}{\sum (x_i - \bar{x})^2}}}$$

has the t-distribution with $n - 2$ degrees of freedom and

$$L_x = \hat{Y}_x - t_{1-\alpha/2} S \sqrt{\frac{1}{n} + \frac{(x - \bar{x})^2}{\sum (x_i - \bar{x})^2}}$$

$$U_x = \hat{Y}_x + t_{1-\alpha/2} S \sqrt{\frac{1}{n} + \frac{(x - \bar{x})^2}{\sum (x_i - \bar{x})^2}}$$

form a $100(1 - \alpha)\%$ confidence interval for $a + bx$, for any given value of x. The observed values of these statistics can be plotted as functions of x, along with the observed regression line $\hat{y}_x$, and it is obvious that they will form a band about the regression line; this is the reason they are referred to as confidence bands. Note that the interval is shortest for $x = \bar{x}$ and that as x moves away from $\bar{x}$, in either direction, the confidence interval for $a + bx$ gets wider. (See Figure 10.2.1.)

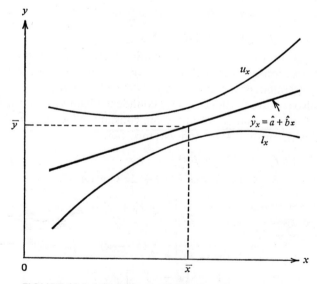

FIGURE 10.2.1 Confidence limits for the mean of **Y**.

EXAMPLE 10.2.1.   For the regression of weights of babies on their ages (Example 10.1.1), we had $n = 10$, $\bar{x} = 3.7$, $\sum (x_i - \bar{x})^2 = 23.1$, $\hat{b} = 2.07$, and $\hat{a} = 7.94$. By assuming that weights of babies of any given age x are normally distributed, we can put confidence intervals about a, b, and $a + bx$. To do this we need to compute s^2 from the data given in Example 10.1.1; we find that

$$s^2 = \tfrac{1}{8}[\sum (y_i - \bar{y})^2 \neq \hat{b}^2 \sum (x_i - \bar{x})^2]$$
$$= \tfrac{1}{8}[111.9 - 98.91] = 1.624$$
$$s = \sqrt{1.624} = 1.274.$$

Then the observed 95% confidence limits for b are $2.07 - (2.306)(1.274)\left(\dfrac{1}{4.806}\right) =$ 1.49 and $2.07 + (2.306)(1.274)\left(\dfrac{1}{4.806}\right) = 2.68$. The estimated standard deviation of our estimate of a is

$$s\sqrt{\frac{\sum x_i^2}{n \sum (x_i - \bar{x})^2}} = (1.274)(.931) = 1.186;$$

thus the observed 95% confidence limits for a are $7.94 - (2.306)(1.186) = 5.190$ and $7.94 + (2.306)(1.186) = 10.689$. The observed 95% confidence bands for the average weight of a baby age x months (for $1 \leq x \leq 6$) are

$$7.94 + 2.07x - 2.306 \sqrt{\frac{1}{10} + \frac{(x - 3.7)^2}{23.1}}$$

and

$$7.94 + 2.07x + 2.306 \sqrt{\frac{1}{10} + \frac{(x - 3.7)^2}{23.1}} ;$$

thus we are 95% sure that the interval (8.52, 11.50) covers the average weight of $x = 1$-month-old babies. Similarly, we are 95% sure that the interval (15.48, 16.96) covers the average weight of 4-month-old babies and that the interval (19.04, 21.68) covers the average weight of 6-month-old babies.

Tests of hypotheses about the values of a and b are easily carried out, based on the distributional result given in Theorem 10.2.2. Suppose that we wanted to test $H_0 : b = b_0$ versus $H_1 : b \neq b_0$, where b_0 is some specified value. The likelihood ratio test criterion (Chapter 8) then says we should reject H_0 if $|\hat{b} - b_0|$ exceeds

$$t_{1-\alpha/2} s \sqrt{\frac{1}{\sum (x_i - \bar{x})^2}} ,$$

where $t_{1-\alpha/2}$ is the $100(1 - \alpha/2)$ percentile of the t-distribution with $n - 2$ degrees of freedom (since we have the two-sided alternative). Similarly, we can easily construct the likelihood ratio test criterion that a takes on any specified value, using the fact that on the basis of a sample of n observations the estimator $\hat{A}$ is normally distributed with mean a and variance that can be independently estimated from the sample values (this also turns out to be a t-test, of course). A frequently tested hypothesis for least-squares models is that $b = 0$; this would test that there is no relation between the mean of Y and the independent variable x (since if the slope is zero, the mean of Y is a constant for all x). The following example is of this type.

EXAMPLE 10.2.2. In Example 10.1.2, we assumed that the cumulative grade point average for students at the end of their freshman year was linearly related to the score x which they made on an entrance exam. We shall now use that data to check the assumption that these two variables are related. Specifically, it would not seem logical that there exists a decreasing relation between y and x (the higher the entrance exam score x, the lower the resulting grade point average); thus let us test $H_0: b = 0$ versus $H_1: b > 0$ with $\alpha = .1$. Since $\hat{B}$ is normally distributed with mean b and standard deviation

$$\sigma \sqrt{\frac{1}{\sum (x_i - \bar{x})^2}},$$

we know that the uniformly most powerful test of this hypothesis versus this one-sided alternative is given by rejecting H_0 only if $\hat{b}$ exceeds

$$t_{.9} s \sqrt{\frac{1}{\sum (x_i - \bar{x})^2}},$$

where $t_{.9}$ is the 90-th percentile of the t-distribution with 6 degrees of freedom (and equals 1.440). From the data given in Example 10.1.2, we find that $\hat{b} = .030$.

$$s \sqrt{\frac{1}{\sum (x_i - \bar{x})^2}} = \sqrt{.0000439} = .0066;$$

since $.030 > (1.44)(.0066) = .0095$, we reject H_0. The probability that we are wrong in so rejecting is .1.

EXAMPLE 10.2.3. A new type of scale has been produced which is designed to give weights accurately (in thousandths of pounds) of objects whose true weights can be as large as half a pound. We have available 5 objects whose true weights are known to be .1, .2, .3, .4, and .5 pounds, respectively. We put each of these objects on the scale twice and record the reading obtained. If we assume that the recorded weight on the scale is a normal random variable Y with mean bx, where x is the true weight, and constant variance σ^2, we can use these 10 weighings to test the hypothesis $H_0: b = 1$ versus $H_1: b \neq 1$, with $\alpha = .05$, for example. If $b \neq 1$, then the scale has a bias in its readings. (The scale would certainly not be of much use if we rejected H_0.) The 10 readings are as recorded below. We find that $\sum x_i = 3.0$,

x_i	y_i
.1	.098
.1	.099
.2	.208
.2	.200
.3	.302
.3	.298
.4	.405
.4	.401
.5	.502
.5	.495

$\sum x_i^2 = 1.1$, $\sum y_i = 3.008$, $\sum y_i^2 = 1.105$, and $\sum x_i y_i = 1.102$. From Exercise 4, Section 10.1, the maximum likelihood (least-squares) estimate of b is

$$\hat{b} = \frac{\sum x_i y_i}{\sum x_i^2} = 1.002;$$

the unbiased estimate of σ^2 is

$$s^2 = \frac{1}{n-1} \sum (y_i - bx_i)^2 = \frac{1}{n-1} [\sum y_i^2 - b^2 \sum x_i^2] = .000066.$$

Then

$$\hat{B} = \frac{\sum x_i Y_i}{\sum x_i^2}, \qquad V(\hat{B}) = \sigma^2 \sum \frac{x_i^2}{(\sum x_i^2)^2} = \frac{\sigma^2}{\sum x_i^2};$$

now

$$\frac{\hat{B} - b}{\sqrt{\dfrac{1}{\sum x_i^2}}}$$

has the t-distribution with 9 degrees of freedom and, using the likelihood ratio criterion for a two-sided test about the mean of a normal, we should reject H_0 if

$$|\hat{b} - 1| \geq 2.262s \sqrt{\frac{1}{\sum x_i^2}}$$

(2.262 is the 97.5 percentile of the t-distribution with 9 degrees of freedom). We find that

$$2.262s \sqrt{\frac{1}{\sum x_i^2}} = .0175$$

and $|\hat{b} - 1| = .002$; thus we accept H_0.

EXERCISE 10.2

1. Assume that the running time to failure of an electron tube is an exponential random variable Y with mean $a + bx$, where x is the number of times the tube has been switched on. Assume that we have a random sample of lives of m tubes $(y_1, y_2, \ldots, y_m)$, each of which has been switched on x_1 times, and a random sample of lives of n tubes $(z_1, z_2, \ldots, z_n)$, each of which has been switched on x_2 times. Compute the maximum likelihood estimates of a and b and compare with the least squares estimates. Why might these be preferred to the least squares estimates?

2. Using the data on average salaries versus education in exercise 3, Section 10.1, test the hypothesis that $b = 0$ versus the alternative that $b > 0$, with $\alpha = .01$.

3. Compute a 90 % confidence interval for a and a 90 % confidence interval for b for the height versus rainfall data of exercise 1, Section 10.1.

4. Assume that the number of sales of a given item in a large department store per week is a Poisson random variable with mean $a + bx$, where x is the amount of money spent on advertising the item during the preceding week. Assume that for m weeks the advertising budget for this item was constant at x_1 dollars and that the weekly numbers of sales made with the budget at this level were $y_1, y_2, \ldots, y_m$. Then some time ago the advertising budget was increased to x_2 dollars; the numbers of sales made during n weeks after the new budget was adopted were $z_1, z_2, \ldots, z_n$. Compute the maximum likelihood estimates of a and b and compare the least squares estimates. Which of these would you prefer and why?

5. For the scale data presented in Example 10.2.3, assume that Y is normal with mean $a + bx$ and constant variance σ^2 and estimate a, b, and σ^2; test the hypothesis that $a = 0$ versus the alternative that $a \neq 0$, with $\alpha = .1$.

6. The question of precision of a scale is related to the σ^2 of the distribution of Y. If σ^2 gets too large, the individual readings would vary a great deal about the true weight, assuming no bias; thus we would not put much faith in the readings. Suppose that it has been decided that the scale in Example 10.2.3 is precise enough if $\sigma^2 \leq .00001$. Use the data presented in the example to test $H_0: \sigma^2 \leq .00001$ versus $H_1: \sigma^2 > .00001$ with $\alpha = .05$.

7. Assume that Y is normal with mean $a + bx$ and variance $x\sigma^2$ for $x > 0$. Given a random sample of n observations of Y, with their associated x values, compute the maximum likelihood estimators of a, b, and σ^2.

8. Assume we have a random sample of n babies of racial group 1, each between 1 and 6 months of age. Thus we have recorded (y_{1i}, x_{1i}), $i = 1, 2, \ldots, n$, where y_{1i} is the weight and x_{1i} the age of the i-th baby. We also have a random sample of m babies from racial group 2, each between 1 and 6 months of age. Their weights, y_{2i}, and ages, x_{2i}, $i = 1, 2, \ldots, m$, are also recorded. We might assume

$$E[Y_{1i}] = a_1 + b_1 x_{1i}, \quad \text{for each } i$$
$$E[Y_{2i}] = a_2 + b_2 x_{2i}, \quad \text{for each } i$$

and that $V(Y_{1i}) = V(Y_{2i}) = \sigma^2$ for all i. It may or may not be the case that $a_1 = a_2$ and/or $b_1 = b_2$.

(a) Assuming all weights are normally distributed, how would you test $H_0 : b_1 = b_2$ versus $H_1 : b_1 \neq b_2$?

(b) Again assuming normality, how would you test $H_0 : a_1 = a_2$ versus $H_1 : a_1 \neq a_2$?

(c) Assuming that $a_1 = a_2$ and $b_1 = b_2$, how would you estimate their common values?

9. Assume in Exercise 8 that the two slopes, b_1 and b_2, are equal but that a_1 and a_2 are not equal. What would you use as estimates for a_1, a_2 and b, the common slope from the two samples?

10. Assume we have a random sample of n values of Y, $Y_{1j}, j = 1, 2, \ldots, n$, each associated with the same value x_1 for x and a random sample of m values of Y, $Y_{2j}, j = 1, 2, \ldots, m$, each associated with the same value x_2 for x (and $x_1 \neq x_2$). We assume $E[Y_{ij}] = a + bx_i, i = 1, 2$, that $V[Y_{ij}] = \sigma^2$ for each i and j, and that the Y_{ij}'s are normally distributed. What are the method of moments estimators for a, b, and σ^2?

11. If $Y_1, Y_2, \ldots, Y_n$ are independent normal random variables, each with variance σ^2 and $E[Y_i] = a + bx_i$, where the x_i's are known constants, verify that the likelihood ratio test of $H_0 : b = b_0$ versus $H_1 : b \neq b_0$ involves the t-statistic as claimed in the text.

10.3. Other Models

We have examined problems of estimation and hypothesis testing when a normal random variable Y has mean $a + bx$, for a given value of x, and variance σ^2. If the mean value of Y is a nonlinear function of x, then the least-squares criteria or the maximum likelihood criteria can still be employed, but the equations to be solved for the estimates may not be as easy to handle. No matter whether the mean of Y is linearly related to x, however, the least squares estimates and the maximum likelihood estimates of the parameters in the mean will be identical. To see that this is the case, suppose that Y is normally distributed with mean $m(x)$ and variance σ^2, where $m(x)$ is some function of x which includes unknown parameters $b_0, b_1, \ldots, b_k$. Then, assuming that $(y_1, x_1), (y_2, x_2), \ldots, (y_n, x_n)$ are the observed values of a random sample of Y, with their associated x values, the least squares estimates of $b_0, b_1, \ldots, b_k$ would be determined by minimizing

$$\sum (y_i - m(x_i))^2$$

simultaneously with respect to $b_0, b_1, \ldots, b_k$. The maximum likelihood estimators of $b_0, b_1, \ldots, b_k$ would be determined by maximizing

$$L(b_0, b_1, \ldots, b_k, \sigma^2) = \left(\frac{1}{2\pi\sigma^2}\right)^{n/2} e^{-\Sigma(y_i - m(x_i))^2/2\sigma^2};$$

L will clearly be maximized, with respect to $b_0, b_1, \ldots, b_k$, by the values that minimize the exponent on e. But these are the least squares values, by definition, so the two will always coincide.

If $E[Y \mid x] = m(x)$ is linear in the unknown parameters, regardless of whether it is linear in x or not, then we have what is called a *linear statistical model*. Examples of linear statistical models are

$$m_1(x) = a + bx$$
$$m_2(x) = b_0 + b_1 x + b_2 x^2$$
$$m_3(x_1, x_2) = b_0 + b_1 x_1 + b_2 x_2 + b_3 x_1 x_2.$$

All of these cases (or any other linear model), employing least squares criteria, are easily handled by using matrix theory which we have not assumed in this book. To see why more advanced mathematical techniques might be called for when treating linear models involving more than two parameters, let us examine the normal (least squares) equations for the quadratic model $m_2(x)$ given above. We assume that $(y_1, x_1), (y_2, x_2), \ldots, (y_n, x_n)$ are the observed values of a random sample of a random variable Y, with their associated x values, that

$$E[Y \mid x] = b_0 + b_1 x + b_2 x^2,$$

and that the variance of Y is σ^2 for any x. The quantity to be minimized, with respect to b_0, b_1, b_2, is

$$Q = \sum (y_i - b_0 - b_1 x_i - b_2 x_i^2)^2.$$

Taking the partial derivatives of Q with respect to b_0, b_1, and b_2 leads to the normal equations

$$n\hat{b}_0 + \hat{b}_1 \sum x_i + \hat{b}_2 \sum x_i^2 = \sum y_i$$
$$\hat{b}_0 \sum x_i + \hat{b}_1 \sum x_i^2 + \hat{b}_2 \sum x_i^3 = \sum x_i y_i$$
$$\hat{b}_0 \sum x_i^2 + \hat{b}_1 \sum x_i^3 + \hat{b}_2 \sum x_i^4 = \sum x_i^2 y_i.$$

You are asked below to show that the solutions to these equations are

$$\hat{b}_0 = \bar{y} - \hat{b}_1 \bar{x} - \hat{b}_2 \sum x_i^2/n$$

$$\hat{b}_1 = \frac{\sum x_i y_i - (\sum x_i^3)(\sum x_i^2 y_i)/\sum x_i^4}{\sum x_i^2 - (\sum x_i^3)^2/\sum x_i^4}$$

$$\hat{b}_2 = \frac{\sum x_i^2 y_i - (\sum x_i^3)(\sum x_i y_i)/\sum x_i^2}{\sum x_i^4 - (\sum x_i^3)^2/\sum x_i^2}.$$

It is easy to see that as the number of parameters increases in a linear statistical model, the number of simultaneous equations to be solved for the estimates also increases; the equations will remain linear in the unknowns, so standard matrix algebra techniques can be used to display and study their solutions. The interested reader may consult the book by Larson and the book by Searle in the references for discussions of this methodology.

As in the above case for $m_2(x)$, the least squares estimators of the unknown parameters will be linear functions of the Y_i's. Thus, if the Y_i's are a random sample of a normal random variable, the least squares estimators of the parameters are linear functions of normal random variables and then are themselves normally distributed. This fact again proves useful in deriving confidence intervals for the unknown parameters and in testing hypotheses about their values, much as we saw in the last section in studying $m_1(x)$.

If $E[Y \mid x] = m(x)$ is a nonlinear function of the parameters, then we have what would be called a nonlinear statistical model. Examples of nonlinear statistical models are

$$m_4(x) = b + b^2 x$$
$$m_5(x) = \cos a + x \sin b$$
$$m_6(x) = a e^{bx};$$

the normal equations are generally nonlinear in the unknowns and thus do not admit easy solutions for the estimates. In the case of $m_4(x)$, above, we would want to minimize

$$\sum (y_i - b - b^2 x_i)^2.$$

By taking the derivative with respect to b and setting it equal to zero, we find the equation we need to solve:

$$\hat{b}^3 \sum x_i^2 + \tfrac{3}{2}\hat{b}^2 \sum x_i + \left(\frac{n}{2} - \sum x_i y_i \right) \hat{b} - \tfrac{1}{2} \sum y_i = 0.$$

Such a cubic equation can be solved by standard methods, of course, but the point is that the normal equations themselves become nonlinear, if $m(x)$ is not linear in the parameters, and this frequently leads to technical difficulties.

For a model such as $m_6(x)$ a simple transformation of the y observations will make it a linear model, since, if

$$y = a e^{bx},$$

then

$$\log_e y = \log_e a + bx$$

where the logarithms are taken to the base e. Thus, using $\log_e y_i$ and x_i observations, the formulas given in Section 10.1 can be applied to estimate $\log_e a$ and b. (The resulting values are not identical with the least squares estimates of a and b in the original model, working directly with the y_i's and x_i's; why is this so?) If then it is assumed that the $\log_e Y_i$ values are normally distributed, the methodology for testing hypotheses about a and b and for constructing confidence intervals for a and b may be applied. In such a case the original random variable Y is said to have a lognormal distribution. Exercise 4 below shows the form of this type of distribution.

The number of nonlinear models that could prove useful in applied research is endless, of course. Before the advent of the modern computer, these nonlinear models were generally avoided, or at most approximated by linear models, because of the technical difficulties both in terms of determining the values of the estimates, given a sample, and in terms of describing the properties of such estimates. Today, the computer allows accurate and rapid solutions to nonlinear equations, and may be used in Monte Carlo studies of the properties of such estimates. Thus, nonlinear models are finding more use in many applied fields.

EXERCISE 10.3

1. Verify the normal equations given above for $m_2(x)$.
2. Verify the solution to the normal equations mentioned in 1.
3. Using the data presented in Exercise 10.1.1, assume that

$$E[Y_i] = b_0 + b_1 x_i + b_2 x_i^2$$

   and compute the least squares estimates for b_0, b_1, and b_2.
4. Assume Y is a continuous random variable with density

$$f_Y(y) = \frac{1}{y \sigma \sqrt{2\pi}} e^{-(\log y - \mu)^2 / 2\sigma^2}, \qquad y > 0$$

$$= 0, \qquad\qquad\qquad y \le 0.$$

   Show that $X = \log Y$ has a normal distribution with mean μ and variance σ^2. Y has the *lognormal* distribution.
5. The gross profits of a new toy company for its first six years in business were .820, 1.355, 2.250, 3.716, 6.090 and 10.035, respectively (each total is divided by \$10,000). Let x denote years in business $(1, 2, \ldots, 6)$ and assume that these gross profit figures are the observed values of a random variable Y whose mean is ae^{bx}. Taking logs of the observed y-values, use least squares to estimate $\log a$ and b. Assuming conditions are unchanged, what you would you estimate their gross profit will be in their seventh year?
6. Assume that the distribution of Y in Exercise 5 is lognormal with mean ae^{bx} and variance σ^2, for each x. Use the data presented in Exercise 5 to estimate σ^2 and test the hypothesis that $b = 1/2$ with $\alpha = .05$. (σ^2 is the variance of $\log Y$.)

References

The following books present more detail and more examples of many of the topics discussed in this text. The probability books listed go deeper into the theory of probability, and the mathematical statistics texts delve more deeply into the theory of statistical methods. The methods books present more practical, applied examples without much indication of the theory behind the techniques employed.

Probability Theory

Barr, D. R. and Zehna, P. W. *Probability* (1971). Brooks/Cole Publishing Co., Monterey, Cal.

Feller, William. *An Introduction to Probability and its Applications*, Vol. 1 (1968, third edition) and Vol. 2 (1966). John Wiley, New York.

Parzen, E. *Modern Probability Theory and its Applications* (1960). John Wiley, New York.

Probability and Mathematical Statistics

Anderson, R. L. and T. A. Bancroft. *Statistical Theory in Research* (1952). McGraw-Hill, New York.

Fisz, Marek. *Probability Theory and Mathematical Statistics* (1963, third edition). John Wiley, New York.

Graybill, F. A. *An Introduction to Linear Statistical Models*, Vol. I (1961). McGraw-Hill, New York.

Larson, Harold J. *Introduction to the Theory of Statistics* (1973). John Wiley, New York.

Rao, C. R. *Linear Statistical Inference and its Applications*, Second edition (1973). John Wiley, New York.

Searle, S. R. *Linear Models* (1971). John Wiley, New York.

Wilks, S. S. *Mathematical Statistics* (1962). John Wiley, New York.

Zehna, Peter W. *Probability Distributions and Statistics* (1970). Allyn and Bacon, Inc., Boston.

Methods of Statistics

Brownlee, K. A. *Statistical Theory and Methodology in Science and Engineering* (1965, second edition). John Wiley, New York.

Ostle, Bernard. *Statistics in Research* (1964, second edition). Iowa State Univ. Press, Ames, Iowa.

Appendix: Tables

TABLE 1
Binomial Distribution Function

$$F_X(t) = \sum_{x=0}^{[t]} \binom{n}{x} p^x (1-p)^{n-x}$$

				p		
n	$[t]$	0.10	0.20	0.30	0.40	0.50
2	0	.8100	.6400	.4900	.3600	.2500
	1	.9900	.9600	.9100	.8400	.7500
3	0	.7290	.5120	.3430	2160	.1250
	1	.9720	.8960	.7840	.6480	.5000
	2	.9990	.9920	.9730	.9360	.8750
4	0	.6561	.4096	.2401	.1296	.0625
	1	.9477	.8192	.6517	.4752	.3125
	2	.9963	.9728	.9163	.8208	.6875
	3	.9999	.9984	.9919	.9744	.9375
5	0	.5905	.3277	.1681	.0778	.0312
	1	.9185	.7373	.5282	.3370	.1875
	2	.9914	.9421	.8369	.6826	.5000
	3	.9995	.9933	.9692	.9130	.8125
	4	1.0000	.9997	.9976	.9898	.9688
6	0	.5314	.2621	.1176	.0467	.0156
	1	.8857	.6554	.4202	.2333	.1094

(*continued*)

TABLE 1

Binomial Distribution Function (*continued*)

n	[t]	0.10	0.20	p 0.30	0.40	0.50
6	2	.9842	.9011	.7443	.5443	.3438
	3	.9987	.9830	.9295	.8208	.6562
	4	.9999	.9984	.9891	.9590	.8906
	5	1.0000	.9999	.9993	.9959	.9844
7	0	.4783	.2097	.0824	.0280	.0078
	1	.8503	.5767	.3294	.1586	.0625
	2	.9743	.8520	.6471	.4199	.2266
	3	.9973	.9667	.8740	.7102	.5000
	4	.9998	.9953	.9712	.9037	.7734
	5	1.0000	.9996	.9962	.9812	.9375
	6	1.0000	1.0000	.9998	.9984	.9922
8	0	.4305	.1678	.0576	.0168	.0039
	1	.8131	.5033	.2553	.1064	.0352
	2	.9619	.7969	.5518	.3154	.1445
	3	.9950	.9437	.8059	.5941	.3633
	4	.9996	.9896	.9420	.8263	.6367
	5	1.0000	.9988	.9887	.9502	.8555
	6	1.0000	.9999	.9987	.9915	.9648
	7	1.0000	1.0000	.9999	.9993	.9961
9	0	.3874	.1342	.0404	.0101	.0020
	·1	.7748	.4362	.1960	.0705	.0195
	2	.9470	.7382	.4628	.2318	.0898
	3	.9917	.9144	.7297	.4826	.2539
	4	.9991	.9804	.9012	.7334	.5000
	5	.9999	.9969	.9747	.9006	.7461
	6	1.0000	.9997	.9957	.9750	.9102
	7	1.0000	1.0000	.9996	.9962	.9805
	8	1.0000	1.0000	1.0000	.9997	.9980
10	0	.3487	.1074	.0282	.0060	.0010
	1	.7361	.3758	.1493	.0464	.0107
	2	.9298	.6778	.3828	.1673	.0547
	3	.9872	.8791	.6496	.3823	.1719
	4	.9984	.9672	.8497	.6331	.3770

TABLE 1
Binomial Distribution Function (*continued*)

n	$[t]$	0.10	0.20	p 0.30	0.40	0.50
10	5	.9999	.9936	.9527	.8338	.6230
	6	1.0000	.9991	.9894	.9452	.8281
	7	1.0000	.9999	.9984	.9877	.9453
	8	1.0000	1.0000	.9999	.9983	.9893
	9	1.0000	1.0000	1.0000	.9999	.9990
11	0	.3138	.0859	.0198	.0036	.0005
	1	.6974	.3221	.1130	.0302	.0059
	2	.9104	.6174	.3127	.1189	.0327
	3	.9815	.8389	.5696	.2963	.1133
	4	.9972	.9496	.7897	.5328	.2744
	5	.9997	.9883	.9218	.7535	.5000
	6	1.0000	.9980	.9784	.9006	.7256
	7	1.0000	.9998	.9957	.9707	.8867
	8	1.0000	1.0000	.9994	.9941	.9673
	9	1.0000	1.0000	1.0000	.9993	.9941
	10	1.0000	1.0000	1.0000	1.0000	.9995
12	0	.2824	.0687	.0138	.0022	.0002
	1	.6590	.2749	.0850	.0196	.0032
	2	.8891	.5583	.2528	.0834	.0193
	3	.9744	.7946	.4925	.2253	.0730
	4	.9957	.9274	.7237	.4382	.1938
	5	.9995	.9806	.8822	.6652	.3872
	6	.9999	.9961	.9614	.8418	.6128
	7	1.0000	.9994	.9905	.9427	.8062
	8	1.0000	.9999	.9983	.9847	.9270
	9	1.0000	1.0000	.9998	.9972	.9807
	10	1.0000	1.0000	1.0000	.9997	.9968
	11	1.0000	1.0000	1.0000	1.0000	.9998
13	0	.2542	.0550	.0097	.0013	.0001
	1	.6213	.2336	.0637	.0126	.0017
	2	.8661	.5017	.2025	.0579	.0112
	3	.9658	.7473	.4206	.1686	.0461
	4	.9935	.9009	.6543	.3530	.1334
	5	.9991	.9700	.8346	.5744	.2905
	6	.9999	.9930	.9376	.7712	.5000
	7	1.0000	.9988	.9818	.9023	.7095

(*continued*)

TABLE 1
Binomial Distribution Function (*continued*)

n	$[t]$	0.10	0.20	p 0.30	0.40	0.50
13	8	1.0000	.9998	.9960	.9679	.8666
	9	1.0000	1.0000	.9993	.9922	.9539
	10	1.0000	1.0000	.9999	.9987	.9888
	11	1.0000	1.0000	1.0000	.9999	.9983
	12	1.0000	1.0000	1.0000	1.0000	.9999
14	0	.2288	.0440	.0068	.0008	.0001
	1	.5846	.1979	.0475	.0081	.0009
	2	.8416	.4481	.1608	.0398	.0065
	3	.9559	.6982	.3552	.1243	.0287
	4	.9908	.8702	.5842	.2793	.0898
	5	.9985	.9561	.7805	.4859	.2120
	6	.9998	.9884	.9067	.6925	.3953
	7	1.0000	.9976	.9685	.8499	.6047
	8	1.0000	.9996	.9917	.9417	.7880
	9	1.0000	1.0000	.9983	.9825	.9102
	10	1.0000	1.0000	.9998	.9961	.9713
	11	1.0000	1.0000	1.0000	.9994	.9935
	12	1.0000	1.0000	1.0000	.9999	.9991
	13	1.0000	1.0000	1.0000	1.0000	.9999
15	0	.2059	.0352	.0047	.0005	.0000
	1	.5490	.1671	.0353	.0052	.0005
	2	.8159	.3980	.1268	.0271	.0037
	3	.9444	.6482	.2969	.0905	.0176
	4	.9873	.8358	.5155	.2173	.0592
	5	.9978	.9389	.7216	.4032	.1509
	6	.9997	.9819	.8689	.6098	.3036
	7	1.0000	.9958	.9500	.7869	.5000
	8	1.0000	.9992	.9848	.9050	.6964
	9	1.0000	.9999	.9963	.9662	.8491
	10	1.0000	1.0000	.9993	.9907	.9408
	11	1.0000	1.0000	.9999	.9981	.9824
	12	1.0000	1.0000	1.0000	.9997	.9963
	13	1.0000	1.0000	1.0000	1.0000	.9995
	14	1.0000	1.0000	1.0000	1.0000	1.0000

TABLE 1
Binomial Distribution Function (*continued*)

n	[t]	0.10	0.20	p 0.30	0.40	0.50
16	0	.1853	.0281	.0033	.0003	.0000
	1	.5147	.1407	.0261	.0033	.0003
	2	.7892	.3518	.0994	.0183	.0021
	3	.9316	.5981	.2459	.0651	.0106
	4	.9830	.7982	.4499	.1666	.0384
	5	.9967	.9183	.6598	.3288	.1051
	6	.9995	.9733	.8247	.5272	.2272
	7	.9999	.9930	.9256	.7161	.4018
	8	1.0000	.9985	.9743	.8577	.5982
	9	1.0000	.9998	.9929	.9417	.7728
	10	1.0000	1.0000	.9984	.9809	.8949
	11	1.0000	1.0000	.9997	.9951	.9616
	12	1.0000	1.0000	1.0000	.9991	.9894
	13	1.0000	1.0000	1.0000	.9999	.9979
	14	1.0000	1.0000	1.0000	1.0000	.9997
	15	1.0000	1.0000	1.0000	1.0000	1.0000
17	0	.1668	.0225	.0023	.0002	.0000
	1	.4818	.1182	.0193	.0021	.0001
	2	.7618	.3096	.0774	.0123	.0012
	3	.9174	.5489	.2019	.0464	.0064
	4	.9779	.7582	.3887	.1260	.0245
	5	.9953	.8943	.5968	.2639	.0717
	6	.9992	.9623	.7752	.4478	.1662
	7	.9999	.9891	.8954	.6405	.3145
	8	1.0000	.9974	.9597	.8011	.5000
	9	1.0000	.9995	.9873	.9081	.6855
	10	1.0000	.9999	.9968	.9652	.8338
	11	1.0000	1.0000	.9993	.9894	.9283
	12	1.0000	1.0000	.9999	.9975	.9755
	13	1.0000	1.0000	1.0000	.9995	.9936
	14	1.0000	1.0000	1.0000	.9999	.9988
	15	1.0000	1.0000	1.0000	1.0000	.9999
	16	1.0000	1.0000	1.0000	1.0000	1.0000

(*continued*)

TABLE 1
Binomial Distribution Function (*continued*)

n	$[t]$	0.10	0.20	p 0.30	0.40	0.50
18	0	.1501	.0180	.0016	.0001	.0000
	1	.4503	.0991	.0142	.0013	.0001
	2	.7338	.2713	.0600	.0082	.0007
	3	.9018	.5010	.1646	.0328	.0038
	4	.9718	.7164	.3327	.0942	.0154
	5	.9936	.8671	.5344	.2088	.0481
	6	.9988	.9487	.7217	.3743	.1189
	7	.9998	.9837	.8593	.5634	.2403
	8	1.0000	.9957	.9404	.7368	.4073
	9	1.0000	.9991	.9790	.8653	.5927
	10	1.0000	.9998	.9939	.9424	.7597
	11	1.0000	1.0000	.9986	.9797	.8811
	12	1.0000	1.0000	.9997	.9942	.9519
	13	1.0000	1.0000	1.0000	.9987	.9846
	14	1.0000	1.0000	1.0000	.9998	.9962
	15	1.0000	1.0000	1.0000	1.0000	.9993
	16	1.0000	1.0000	1.0000	1.0000	.9999
	17	1.0000	1.0000	1.0000	1.0000	1.0000
19	0	.1351	.0144	.0011	.0001	.0000
	1	.4203	.0829	.0104	.0008	.0000
	2	.7054	.2369	.0462	.0055	.0004
	3	.8850	.4551	.1332	.0230	.0022
	4	.9648	.6733	.2822	.0696	.0096
	5	.9914	.8369	.4739	.1629	.0318
	6	.9983	.9324	.6655	.3081	.0835
	7	.9997	.9767	.8180	.4878	.1796
	8	1.0000	.9933	.9161	.6675	.3238
	9	1.0000	.9984	.9674	.8139	.5000
	10	1.0000	.9997	.9895	.9115	.6762
	11	1.0000	1.0000	.9972	.9648	.8204
	12	1.0000	1.0000	.9994	.9884	.9165
	13	1.0000	1.0000	.9999	.9969	.9682
	14	1.0000	1.0000	1.0000	.9994	.9904
	15	1.0000	1.0000	1.0000	.9999	.9978
	16	1.0000	1.0000	1.0000	1.0000	.9996
	17	1.0000	1.0000	1.0000	1.0000	1.0000

TABLE 1
Binomial Distribution Function (*continued*)

n	[t]	0.10	0.20	p 0.30	0.40	0.50
20	0	.1216	.0115	.0008	.0000	.0000
	1	.3917	.0692	.0076	.0005	.0000
	2	.6769	.2061	.0355	.0036	.0002
	3	.8670	.4114	.1071	.0160	.0013
	4	.9568	.6296	.2375	.0510	.0059
	5	.9887	.8042	.4164	.1256	.0207
	6	.9976	.9133	.6080	.2500	.0577
	7	.9996	.9679	.7723	.4159	.1316
	8	.9999	.9900	.8867	.5956	.2517
	9	1.0000	.9974	.9520	.7553	.4119
	10	1.0000	.9994	.9829	.8725	.5881
	11	1.0000	.9999	.9949	.9435	.7483
	12	1.0000	1.0000	.9987	.9790	.8684
	13	1.0000	1.0000	.9997	.9935	.9423
	14	1.0000	1.0000	1.0000	.9984	.9793
	15	1.0000	1.0000	1.0000	.9997	.9941
	16	1.0000	1.0000	1.0000	1.0000	.9987
	17	1.0000	1.0000	1.0000	1.0000	.9998
	18	1.0000	1.0000	1.0000	1.0000	1.0000

TABLE 2
Poisson Distribution Function

$$F_X(t) = \sum_{x=0}^{[t]} \frac{\lambda^x}{x!} e^{-\lambda}$$

[t]	.50	1.0	2.0	3.0	4.0	5.0	6.0	7.0	8.0	9.0
0	.607	.368	.135	.050	.018	.007	.002	.001	.000	.000
1	.910	.736	.406	.199	.092	.040	.017	.007	.003	.001
2	.986	.920	.677	.423	.238	.125	.062	.030	.014	.006
3	.998	.981	.857	.647	.433	.265	.151	.082	.042	.021
4	1.000	.996	.947	.815	.629	.440	.285	.173	.100	.055
5	1.000	.999	.983	.961	.785	.616	.446	.301	.191	.116
6	1.000	1.000	.995	.966	.889	.762	.606	.450	.313	.207
7	1.000	1.000	.999	.988	.949	.867	.744	.599	.453	.324
8	1.000	1.000	1.000	.996	.979	.932	.847	.729	.593	.456
9	1.000	1.000	1.000	.999	.992	.968	.916	.830	.717	.587
10	1.000	1.000	1.000	1.000	.997	.986	.957	.901	.816	.706
11	1.000	1.000	1.000	1.000	.999	.995	.980	.947	.888	.803
12	1.000	1.000	1.000	1.000	1.000	.998	.991	.973	.936	.876
13	1.000	1.000	1.000	1.000	1.000	.999	.996	.987	.966	.926
14	1.000	1.000	1.000	1.000	1.000	1.000	.999	.994	.983	.959
15	1.000	1.000	1.000	1.000	1.000	1.000	.999	.998	.992	.978
16	1.000	1.000	1.000	1.000	1.000	1.000	1.000	.999	.996	.989
17	1.000	1.000	1.000	1.000	1.000	1.000	1.000	1.000	.998	.995
18	1.000	1.000	1.000	1.000	1.000	1.000	1.000	1.000	.999	.998
19	1.000	1.000	1.000	1.000	1.000	1.000	1.000	1.000	1.000	.999
20	1.000	1.000	1.000	1.000	1.000	1.000	1.000	1.000	1.000	1.000

TABLE 2
Poisson Distribution Function (*continued*)

[r]	λ					
	10.0	11.0	12.0	13.0	14.0	15.0
2	.003	.001	.001	.000	.000	.000
3	.010	.005	.002	.001	.000	.000
4	.029	.015	.008	.004	.002	.001
5	.067	.038	.020	.011	.006	.003
6	.130	.079	.046	.026	.014	.008
7	.220	.143	.090	.054	.032	.018
8	.333	.232	.155	.100	.062	.037
9	.458	.341	.242	.166	.109	.070
10	.583	.460	.347	.252	.176	.118
11	.697	.579	.462	.353	.260	.185
12	.792	.689	.576	.463	.358	.268
13	.864	.781	.682	.573	.464	.363
14	.917	.854	.772	.675	.570	.466
15	.951	.907	.844	.764	.669	.568
16	.973	.944	.899	.835	.756	.664
17	.986	.968	.937	.890	.827	.749
18	.993	.982	.963	.930	.883	.819
19	.997	.991	.979	.957	.923	.875
20	.998	.995	.988	.975	.952	.917
21	.999	.998	.994	.986	.971	.947
22	1.000	.999	.997	.992	.983	.967
23	1.000	1.000	.999	.996	.991	.981
24	1.000	1.000	.999	.998	.995	.989
25	1.000	1.000	1.000	.999	.997	.994
26	1.000	1.000	1.000	1.000	.999	.997
27	1.000	1.000	1.000	1.000	.999	.998
28	1.000	1.000	1.000	1.000	1.000	.999
29	1.000	1.000	1.000	1.000	1.000	1.000

Abridged with permission from E. C. Molina, *Poisson's Exponential Binomial Limit*, D. Van Nostrand, 1949.

TABLE 3
Standard Normal Distribution Function

$$N_Z(t) = \int_{-\infty}^{t} \frac{1}{\sqrt{2\pi}} e^{-x^2/2} \, dx$$

t	0	1	2	3	4	5	6	7	8	9
−3.	.0013									
−2.9	.0019	.0018	.0017	.0017	.0016	.0016	.0015	.0015	.0014	.0014
−2.8	.0026	.0025	.0024	.0023	.0023	.0022	.0021	.0021	.0020	.0019
−2.7	.0035	.0034	.0033	.0032	.0031	.0030	.0029	.0028	.0027	.0026
−2.6	.0047	.0045	.0044	.0043	.0041	.0040	.0039	.0038	.0037	.0036
−2.5	.0062	.0060	.0059	.0057	.0055	.0054	.0052	.0051	.0049	.0048
−2.4	.0082	.0080	.0078	.0075	.0073	.0071	.0069	.0068	.0066	.0064
−2.3	.0107	.0104	.0102	.0099	.0096	.0094	.0091	.0089	.0087	.0084
−2.2	.0139	.0136	.0132	.0129	.0125	.0122	.0119	.0116	.0113	.0110
−2.1	.0179	.0174	.0170	.0166	.0162	.0158	.0154	.0150	.0146	.0143
−2.0	.0227	.0222	.0217	.0212	.0207	.0202	.0197	.0192	.0188	.0183
−1.9	.0287	.0281	.0274	.0268	.0262	.0256	.0250	.0244	.0239	.0233
−1.8	.0359	.0351	.0344	.0336	.0329	.0322	.0314	.0307	.0300	.0294
−1.7	.0446	.0436	.0427	.0418	.0409	.0401	.0392	.0384	.0375	.0367
−1.6	.0548	.0537	.0526	.0516	.0505	.0495	.0485	.0475	.0465	.0455
−1.5	.0668	.0655	.0643	.0630	.0618	.0606	.0594	.0582	.0571	.0559
−1.4	.0808	.0793	.0778	.0764	.0749	.0735	.0721	.0708	.0694	.0681
−1.3	.0968	.0951	.0934	.0918	.0901	.0885	.0869	.0853	.0838	.0823
−1.2	.1151	.1131	.1112	.1093	.1075	.1056·	.1038	.1020	.1003	.0985
−1.1	.1357	.1335	.1314	.1292	.1271	.1251	.1230	.1210	.1190	.1170
−1.0	.1587	.1562	.1539	.1515	.1492	.1469	.1446	.1423	.1401	.1379
−.9	.1841	.1814	.1788	.1762	.1736	.1711	.1685	.1660	.1635	.1611
−.8	.2119	.2090	.2061	.2033	.2005	.1977	.1949	.1921	.1894	.1867
−.7	.2420	.2389	.2358	.2326	.2297	.2266	.2236	.2206	.2177	.2148
−.6	.2743	.2709	.2676	.2643	.2611	.2578	.2546	.2514	.2483	.2451
−.5	.3085	.3050	.3015	.2981	.2946	.2912	.2877	.2843	.2810	.2776
−.4	.3446	.3409	.3372	.3336	.3300	.3264	.3228	.3192	.3156	.3121
−.3	.3821	.3783	.3745	.3707	.3669	.3632	.3594	.3557	.3520	.3483
−.2	.4207	.4168	.4129	.4090	.4052	.4013	.3974	.3936	.3897	.3859
−.1	.4602	.4562	.4522	.4483	.4443	.4404	.4364	.4325	.4286	.4247
−.0	.5000	.4960	.4920	.4880	.4840	.4801	.4761	.4721	.4681	.4641

TABLE 3
Standard Normal Distribution Function (*continued*)

t	0	1	2	3	4	5	6	7	8	9
.0	.5000	.5040	.5080	.5120	.5160	.5199	.5239	.5279	.5319	.5359
.1	.5398	.5438	.5478	.5517	.5557	.5596	.5636	.5675	.5714	.5753
.2	.5793	.5832	.5871	.5910	.5948	.5987	.6026	.6064	.6103	.6141
.3	.6179	.6217	.6255	.6293	.6331	.6368	.6406	.6443	.6480	.6517
.4	.6554	.6591	.6628	.6664	.6700	.6736	.6772	.6808	.6844	.6879
.5	.6915	.6950	.6985	.7019	.7054	.7088	.7123	.7157	.7190	.7224
.6	.7257	.7291	.7324	.7357	.7389	.7422	.7454	.7486	.7517	.7549
.7	.7580	.7611	.7642	.7673	.7704	.7734	.7764	.7794	.7823	.7852
.8	.7881	.7910	.7939	.7967	.7995	.8023	.8051	.8079	.8106	.8133
.9	.8159	.8186	.8212	.8238	.8264	.8289	.8315	.8340	.8365	.8389
1.0	.8413	.8438	.8461	.8485	.8508	.8531	.8554	.8577	.8599	.8621
1.1	.8643	.8665	.8686	.8708	.8729	.8749	.8770	.8790	.8810	.8830
1.2	.8849	.8869	.8888	.8907	.8925	.8944	.8962	.8980	.8997	.9015
1.3	.9032	.9049	.9066	.9082	.9099	.9115	.9131	.9147	.9162	.9177
1.4	.9192	.9207	.9222	.9236	.9251	.9265	.9279	.9292	.9306	.9319
1.5	.9332	.9345	.9357	.9370	.9382	.9394	.9406	.9418	.9429	.9441
1.6	.9452	.9463	.9474	.9484	.9495	.9505	.9515	.9525	.9535	.9545
1.7	.9554	.9564	.9573	.9582	.9591	.9599	.9608	.9616	.9625	.9633
1.8	.9641	.9649	.9656	.9664	.9671	.9678	.9686	.9693	.9700	.9706
1.9	.9713	.9719	.9726	.9732	.9738	.9744	.9750	.9756	.9761	.9767
2.0	.9773	.9778	.9783	.9788	.9793	.9798	.9803	.9808	.9812	.9817
2.1	.9821	.9826	.9830	.9834	.9838	.9842	.9846	.9850	.9854	.9857
2.2	.9861	.9864	.9868	.9871	.9875	.9878	.9881	.9884	.9887	.9890
2.3	.9893	.9896	.9898	.9901	.9904	.9906	.9909	.9911	.9913	.9916
2.4	.9918	.9920	.9922	.9925	.9927	.9929	.9931	.9932	.9934	.9936
2.5	.9938	.9940	.9941	.9943	.9945	.9946	.9948	.9949	.9951	.9952
2.6	.9953	.9955	.9956	.9957	.9959	.9960	.9961	.9962	.9963	.9964
2.7	.9965	.9966	.9967	.9968	.9969	.9970	.9971	.9972	.9973	.9974
2.8	.9974	.9975	.9976	.9977	.9977	.9978	.9979	.9979	.9980	.9981
2.9	.9981	.9982	.9982	.9983	.9984	.9984	.9985	.9985	.9986	.9986
3.	.9987									

TABLE 4

χ^2 Distribution Function

$$F_{\chi^2}(t) = \int_0^t \frac{1}{2^{d/2} \, \Gamma(d/2)} x^{(d-2)/2} e^{-x/2} \, dx = 1 - \alpha$$

Degrees of freedom = d

$1 - \alpha$

d	.005	.010	.025	.050	.100	.250	.500	.750	.900	.950	.975	.990	.995
1	.0000393	.000157	.000982	.00393	.0158	.102	.455	1.32	2.71	3.84	5.02	6.63	7.88
2	.0100	.0201	.0506	.103	.211	.575	1.39	2.77	4.61	5.99	7.38	9.21	10.6
3	.0717	.115	.216	.352	.584	1.21	2.37	4.11	6.25	7.81	9.35	11.3	12.8
4	.207	.297	.484	.711	1.06	1.92	3.36	5.39	7.78	9.49	11.1	13.3	14.9
5	.412	.554	.831	1.15	1.61	2.67	4.35	6.63	9.24	11.1	12.8	15.1	16.7
6	.676	.872	1.24	1.64	2.20	3.45	5.35	7.84	10.6	12.6	14.4	16.8	18.5
7	.989	1.24	1.69	2.17	2.83	4.25	6.35	9.04	12.0	14.1	16.0	18.5	20.3
8	1.34	1.65	2.18	2.73	3.49	5.07	7.34	10.2	13.4	15.5	17.5	20.1	22.0
9	1.73	2.09	2.70	3.33	4.17	5.90	8.34	11.4	14.7	16.9	19.0	21.7	23.6
10	2.16	2.56	3.25	3.94	4.87	6.74	9.34	12.5	16.0	18.3	20.5	23.2	25.2
11	2.60	3.05	3.82	4.57	5.58	7.58	10.3	13.7	17.3	19.7	21.9	24.7	26.8
12	3.07	3.57	4.40	5.23	6.30	8.44	11.3	14.8	18.5	21.0	23.3	26.2	28.3
13	3.57	4.11	5.01	5.89	7.04	9.30	12.3	16.0	19.8	22.4	24.7	27.7	29.8
14	4.07	4.66	5.63	6.57	7.79	10.2	13.3	17.1	21.1	23.7	26.1	29.1	31.3
15	4.60	5.23	6.26	7.26	8.55	11.0	14.3	18.2	22.3	25.0	27.5	30.6	32.8

df													
16	5.14	5.81	6.91	7.96	9.31	11.9	15.3	19.4	23.5	26.3	28.8	32.0	34.3
17	5.70	6.41	7.56	8.67	10.1	12.8	16.3	20.5	24.8	27.6	30.2	33.4	35.7
18	6.26	7.01	8.23	9.39	10.9	13.7	17.3	21.6	26.0	28.9	31.5	34.8	37.2
19	6.84	7.63	8.91	10.1	11.7	14.6	18.3	22.7	27.2	30.1	32.9	36.2	38.6
20	7.43	8.26	9.59	10.9	12.4	15.5	19.3	23.8	28.4	31.4	34.2	37.6	40.0
21	8.03	8.90	10.3	11.6	13.2	16.3	20.3	24.9	29.6	32.7	35.5	38.9	41.4
22	8.64	9.54	11.0	12.3	14.0	17.2	21.3	26.0	30.8	33.9	36.8	40.3	42.8
23	9.26	10.2	11.7	13.1	14.8	18.1	22.3	27.1	32.0	35.2	38.1	41.6	44.2
24	9.89	10.9	12.4	13.8	15.7	19.0	23.3	28.2	33.2	36.4	39.4	43.0	45.6
25	10.5	11.5	13.1	14.6	16.5	19.9	24.3	29.3	34.4	37.7	40.6	44.3	46.9
26	11.2	12.2	13.8	15.4	17.3	20.8	25.3	30.4	35.6	38.9	41.9	45.6	48.3
27	11.8	12.9	14.6	16.2	18.1	21.7	26.3	31.5	36.7	40.1	43.2	47.0	49.6
28	12.5	13.6	15.3	16.9	18.9	22.7	27.3	32.6	37.9	41.3	44.5	48.3	51.0
29	13.1	14.3	16.0	17.7	19.8	23.6	28.3	33.7	39.1	42.6	45.7	49.6	52.3
30	13.8	15.0	16.8	18.5	20.6	24.5	29.3	34.8	40.3	43.8	47.0	50.9	53.7

Abridged with permission from Catherine M. Thompson, "Tables of percentage points of the incomplete Beta function and of the chi-square distribution," *Biometrika*, **32**, 1941.

TABLE 5
Student's t Distribution Function

$$F_t(x) = \int_{-\infty}^{x} \frac{\Gamma((d+1)/2)}{\sqrt{d\pi}\,\Gamma(d/2)} (1 + u^2)^{-(d+1)/2}\, du = 1 - \alpha$$

| | | | | Degrees of freedom $= d$ | | | | |
| | | | | $1 - \alpha$ | | | | |
d	.60	.75	.90	.95	.975	.99	.995	.9995
1	.325	1.000	3.078	6.314	12.706	31.821	63.657	636.619
2	.289	.816	1.886	2.920	4.303	6.965	9.925	31.598
3	.277	.765	1.638	2.353	3.182	4.541	5.841	12.941
4	.271	.741	1.533	2.132	2.776	3.747	4.604	8.610
5	.267	.727	1.476	2.015	2.571	3.365	4.032	6.859
6	.265	.718	1.440	1.943	2.447	3.143	3.707	5.959
7	.263	.711	1.415	1.895	2.365	2.998	3.499	5.405
8	.262	.706	1.397	1.860	2.306	2.896	3.355	5.041
9	.261	.703	1.383	1.833	2.262	2.821	3.250	4.781
10	.260	.700	1.372	1.812	2.228	2.764	3.169	4.587
11	.260	.697	1.363	1.796	2.201	2.718	3.106	4.437
12	.259	.695	1.356	1.782	2.179	2.681	3.055	4.318
13	.259	.694	1.350	1.771	2.160	2.650	3.012	4.221
14	.258	.692	1.345	1.761	2.145	2.624	2.977	4.140
15	.258	.691	1.341	1.753	2.131	2.602	2.947	4.073
16	.258	.690	1.337	1.746	2.120	2.583	2.921	4.015
17	.257	.689	1.333	1.740	2.110	2.567	2.898	3.965
18	.257	.688	1.330	1.734	2.101	2.552	2.878	3.922
19	.257	.688	1.328	1.729	2.093	2.539	2.861	3.883
20	.257	.687	1.325	1.725	2.086	2.528	2.845	3.850
21	.257	.686	1.323	1.721	2.080	2.518	2.831	3.819
22	.256	.686	1.321	1.717	2.074	2.508	2.819	3.792
23	.256	.685	1.319	1.714	2.069	2.500	2.807	3.767
24	.256	.685	1.318	1.711	2.064	2.492	2.797	3.745
25	.256	.684	1.316	1.708	2.060	2.485	2.787	3.725
26	.256	.684	1.315	1.706	2.056	2.479	2.779	3.707
27	.256	.684	1.314	1.703	2.052	2.473	2.771	3.690
28	.256	.683	1.313	1.701	2.048	2.467	2.763	3.674
29	.256	.683	1.311	1.699	2.045	2.462	2.756	3.659
30	.256	.683	1.310	1.697	2.042	2.457	2.750	3.646
40	.255	.681	1.303	1.684	2.021	2.423	2.704	3.551
60	.254	.679	1.296	1.671	2.000	2.390	2.660	3.460
120	.254	.677	1.289	1.658	1.980	2.358	2.617	3.373
∞	.253	.674	1.282	1.645	1.960	2.326	2.576	3.291

Abridged with permission from R. A. Fisher and Frank Yates, *Statistical Tables*, Oliver and Boyd, Ltd., Edinburgh, 1938.

Answers to Exercises

Chapter 1

EXERCISE 1.1

5. (a) False.   (b) False.   (c) True.   (d) True.   (e) True.
   (f) True.   (g) False.   (h) True.   (i) True.   (j) False.
   (k) False.   (l) False.   (m) False.   (n) False.

6. (a) True.   (b) False.   (c) False.   (d) True.   (e) False.
   (f) True.   (g) False.

7. Yes, no, no.

8. (a) Definition of family must be specified.   (b) Element.

9. No, How are partially assembled autos counted?

10. No.

EXERCISE 1.2

2. $\{x : 0 \leq x \leq 1\}$, $\{0, \frac{1}{2}, 1\}$, B, $\varnothing$, $\varnothing$, C.

4. $F \subset E$.   5. $F \subset E$.

6. (a) B.   (b) A.
   (c) $\{x : x = 0, 1, 2, \ldots, 10\}$.   (d) $\{1, 2, 3, 4, 5, 6\}$.
   (e) $\{0, 1, 2, 3, 4, 5, 6, 7, 8, 9, 10, 20, 30\}$.
   (f) $\{10\}$.   (g) $\{x : x = 0 \text{ or } 1 \leq x \leq 10\}$.
   (h) $\{1, 2, 3, 4, 5, 6\}$.   (i) $\{x : x = 0, 20, 30 \text{ or } 1 \leq x \leq 10\}$.
   (j) $\{10\}$.   (k) $\{0, 1, 2, 3, 4, 5, 6, 10, 20, 30\}$.
   (l) $\{0\}$.   (m) $\{x : x = 0 \text{ or } 1 \leq x \leq 10\}$.

(n) $\{1, 2, 3, 4, 5, 6\}$. (o) same as (e).

(p) same as (n). (q) $\{0, 1, 2, 3, 4, 5, 6, 10\}$.

(r) B.

8. Safe crackers in prison. 9. All sociologists in the U.S.

10. $a = 1, b = c = 0$, plus various other 0, 1 assignments.

11. County, state.

12. $\{4, 5, 6, \ldots, n\}, \{1, 4, 5, 6, \ldots, n - 1\}, \{1, 4, 5, 6, \ldots, n\}$, same.

13. No, No.

18. $A \times B = \{(1, 2), \ (1, 1), \ (2, 2), \ (2, 1)\}, \ A \times C = \{(1, 10), \ (1, 12), \ (2, 10),$ $(2, 12)\} = B \times C, \ B \times A = A \times B, \ C \times A = C \times B = \{(10, 1), \ (10, 2),$ $(12, 1), \ (12, 2)\}, \ A \times A = A \times B = B \times B, \ C \times C = \{(10, 10), \ (10, 12),$ $(12, 10), \ (12, 12)\}, \ A \times B \times C = \{(1, 2, 10), (1, 2, 12), (1, 1, 10), (1, 1,12),$ $(2, 2, 10), \ (2, 2, 12), \ (2, 1, 10), \ (2, 1, 12)\}, \ C \times B \times A = C \times A \times B =$ $\{(10, 2, 1), \ (10, 2, 2), \ (10, 1, 1), \ (10, 1, 2), \ (12, 2, 1), \ (12, 2, 2), \ (12, 1, 1),$ $(12, 1, 2)\}$.

19. $A = B$. 20. No. 21. No.

22. $\varnothing$. 26. $U \times U$.

28. $(\bar{A} \times \bar{B}) \cup (\bar{A} \times B) \cup (A \times \bar{B})$.

EXERCISE 1.3

2. $R_g = \{(2, 3, 4, \ldots, 12\}$. 3. $R_h = \{(0, 1, 2, 3\}, 1, 3$.

4. $\{\frac{1}{4}, \frac{1}{3}, \frac{1}{2}, \frac{2}{3}, \frac{3}{4}, 1, \frac{4}{3}, \frac{3}{2}, 2, 3, 4\}$. 5. $\{0, \pm 1, \pm 2, \pm 3, \pm 4, \pm 5\}$.

6. $\{0, 1, 2, 3, 4, 5\}$. 7. $\{0, 1\}, 4$.

8. $\{0, \frac{1}{5}, \frac{2}{5}, \frac{3}{5}, \frac{4}{5}, 1\}$. 9. $\{\frac{1}{4}, \frac{1}{3}, \frac{1}{2}, 1, 2\}$.

Chapter 2

EXERCISE 2.1

1. $\{x : x = 1, 2, \ldots, 10\}$.

2. $\{(x, y) : x = 1, 2, \ldots, 10; \ y = 1, 2, \ldots, 10\}$.

3. $\{(x, y) : x = 1, 2, \ldots, 10; \ y = 1, 2, \ldots, 10; \ x \neq y\}$.

4. White: $1, 2, 3, 4$; Red: $5, 6, \ldots, 10$; $A = \{1, 2, 3, 4\}, B = \{5, 6, \ldots, 10\}$.

5. Same numbering as 4;
$C = \{(x, y) : x = 1, 2, 3, 4; \ y = 1, 2, \ldots, 10\}$;
$D = \{(x, y) : x = 1, 2, \ldots, 10; \ y = 1, 2, 3, 4\}$;
$E = \{(x, y) : x = 1, 2, 3, 4; \ y = 1, 2, 3, 4\}$; Yes.

6. $\{(x, y): x = a, b, c, d, e; \ y = a, b, c, d, e\}$.

7. $S = \{(x, y): x = \text{bald, brown, black}; \ y = \text{blue, brown}\}$;
   $A = \{(\text{bald, blue}), (\text{bald, brown})\}$; $B = \{(\text{bald, blue}), (\text{brown, blue}),$
   $(\text{black, blue})\}$; $C = \{(\text{brown, brown})\}$.

8. $S = \{(\text{Marie, Sandy}), (\text{Marie, Tina}), (\text{Sandy, Tina}), (\text{Sandy, Marie}),$
        $(\text{Tina, Marie}), (\text{Tina, Sandy})\}$;
   $A = \{(\text{Marie, Sandy}), (\text{Marie, Tina})\}$;
   $B = \{(\text{Sandy, Marie}), (\text{Tina, Marie})\}$; $\overline{A \cup B}$.

9. $S = \{(x_1, x_2, x_3): x_i = \text{red, blue, black, green}; \ i = 1, 2, 3\}$;
   $A = \{(\text{red, red, red,})\}$;
   $B = \{(x_1, x_2, x_3): \text{ exactly one } x_i \text{ is red, one is green and one is blue}\}$,
   $C = \{(x_1, x_2, x_3): x_1 \neq x_2, x_1 \neq x_3, x_2 \neq x_3\}$;
   $D = \varnothing$.

10. $S = \{(x_1, x_2, x_3, x_4): x_i = 1, 2, 3, \ i = 1, 2, 3, 4\}$;
    $A = \{(1, 1, 1, 1)\}$;
    $B = \{(x_1, x_2, x_3, x_4): \ 2 \ x_i\text{'s are } 1, 2 \ x_i\text{'s are } 2\}$;
    $C = \{(x_1, x_2, x_3, x_4): \text{ three different numbers occur in } x_1, x_2, x_3, x_4\}$.

11. $S = \{(x_1, x_2, x_3): x_i = 1, 2, \ldots, 9; \ i = 1, 2, 3; \text{ no two } x_i\text{'s are equal}\}$;
    $A = \{(5, x_2, x_3): x_i = 1, 2, \ldots, 9; \ i = 2, 3; \ x_2 \neq x_3, x_2 \neq 5, x_3 \neq 5\}$;
    $B = \{(x_1, x_2, 5): x_i = 1, 2, \ldots, 9; \ i = 1, 2; \ x_1 \neq x_2, x_1 \neq 5, x_2 \neq 5\}$;
    $C = \{(x_1, x_2, x_3): (x_1, x_2, x_3) \in S \text{ and } x_i \neq 5, i = 1, 2, 3\}$.

12. $S = \{(x_1, x_2): \ 0 \leq x_1 \leq 1600, 0 \leq x_2 \leq 1600\}$;
    $A = \{(x_1, x_2): \ 0 \leq x_1 \leq 1000, 0 \leq x_2 \leq 1000\}$;
    $B = \{(x_1, x_2): \ 1000 \leq x_1 \leq 1600, 1000 \leq x_2 \leq 1600\}$;
    $C = \{(x_1, x_2): \ x_1 = 1000 \text{ and } 1000 \leq x_2 \leq 1600 \text{ or } x_2 = 1000 \text{ and}$
         $1000 \leq x_1 \leq 1600\}$;
    $D = \{(x_1, x_2): \ x_1 = 1000 \text{ and } 0 \leq x_2 \leq 1000 \text{ or } x_2 = 1000 \text{ and}$
         $0 \leq x_1 \leq 1000\}$.

13. $S = \{x: x = 0, 1, 2, \ldots, 30\}$; $A = \{x: x = 5, 6, \ldots, 30\}$;
    $B = \{x: x = 0, 1, 2, \ldots, 5\}$; $C = A \cap B$.

14. $S = \{(x_1, x_2): x_i = 0, 1, 2, \ldots, 30, i = 1, 2\}$;
    $A = \{(x_1, x_2): x_1 = 5, 6, \ldots, 30; \ x_2 = 0, 1, 2, \ldots, 30\}$;
    $B = \{(x_1, x_2): x_1 = 0, 1, 2, \ldots, 30; \ x_2 = 5, 6, \ldots, 30\}$;
    $C = A \cap B$.

15. $d = $ defective, $n = $ nondefective,
    $S = \{(x_1, x_2, \ldots, x_{50}): x_i = n \text{ or } d, i = 1, 2, \ldots, 50\}$;
    $A = \{(n, n, \ldots, n)\}$; $B = \{(d, d, \ldots, d)\}$;
    $C = \{(d, d, x_3, x_4, \ldots, x_{50}): x_i = n \text{ or } d, i = 3, 4, \ldots, 50\}$.

EXERCISE 2.2

1. (a) $\frac{1}{3}$ (b) $\frac{2}{3}$ (c) $\frac{2}{3}$ (d) $\frac{1}{3}$ (e) 1 (f) $\frac{2}{3}$.

2. (a) $\frac{3}{10}$ (b) $\frac{7}{10}$ (c) $\frac{1}{2}$ (d) $\frac{3}{10}$ (e) 1 (f) $\frac{1}{2}$.

3. (a) 0   (b) 1   (c) 0   (d) 0   (e) 1   (f) 0, No.

4. $P(B) = 0, \frac{1}{2}, 1$.

7. (a) $\frac{1}{2}$ (b) $\frac{1}{2}$ (c) $\frac{1}{3}$ (d) $\frac{1}{3}$ (e) $\frac{2}{3}$ (f) $\frac{1}{6}$ (g) $\frac{1}{6}$ (h) $\frac{5}{6}$.

8. (a) $\frac{7}{12}$ (b) $\frac{3}{4}$ (c) $\frac{1}{12}$ (d) $\frac{1}{4}$ (e) $\frac{5}{12}$ (f) $\frac{3}{4}$.

9. No.   10. $P(A) + P(B) = 1$.   12. (a) $\frac{1}{2}$ (b) $\frac{1}{2}$.

13. (a) 1   (b) 1.   15. P(person 1 wins) $\geq \frac{1}{2}$.

EXERCISE 2.3

1. $\frac{1}{2}$.   2. $\frac{1}{2}, \frac{1}{4}, \frac{1}{13}, \frac{1}{52}$.   3. $\frac{1}{5}, \frac{1}{5}, \frac{1}{25}$.

4.
sum	2	3	4	5	6	7	8	9	10	11	12
36p	1	2	3	4	5	6	5	4	3	2	1

5.
sum	3	4	5	6
8p	1	3	3	1

6. $\frac{18}{40}, \frac{6}{40}, \frac{7}{40}, \frac{15}{40}, \frac{17}{40}, \frac{4}{40}$,

7.
number	1	2	3	4	5	6
21p	1	2	3	4	5	6

(a) $\frac{12}{21}$ (b) $\frac{11}{21}$.

8.
number	1	2	3	4	5	6
147p	60	30	20	15	12	10

(a) $\frac{55}{147}$ (b) $\frac{22}{147}$.

9. Doug: $\frac{8}{27}$; Joe: $\frac{2}{27}$; Hugh: $\frac{13}{27}$; Ray: $\frac{4}{27}$.

10. Doug: $\frac{9}{27}$; Joe: $\frac{18}{27}$; Hugh: $\frac{9}{27}$; Ray: $\frac{18}{27}$.

11. $\frac{1}{8}, \frac{1}{8}$.

12. (a) 0   (b) Elements 2 through n will not occur.

EXERCISE 2.4

1. $3! = 6$.   2. 6.   3. $6! = 720$.

4. $3^3 = 27$.   5. $2^4 = 16$.

6. Identical awards: $\binom{4}{3}$; Different awards: $_4P_3$ 7. $2^3 = 8$.

8. No.   9. $\binom{28}{5} = 98,280$.   10. $\binom{8}{2} = 28$.

11. $\binom{10}{3} = 120$.   12. $\binom{20}{5} = 15,504$.   13. $\binom{15}{2} = 105$.

14. $\binom{15}{3} = 455, \binom{15}{k}$.   15. $\binom{9}{3} = 84$.   16. 7, $\binom{7}{3} = 35$.

17. $\binom{2}{1}\binom{3}{1}\binom{4}{1} = 24.$

18. (a) $(26)^3(10)^3 = 17,576,000.$ (b) 17,380,000.

(c) $\binom{6}{3}(26)^3(10)^3 = (35,152)10^4.$

19. (a) $\binom{100}{10}$ (b) $\sum_{k=6}^{10}\binom{60}{k}\binom{40}{10-k} \doteq (.633)\binom{100}{10}$

(c) $\sum_{k=0}^{5}\binom{60}{k}\binom{40}{10-k} \doteq (.367)\binom{100}{10}.$

21. (a) $\binom{3}{1}4 = 12$ (b) $\binom{3}{2}4 \cdot 3 = 36$ (c) $4 \cdot 3 \cdot 2 = 24$

(d) 73.

22. (a) $\binom{10}{3 \ \ 3 \ \ 2 \ \ 1 \ \ 1} = 50,400$ (b) $\binom{11}{4 \ \ 4 \ \ 2 \ \ 1} = 34,650.$

23. $\binom{20}{8 \ \ 7 \ \ 5} = 99,768,240.$

24. (a) $\binom{19}{7 \ \ 7 \ \ 5} = 39,907,296$ (b) $\binom{18}{6 \ \ 7 \ \ 5} = 14,702,688.$

25. $\binom{50}{20 \ \ 20 \ \ 10}.$

26. (a) 6   (b) 12   (c) 6
    (d) 25.

EXERCISE 2.5

1. $\binom{30}{12}.$ 2. $\binom{6}{1}.$ 3. 2.

4. $\binom{20}{15}, \binom{20}{10}.$ 5. $\binom{9}{2}\big/\binom{10}{3} = \frac{3}{10}, \binom{8}{1}\big/\binom{10}{3} = \frac{6}{90}.$

6. $\frac{1}{50}, 1\big/\binom{100}{2}.$ 7. $\frac{1}{2^{30}}, \frac{1}{2^{29}}.$ 8. $\frac{1}{2}, \frac{1}{3}.$

9. $\frac{1}{2^5}, \frac{6}{2^5}.$ 10. (a) $\binom{13}{2}\binom{4}{2}\binom{4}{2}\binom{44}{1}\big/\binom{52}{5}.$

(b) $13 \cdot 12 \cdot \binom{4}{3}\binom{4}{2}\big/\binom{52}{5}$ (c) $4\binom{13}{5}\big/\binom{52}{5}$

(d) $\binom{4}{1}^5 \cdot 10\big/\binom{52}{5}.$

11. $n \leq 12$, Prob $= 1 - {}_{12}P_n/(12)^n$; $n > 12$, Prob $= 1$.

n	3	4	5	6
Prob	.236	.427	.618	.777

12. $\sum_{k=8}^{10} \binom{10}{k} / 2^{10} = 56/1024$.   13. $\frac{1}{5}, \frac{1}{45}$.   14. $\frac{1}{20}, 0, \frac{9}{20}$.

15. $\frac{36}{70}$.   16. (a) .020 (b) .0035 (c) .0087.

17. $\frac{1}{16}$.   18. $\frac{6}{64}$.   19. $\frac{11}{1024} = .011$.

EXERCISE 2.6

1. $\frac{1}{3}, i = 1, 2, 3, 4$; $\frac{1}{4}, i = 1, 2, 3, 4$.   2. $\frac{1}{4}, i = 1, 2, 3, 4$; $\frac{1}{4}, i = 1, 2, 3, 4$.

3. $\frac{1}{2}$.   4. $\frac{3}{4}, \frac{1}{2}$.   5. $\frac{5}{12}, \frac{7}{12}$.

6. $\frac{1}{4}, \frac{3}{4}$.

7. $\binom{26}{5} / \binom{52}{5} = 253/9996$, $\binom{13}{5} / \binom{52}{5} = 33/66,640$.

8. $\frac{1}{3}$.   9. $\frac{41}{63}$.   10. $\frac{35}{41}$.

11. .087.   12. .199, $\frac{1}{199}$.   13. .807.

14. $\frac{6}{21}, \frac{5}{21}, \frac{4}{21}, \frac{3}{21}, \frac{2}{21}, \frac{1}{21}$.   15. (a) $\frac{1}{3}$ (b) No.

16. (a) .3024 (b) .216, .18, .1.   17. (a) $\frac{37}{256}$ (b) $\frac{29}{128}$ (c) $\frac{22}{64}$.

18. (a) $\frac{1}{25}$ (b) $\frac{1}{10}$.

EXERCISE 2.7

1. Yes.   2. Yes.   3. No.

4. Yes.   5. No.

8. $(.7)^{15}$, $\binom{15}{14}(.7)^{14}(.3)$, independent shots.   9. (a) .07 (b) .24 (c) .504.

10. (a) $S_i = \{n, d\}$, $S = S_1 \times S_2 \times \cdots \times S_{10}$ (b) $(.9)^{10} = .349$
    (c) .736.

11. $\binom{990}{20} / \binom{1000}{20}$, $\binom{10}{1}\binom{990}{19} / \binom{1000}{20}$.   12. $1 - (.9)^{16} = .815$.

13. (a) .007 (b) .0443.

15.

0	1	2
$(1 - p_1)(1 - p_2)$	$p_1(1 - p_2) + (1 - p_1)p_2$	$p_1 p_2$

16.

0	1	2	3	4	5
.0839	.3497	.3996	.1499	.0166	.0003

17.

0	1	2	3	4	5
.1317	.3292	.3292	.1646	.0412	.0041

EXERCISE 2.8

1. (a) $\left(\frac{5}{6}\right)^9\left(\frac{1}{6}\right)$ = .032 (b) .421 (c) $\frac{6}{11}$ = .545.   2. (a) $\frac{5}{36}$ (b) $\frac{5}{11}$.

3. (a) .01 (b) $\frac{1}{111}$.   4. (a) $(.9)^4$ = .66 (b) $(.9)^{10}$ = .35.

5. (a) $\frac{3}{10}$ (b) $\frac{2}{10}$.   6. $\frac{1}{12}, \frac{1}{12}, \frac{5}{12}$.   7. $1/\sqrt{2}$.

8. $\frac{2}{3}$.   9. (a) $\frac{3}{4}$ (b) $\frac{1}{5}$.   10. (a) $\frac{1}{2}$ (b) $\frac{1}{3}$.

11. (a) $n/2^{n-1}$ (b) $(2^{n-1} - n)/(2^n - n)$, no.   12. (b) $1/(b - a)$.

13. $P(A) = .32$, $P(B) = .24$.   14. $P(A) = \frac{39}{50}$, $P(B) = \frac{433}{450}$, $P(C) = \frac{146}{225}$.

Chapter 3

EXERCISE 3.1

1.

y	1	2	3	4
$p_Y(y)$	$\frac{1}{4}$	$\frac{1}{4}$	$\frac{1}{4}$	$\frac{1}{4}$

2.

z	3	4	5	6	7
$p_Z(z)$	$\frac{1}{6}$	$\frac{1}{6}$	$\frac{2}{6}$	$\frac{1}{6}$	$\frac{1}{6}$

3.

z	2	3	4	5	6	7	8
$p_Z(z)$	$\frac{1}{16}$	$\frac{2}{16}$	$\frac{3}{16}$	$\frac{4}{16}$	$\frac{3}{16}$	$\frac{2}{16}$	$\frac{1}{16}$

4.

x	2	5	8	10	13	17	18	20	25	32
$p_X(x)$	$\frac{1}{16}$	$\frac{2}{16}$	$\frac{1}{16}$	$\frac{2}{16}$	$\frac{2}{16}$	$\frac{2}{16}$	$\frac{1}{16}$	$\frac{2}{16}$	$\frac{2}{16}$	$\frac{1}{16}$

5.

x	19	$19\frac{1}{2}$	20	$20\frac{1}{2}$	$21\frac{1}{2}$	22	$22\frac{1}{2}$	23	$23\frac{1}{2}$	25
$p_X(x)$	$\frac{3}{45}$	$\frac{12}{45}$	$\frac{9}{45}$	$\frac{4}{45}$	$\frac{3}{45}$	$\frac{4}{45}$	$\frac{4}{45}$	$\frac{4}{45}$	$\frac{1}{45}$	$\frac{1}{45}$

6. $p_X(x) = \frac{1}{4}$, $x = 1, 2, 3, 4.$

7. $p_Y(y) = \dbinom{26}{y}\dbinom{26}{5-y}\Big/\dbinom{52}{5}$ $y = 0, 1, 2, 3, 4, 5.$

8. $p_Z(z) = \dbinom{13}{z}\dbinom{39}{5-z}\Big/\dbinom{52}{5}$, $z = 0, 1, 2, 3, 4, 5.$

EXERCISE 3.2

1. $p_X(x) = \frac{1}{3}$ at $x = -3, -1, 0.$ 2. $p_Z(z) = \frac{1}{2}$ at $z = -2, 0;\ \frac{1}{2}.$

3. $p_W(w) = \frac{1}{3}$ at 3, 6, $\frac{1}{6}$ at 4, 5; $\frac{1}{3}.$

4.

y	0	5	7	100	102
$p_Y(y)$	$\frac{1}{4}$	$\frac{1}{12}$	$\frac{1}{6}$	$\frac{1}{3}$	$\frac{1}{6}$

, $P(Y \leq 100) = \frac{5}{6}$

5. $F_Z(x) = 0,\ x < 0$
$= \frac{1}{3},\ 0 \leq x < 1$
$= \frac{2}{3},\ 1 \leq x < 2$
$= 1,\ x \geq 2.$

6. $F_U(t) = 0,\quad t < -3$
$= \frac{1}{2},\ -3 \leq t < 0$
$= \frac{2}{3},\quad 0 \leq t < 4$
$= 1,\quad t \geq 4.$

7. $f_X(t) = \frac{1}{2},\ -1 < t < 1,\ P(-\frac{1}{2} \leq x \leq \frac{1}{2}) = \frac{1}{2}.$

8. $f_Y(t) = 1/2(t)^{1/2},\ 0 < t < 1,\ P(\frac{1}{4} < Y < \frac{3}{4}) = \frac{1}{2}(3)^{1/2} - \frac{1}{2}.$

9. $f_Z(t) = 2t,\qquad 0 < t < \frac{1}{2}$
$= 6(1 - t),\ \frac{1}{2} < t < 1.$

10. No.

12. $F_X(t) = 0,\qquad t < 99$
$= t - 99,\ 99 \leq t \leq 100$
$= 1,\qquad 100 < t.$

13. $F_Y(t) = 0,\qquad t < 0$
$= 1 - (1 - t)^2,\ 0 \leq t \leq 1$
$= 1,\qquad t > 1.$

14. $F_Z(t) = 0,\qquad t < 0$
$= 1 - e^{-10t},\ t \geq 0.$

15. $f_X(x) = 1/(b - a),\qquad a < x < b$
$F_X(t) = 0,\qquad t < a$
$= (t - a)/(b - a),\ a \leq t \leq b$
$= 1,\qquad t < b.$

16. $f_Y(y) = \frac{1}{30},\ 0 < y < 30$
$F_Y(t) = 0,\quad t < 0$
$= \dfrac{t}{30},\ 0 \leq t \leq 30$
$= 1,\quad t > 30.$

17. (a) $f_X(x) = -11.04 + .16x,$ $69 < x < 71.5$
            $= 11.84 - .16x,$ $71.5 < x < 74$
     $F_X(t) = 0,$ $t < 69$
            $= .08(t - 69)^2,$ $69 < t < 71.5$
            $= .1 - .08(t - 74)^2,$ $71.5 \le t \le 74$
            $= 1,$ $t > 74.$

18. $f_Y(y) = (15 - x)/225,$ $0 < x < 15$
           $= (x - 15)/225,$ $15 < x < 30$
     $F_Y(t) = 0,$ $t < 0$
           $= \frac{1}{2} - (15 - t)^2/450,$ $0 < t \le 15$
           $= \frac{1}{2} + (t - 15)^2/450,$ $15 < t \le 30$
           $= 1,$ $t > 30.$

EXERCISE 3.3

1. $\dfrac{15}{2}, 85, \dfrac{15}{64}, \dfrac{139}{2}, \dfrac{115}{4}, \dfrac{\sqrt{115}}{2}.$ 2. $0, \dfrac{1}{3}, 2, 7, \dfrac{1}{3}, \dfrac{1}{\sqrt{3}}.$

3. $\dfrac{1}{3}, \dfrac{1}{6}, \dfrac{641}{6}, 2, \dfrac{1}{18}, \dfrac{\sqrt{2}}{6}.$ 5. 10, 20, 55, 99, 47, 80, 50.

6. $e^{.5}, e^{.75} - e^{.25}, e - 1, (1 - e)(e - 3)/2.$

7. $\log_2 1.5, \log_2 1.75 - \log_2 1.25, 2 - (\log_e 2)^{-1}, 6/\log_e 2 - 2(1 + 1/(\log_e 2)^2).$

8. 2, 2, 3, 0, 1, 2.   10. $5'8\frac{1}{2}''.$

11. $-.70, -70, 899.91, 89,100.$ 13. 7, $\frac{59}{6}.$

14. 3.22, 42.15.

15. (a) $2^{-x}, x = 1, 2, 3, \ldots.$ (b) ∞ (c) Impossible to pay enough.

16. 2.

EXERCISE 3.4

1. $m_1 = m_2, m_3 = \frac{1}{6}$ 2. $(e^{40t} - e^{30t})/10t, (e^{5t} - e^{-5t})/10t.$

3. $.0001/(.0001 - t), 10^4, 10^8.$ 4. $(1 + t)^3/8, \frac{3}{2}, \frac{3}{4}.$

5. $t(1 - t^n)/n(1 - t).$ 6. $m_1, m_2 - m_1, m_3 - 3m_2 + 2m_1.$

7. $\xi_1, \xi_2 + \xi_1, \xi_3 + 3\xi_2 + \xi_1.$ 8. $(.9t)/(1 - .1t), \frac{10}{9}, \frac{10}{81}.$

11. $p_X(x) = 1/(n + 1)$ at $x = 0, 1, 2, \ldots, n;$ $[1 - e^{(n+1)t}]/(n + 1)[1 - e]^t.$

12. $p_X(x) = \dbinom{n}{x} p^x (1 - p)^{n-x}, x = 0, 1, 2, \ldots, n$

     $m_X(t) = [1 - p(1 - e^t)]^n.$

13. $m_X(t) = pe^t/[1 - (1 - p)e^t].$ 15. $c_X(t) = n \log [1 - p(1 - e^t)].$

16. $c_X(t) = t + \log p - \log [1 - (1 - p)e^t].$

EXERCISE 3.5

1. $1 - F_X\left(\dfrac{t-a}{b}\right) + p_X\left(\dfrac{t-a}{b}\right).$

2. $F_Y(t) = 0,\ t < a$
   $= 1,\ t \geq a.$

3. $F_Y(t) = 0,\qquad\qquad t < 13$
   $= (t-13)/4,\quad 13 \leq t \leq 17$
   $= 1,\qquad\qquad t > 17.$
   $f_Y(t) = \tfrac{1}{4},\quad 13 < t < 17$

4. $F_Z(t) = 0,\qquad\qquad t < -1$
   $= (t+1)^3,\quad -1 \leq t \leq 0$
   $= 1,\qquad\qquad t > 0.$
   $f_Z(t) = 3(t+1)^2,\quad -1 < t < 0$

5. $F_U(t) = 0,\qquad t < -120$
   $= \tfrac{1}{4},\qquad -120 \leq t < -50$
   $= \tfrac{3}{4},\qquad -50 \leq t < 20.$
   $= 1,\qquad t \geq 20.$
   $p_U(t) = \tfrac{1}{4},\quad t = -120$
   $= \tfrac{1}{2},\quad t = -50$
   $= \tfrac{1}{4},\quad t = 20$

6. $F_X(t) = 0,\qquad\qquad t < -7$
   $= 1 - e^{-(t+7)/2},\quad t \geq -7.$
   $f_X(t) = \tfrac{1}{2}e^{-(t+7)/2},\quad t > -7$

7. $F_{(X-\mu)/\sigma}(t) = 0,\qquad\qquad t < -\sqrt{3}$
   $= t\sqrt{3}/6 + \tfrac{1}{2},\quad -\sqrt{3} \leq t \leq \sqrt{3}$
   $= 1,\qquad\qquad t > \sqrt{3}$
   $f_{(X-\mu)/\sigma}(t) = \sqrt{3}/6,\ -\sqrt{3} < t < \sqrt{3}.$

8. $F_{(Z-\mu)/\sigma}(t) = 0\qquad\qquad t < -1$
   $= 1 - e^{-(t+1)},\quad t \geq -1.$
   $f_{(Z-\mu)/\sigma}(t) = e^{-(t+1)},\quad t > -1$

9. $F_{(U-\mu)/\sigma}(t) = 0,\qquad t < -\sqrt{3/2}$
   $= \tfrac{1}{3},\qquad -\sqrt{3/2} \leq t < 0$
   $= \tfrac{2}{3},\qquad 0 \leq t < \sqrt{3/2}.$
   $= 1,\qquad t > \sqrt{3/2}.$
   $p_{(U-\mu)/\sigma}(t) = \tfrac{1}{3},\quad t = 0,\ \pm\sqrt{3/2}$

10. $F_Z(t) = 0,\ t < 0$
    $= \sqrt{t},\ 0 \leq t \leq 1$
    $= 1,\quad t > 1.$
    $f_Z(t) = \tfrac{1}{2}\sqrt{t},\ 0 < t < 1$

11. Same as 10.

14. $F_Z(t) = 0,\ t < 0$
    $= t,\ 0 \leq t \leq 1$
    $= 1,\ t > 1.$
    $f_Z(t) = 1,\ 0 < t < 1$
    $= 0,\ t \leq 0\ \text{or}\ t \geq 1$

15. $F_Y(t) = 1 - e^{-t},\ t \geq 0$
    $= 0,\qquad t < 0.$
    $f_Y(t) = e^{-t},\ t > 0$
    $= 0,\quad t \leq 0$

17. $f_X(x) = \tfrac{5}{2}\sqrt{x\pi},\ \pi(r - .1)^2 < x < \pi(r + .1)^2,\ \tfrac{1}{2}.$

18. $f_X(x) = \dfrac{5}{2r}\sqrt{x\pi}$, $\pi(.9r)^2 < x < \pi(1.1r)^2$, $\frac{1}{2}$.

19. $F_X(t^2)$.   20. $2tf_X(t)$.   21. $\sqrt{t_k}$.   22. t_k^2.

24. (a) 100.5, $\frac{1}{12}$ (b) 10,100.33, 3280   (c) 10,100.25, 3366.75.

Chapter 4

EXERCISE 4.1

1. $\frac{5}{6}, \frac{25}{36}, \frac{4650}{7776}, \frac{1526}{7776}$.   2. 16, $4/\sqrt{5}$, .2182, .0867, .0692.

3. .246, 5.   4. .3487, .1.   7. $n = 18, p = \frac{1}{3}$.

8. 99, No.   10. $-.0965$, $-.965$.   11. .598, .196.

12. .590, .00001.   13. 1, .358.   14. 3.58, 3.5×10^{-5}.

15. Binomial, $n = 10$, $p = \frac{1}{2}$ for right and left, $11/2^9$.

16. Profit $= 1000 - 300X$, $X =$ number that don't pay, X is binomial $n = 10$, $p = .1$, 700.

EXERCISE 4.2

1. $\frac{3}{4}, \frac{7}{8}$.   2. $(\frac{20}{38})^{k-1}(\frac{18}{38}), \frac{38}{18}$.   3. $(\frac{36}{38})^{k-1}(\frac{2}{38})$, 19, 342.

4. $(\frac{9}{10})^{k-1}(\frac{1}{10})$, 10.   5. $\frac{1}{10}$ for $k = 1, 2, \ldots, 10$, $5\frac{1}{2}$.

6. $\dbinom{3}{k}\dbinom{2}{2-k}\Big/\dbinom{5}{2}$.   7. $\dbinom{26}{k}\dbinom{26}{13-k}\Big/\dbinom{52}{13}$, $\frac{13}{2}, \frac{169}{68}$.

8. $\dbinom{8}{k}\dbinom{2}{5-k}\Big/\dbinom{10}{5}$, $\dbinom{2}{k}\dbinom{8}{5-k}\Big/\dbinom{10}{5}$.   9. \$5.

10. $\frac{22}{35}$.   11. $\dbinom{k-1}{k-r}q^{k-r}p^r$, for $k = r, r+1, r+2, \ldots$.

12. $\dbinom{k-1}{k-3}(.999)^{k-3}(.001)^3$, $k = 3, 4, \ldots$.   14. (b) $\frac{1}{3}$ (c) No.

17. $\dbinom{95}{n}\Big/\dbinom{100}{n}$, .8119, .7696.   18. .9721.

19. (a) binomial, $n = M, p$; (b) np.

EXERCISE 4.3

1. .368, .264.   2. .0003, .135, .0003.   3. .632, $(.632)^3$.

4. 6, $\sqrt{6}$, .607.   7. exp $[\lambda s(e^t - 1)]$.   8. .607, .303, .090.

9. $e^{-.02} = .980$, $(.999)^{20} = .980$.   10. .050, .224, 149.

11. $\lambda s(e^t - 1)$.   13. .180, .323, .857, .002.

14. (a) Binomial, $n = 3000$, $p = \dfrac{1}{2n}$

 (b) $(5999/6000)^{3000} \doteq e^{-1/2} = .607$

 (c) $(.607)^{16} \doteq .0003$.

15. $(.999)^5 \doteq e^{-.005} = .995$.

EXERCISE 4.4

1. $1/2\sqrt{t}$, $2 - \sqrt{2}$.   2. $\dfrac{2t}{3}$.   3. $(e^{tb} - e^{ta})/t(b - a)$.

4. $(t^b - t^a)/(b - a)\log t$.   5. .316, .422.   6. $\lambda = \log_e 2 = .693$.

7. .2865, .5654, 0.   8. both .1882.   9. $\sqrt{3}/2$.

10. $-\log(1 - p)$.   11. $\lambda^2 s e^{-\lambda s}$.   13. $\dfrac{\lambda^r s^{r-1}}{\Gamma(r)} e^{-\lambda s}$.

14. $\log \lambda - \log(\lambda - t)$.   16. $r!/\lambda^r$.   17. $e^{-\lambda t}$.

18. $q^{[t]}$, where $[t]$ is the integer part of t.

EXERCISE 4.5

1. .3085, .9987.   2. .3085, .8413, .1574.   3. $2N_Z(t) - 1$ for $t > 0$.

4. .3174, .3830.   5. 0, 1.35.   6. μ, 1.35σ.

8. .456, 1.22.   9. $\exp(-t/2)/\Gamma(1/2)\sqrt{2t}$. 10. $\exp(t^2/2)$.

11. $2n_Z(t)$, $t > 0$.   12. (a) 6 feet, 3.94 inches (b) 5 feet, 9 inches.

13. (a) 628   (b) 432.5, 500, 567.5   (c) .8023.

15. (a) $\mu = 30.69$, $\sigma = 15.09$ (b) $\mu - .253\sigma = 26.87$.

16. $\chi_k^2 = z_{(k+1)/2}^2$, k-th percentile of $\chi^2 = \chi_k^2$.

17. k-th percentile of folded normal $= u_k = z_{(k+1)/2}$. 18. $\chi^2 = $ (folded normal)^2.

Chapter 5

EXERCISE 5.1

1.

	x		
	2	3	4
y $\begin{array}{c} -1 \\ 0 \\ 1 \end{array}$	$\begin{array}{c} 0 \\ \frac{1}{4} \\ 0 \end{array}$	$\begin{array}{c} \frac{1}{4} \\ 0 \\ \frac{1}{4} \end{array}$	$\begin{array}{c} 0 \\ \frac{1}{4} \\ 0 \end{array}$	$\begin{array}{c} \frac{1}{4} \\ \frac{1}{2} \\ \frac{1}{4} \end{array}$
	$\frac{1}{4}$	$\frac{1}{2}$	$\frac{1}{4}$	

2.

x

	2	3	4	
2	$\frac{1}{8}$	$\frac{1}{8}$	0	$\frac{1}{4}$
y 3	$\frac{1}{8}$	$\frac{1}{4}$	$\frac{1}{8}$	$\frac{1}{2}$
4	0	$\frac{1}{8}$	$\frac{1}{8}$	$\frac{1}{4}$
	$\frac{1}{4}$	$\frac{1}{2}$	$\frac{1}{4}$	

3. Divide entries by 2652, $P(X > Y) = \frac{228}{2652}$

x

	0	1	2	
0	1260	216	6	1482
y 1	864	144	6	1014
2	132	24	0	156
	2256	384	12	

4. Divide entries by 90, $P(X \le Y) = \frac{84}{90}$

x

	0	1	2	
0	0	0	6	6
y 1	0	42	0	42
2	42	0	0	42
	42	42	6	

5. $f_{X,Y}(x, y) = 1/\pi$, $x^2 + y^2 \le 1$, $f_X(t) = f_Y(t) = 2(1 - t^2)^{1/2}/\pi$.
6. $2/\log_e 2$. 7. 9/40.
11. (a) $f_X(x) = 1, 0 < x < 1, f_Y(y) = -\log_e y, 0 < y < 1$.
12. (a) $a = \sqrt{2}$ (b) $f_X(x) = f_Y(x) = 1/2\sqrt{2}, -\sqrt{2} < x < \sqrt{2}$.
13. (a) $a = 1$ (b) $f_X(x) = f_Y(x) = x + 1, -1 < x < 0$.
   $= 1 - x, 0 < x < 1$.
14. $\lambda^2 e^{-\lambda(t_1 + t_2)}, t_1 > 0, t_2 > 0$.

EXERCISE 5.2

1.

x

	-1	1	
-1	$\frac{1}{4}$	$\frac{1}{4}$	$\frac{1}{2}$
y 1	$\frac{1}{4}$	$\frac{1}{4}$	$\frac{1}{2}$
	$\frac{1}{2}$	$\frac{1}{2}$	

, $E\left[\dfrac{X}{Y}\right] = 0.$

2. 62.5, 3, $195\frac{5}{6}$.　　　　3. 260.42, $\frac{1}{3}$, $8\frac{1}{3}$, .894.

4. 1, 1, $(n + 1)(2n + 1)/3$, $(n + 1)/2$.　　　　5. $\frac{1}{4}, \frac{1}{4}, \frac{1}{20}, -\frac{1}{80}$.

7. .9923.　　　　　　　　　8. .9971.　　　　　　　　9. $E[t_1^X t_2^Y]$.

10. (a) 0 (b) 0.　　　　　　11. $a/3$, $2(a^2 + 3)/9$, $b/3$, $2(b^2 + 3)/9$, $2(ab + 3)/9$.

12. $p_{X,Y}(x, y) = \binom{3}{x} \Big/ 2^3$, at $(x, y) = (3, 0), (2, 1), (1, 2), (0, 3)$, $\rho = -1$.

14. (a) $1/(1 - 2a)$ (b) 0, 1.　　　　　　15. $\frac{1}{2}$.

EXERCISE 5.3

2. $(x + 1)/2$, $(n + y)/2$.　　4. 0, $\frac{1}{2}(1 - \sqrt{|y|})$.　　　　5. No.

6. μ_Y, No, No.　　　　　8. X or Y is a constant.

9. $\binom{n}{x}\binom{n}{y} p^{x+y}(1 - p)^{2n-x-y}$.　　10. $\dfrac{2^{x+y}}{x!y!} e^{-4}$.

11. $\exp [t_1\mu_X + t_2\mu_Y + t_1^2\sigma_X^2/2 + t_2^2\sigma_Y^2/2]$.　　　12. $\frac{1}{2}$, 0.

15. Yes.　　　　　　　　16. No.

17. 0, 0, $(a_1^2 + a_2^2)\sigma^2$, $(b_1^2 + b_2^2)\sigma^2$, $(a_1b_1 + a_2b_2)/\sqrt{(a_1^2 + a_2^2)(b_1^2 + b_2^2)}$.

18. No.　　　　　　　　19. $\lambda e^{-\lambda(x-a)}$, $x > a$, $\mu = (1 + \lambda a)/\lambda$, $\sigma^2 = 1/\lambda^2$.

EXERCISE 5.4

1. .5881, .0207.　　　2. (a) $\dfrac{(\lambda s)^{\Sigma x_i}}{\Pi x_i!} e^{-3\lambda s}$　　(b) $e^{-s\Sigma \lambda_i} \Pi \dfrac{(\lambda_i s)^{x_i}}{x_i!}$.

3. .8686.　　　　　　11. (a) $(.1587)^2(.8413)(.6826) = .0145$

　　　　　　　　　　　(b) $\binom{4}{2} (.1587)^2(.6826)^2 = .0704$.

12. (a) $\dfrac{7^7}{7!} e^{-7} = .149$

　　(b) $e^{-7} \Big/ \dfrac{7^7}{7!} e^{-7} = \dfrac{7!}{7^7} = .006$.

13. (a) Normal, $\mu = 288$ ounces, $\sigma = .4898$ ounces (b) .4793.

14. No.　　　　　　15. (a) .0207 (b) .4443.

16. (a) Normal, $\mu_Y = \Sigma \mu_i$, $\sigma_Y^2 = \Sigma \sigma_i^2$ (b) Probably not.

EXERCISE 5.5

1.

k	1.25	1.50	1.75	2
$P(|X - .5| < .289k)$	.722	.866	1	1

.

2. $P(|X - \mu_X| > k\sigma_X) = 1 - 1/k^2$.　　4. $\mu = 0$, $\sigma^2 = 1/n$.

5. $\bar{X}$.　　　　　　　　6. $1/\bar{X}$.

8. $n \geq 1/(.1)(4)(.05)^2 = 1000$, $n \geq 1/(.1)(4)(.01)^2 = 25,000$.

EXERCISE 5.6

1. .8643, .6920
2. .0150.
3. 0.
4. .3669.
5. .5558.
6. .0262.
7. .7190.
8. $(1.28)^2/4(.1)^2 \doteq 41.$
9. .5222.
10. .9164.
11. .51.
12. (a) .3085 (b) .9236.
13. .5230.
14. .9332.
15. (a) Negative binomial, $p = .05, r = 50$ (b) .9265.

Chapter 6

EXERCISE 6.1

1. .2146.
2. $\lambda^n e^{-\lambda \Sigma x_i}, x_i > 0, i = 1, 2, \ldots, n.$
3. $1, 0 < x_i < 1, i = 1, 2, \ldots, 5.$
4. $(2\pi\sigma^2)^{-10} e^{-\Sigma(x_i-\mu)^2/2\sigma^2}, 1 - N_Z([2000 - 10\mu]/\sigma\sqrt{5}).$
5. $(2\pi\sigma^2)^{-5} e^{-\Sigma(x_i-\mu)^2/2\sigma^2}, N_Z(0) = \frac{1}{2}.$
6. .7948, .2052.
7. $5^{-\Sigma x_i}.$
8. $p^{\Sigma x_i}(1 - p)^{50-\Sigma x_i} \Pi \binom{10}{x_i}.$
9. $p^{10} q^{\Sigma x_i - 10}.$
10. (a) 50 (b) No.
11. (a) Yes. (b) No.
12. (a) $\binom{15}{12} p^{12} q^3.$ (b) $\dfrac{12}{15} = .8.$ (c) Yes.
13. .3174.

EXERCISE 6.2

1. $118\frac{1}{3}$, $14{,}024\frac{1}{3}$, $1{,}664{,}602\frac{1}{3}$, $118\frac{1}{3}$, 25.87, 119, 120, 15.
2. 9.79, .047, and 9.6, 9.6, 9.6, 9.7, 9.8, 9.8, 10, 10.2.
3. 163.6, 146.3, 12.1, 162, 30.
4. 3.175, .63, 2.925, .88.
5. 56.2, 876.8, 55, 38, 46, 55, 70, 72
6. 0, 20/32
7. .6513, .000977.
8. .9990, $(.1)^{10}.$
10. For $X_{(1)}$: $n\, n_Z\left(\dfrac{t-\mu}{\sigma}\right)\left[1 - N_Z\left(\dfrac{t-\mu}{\sigma}\right)\right]^{n-1};$

For $X_{(n)}$: $n\, n_Z\left(\dfrac{t-\mu}{\sigma}\right)\left[N_Z\left(\dfrac{t-\mu}{\sigma}\right)\right]^{n-1}.$

11. $n/(n + 1), 1/(n + 1).$ 12. (a) standard normal (b) No.   13. (a) .3281 (b) .1856.
14. $\sigma^2, \sigma\Gamma(n/2)/\Gamma([n - 1]/2)\sqrt{2(n - 1)}.$

15. Bernoulli with parameter p. (b) All equal p. (c) Both are normal, $\mu = p$, $\sigma = \sqrt{pq/n}$.

16. (a) $1 - [1 - t/10]^5, 0 < t < 10$ (b) $\frac{1}{2}[1 - t/10]^4, 0 < t < 10$ (c) $10/6$.

17. $t^5/10^5, 0 < t < 10$; $50/6$.   18. $\sum_{k=3}^{5} \binom{5}{k} \left(\frac{t}{10}\right)^k \left(1 - \frac{t}{10}\right)^{5-k}, 0 < t < 10; \frac{1}{2}$.

19. .94.   20. .90.   21. .95.

24. χ^2, $2n$ degrees of freedom.   26. $n \binom{n-1}{(n-1)/2} t^{(n-1)/2}(1 - t)^{(n-1)/2}$.

Chapter 7

EXERCISE 7.1

1. .62.   2. 3, 1.5.   3. $63\frac{1}{3}$.
4. $(\bar{X})^{-1}$.   5. .192.   6. $(\bar{X})^{-1}$.
7. $\bar{X}$.   8. $\bar{X}/n$.   9. $\sum (x_i - 5)^2/n$.
10. $\bar{X} - 19/8$.   11. $\tilde{p} = \bar{X}/\tilde{n}$, $\tilde{n} = \bar{X}^2/(\bar{X} + \bar{X}^2 - M_2)$.
12. $\sum (X_i - 10)^2/n$.   13. 15, $8\frac{1}{4}$.   14. $\bar{X}$.
15. 1.5.   16. $(\bar{X})^{-1}$.
17. Any value between $x_{(n)} - \frac{1}{2}, x_{(1)} + \frac{1}{2}$.   18. $(\bar{X})^{-1}$.
19. .000914.   20. $\bar{X}/n$.   21. .405.
22. .405.   23. $\bar{X}$.
24. $[-\bar{X} \pm \sqrt{\bar{X}^2 - 4M_2}]/2$, whichever root is appropriate.
25. (a) $\bar{X}, \sum (X_i - \bar{X})^2/n_1, \bar{Y}, \sum (Y_i - \bar{Y})^2/n_2$ (b) same as (a).
26. $\bar{Z} = (n_1\bar{X} + n_2\bar{Y})/(n_1 + n_2), [\sum (X_i - \bar{Z})^2 + \sum (Y_i - \bar{Z})^2]/[n_1 + n_2]$.
27. $\bar{X}, \bar{Y}, [\sum (X_i - \bar{X})^2 + \sum (Y_i - \bar{Y})^2]/(n_1 + n_2)$ 28. $\sqrt{3M_2}$, $\max |X_i|$.
29. $\bar{X} - 1, X_{(1)}$.

EXERCISE 7.2

1. $\frac{1}{6}$.   2. $2\sigma^4/3$, No.   3. $n_1/(n_1 + n_2)$.
4. Yes.   5. .407.   6. .9671, .9319.
7. Normal, $\mu = \lambda, \sigma = \lambda/\sqrt{n}$.
8. $E[\hat{\Gamma}] = n\gamma/(n + 1)$, $V[\hat{\Gamma}] = n\gamma^2/(n + 2)(n + 1)^2$ $E[\tilde{\Gamma}] = \gamma$, $V[\tilde{\Gamma}] = \gamma^2/12n$, $\tilde{\Gamma}$, Yes.
9. nT_1, $\bar{T}$.   10. $2\sigma^4/n$.
11. (b) $a = V[\Theta_1]/(V[\Theta_1] + V[\Theta_2])$.   12. $\bar{X}$.

13. Normal, $\mu = \lambda$, $\sigma = \lambda/\sqrt{n}$.

14. (a) $\hat{\lambda}_1 = 10/13$, $\hat{\lambda}_2 = \dfrac{21}{26}$ (b) .78799 (c) .78846.   17. $(X_1 + X_2)/3$.

EXERCISE 7.3

1. (158.2, 172.2).   2. $(n-1)S^2/\chi_\alpha^2$, $n-1$ degrees of freedom.

3. (a) binomial, $n = 100$, $p = .9$ (b) .1350.

4. $\bar{X} - \bar{Y} \pm t_{1-\alpha/2}S\sqrt{\dfrac{1}{n_1} + \dfrac{1}{n_2}}$, $S^2 = [(n_1 - 1)S_X^2 + (n_2 - 1)S_Y^2]/(n_1 + n_2 - 2)$,

   t distribution has $n_1 + n_2 - 2$ degrees of freedom.

5. $(-.1, 10.5)$.   6. (a) .372 (b) lower limit $= .317$.

7. (a) Length $= L = k \sum (X_i - \bar{X})^2$, $\dfrac{L}{k\sigma^2}$ is χ^2, $n-1$ degrees of freedom.

   (b) $n = 20$, $\alpha = .05$, $k = (1/8.91 - 1/32.9) = .082$

   $$P(L \le \sigma^2) = P\left(\dfrac{L}{k\sigma^2} \le \dfrac{1}{.082}\right) \doteq .13.$$

8. (a) Length $= L = 2t_{1-\alpha/2}S/\sqrt{n} = jS$; $\dfrac{(n-1)L^2}{j^2\sigma^2}$ is χ^2, $n-1$ degrees of freedom.

   (b) $P(L < \sigma) = P\left(\dfrac{(n-1)L^2}{j^2\sigma^2} < \dfrac{n-1}{j^2}\right) \doteq .71$ for $\alpha = .05$.

9. 16, 62.

10. (a) $1 - N_Z(t - \mu)$ (b) Use confidence interval for μ.

    (c) Lower $= 1 - N_Z(10 - 10.48) = .6700$

    (c) Upper $= 1 - N_Z(10 - 11.72) = .9573$.

11. $100(1 - \alpha)\%$ two-sided interval: $(X_{(n)}/\sqrt[n]{1 - \alpha/2}, X_{(n)}/\sqrt[n]{\alpha/2})$.

12. $s < z_{1-\alpha/2}/t_{1-\alpha/2}$, procedure of example 7.3.2.   13. $\bar{X} \pm z_{1-\alpha/2}\sqrt{\bar{X}/n}$.

14. $(\sum X_i^2/\chi_{1-\alpha}^2, \sum X_i^2/\chi_\alpha^2)$ where χ^2 has n degrees of freedom.

Chapter 8

EXERCISE 8.1

1. $\bar{X} > c$.   2. (a) $\mu > 2$ (b) $\mu < 2$ (c) None.

3. $\alpha = .0250$, $Q(115) = .7054$.   4. $\alpha = .05$.

6. $\alpha = .01$, $Q(\theta) = 1 - .99/\theta$.   7. $Q(\lambda) = 1 - e^{-5.13\lambda}$.

8. $\alpha = .01$, $Q(\theta) = 1 - .99/\theta^{10}$.

9. $\alpha = .01$, $Q(\theta) = \{1 - (1 - [.01]^{1/10})\}/\theta^{10}$.

10. (a) $Q(\mu_1) = 1 - N_Z(\sqrt{n}[c - \mu_1])$ (b) $Q'(\mu_1) = n_Z(\sqrt{n}[c - \mu_1])/2\sqrt{n}$.

11. $Q'(\theta) = .99/\theta^2$.

EXERCISE 8.2

1. $\alpha = 1/256$, $\beta = 175/256$.   2. $\alpha = .0002$, $\beta = .49$.

5. $\beta \doteq .0001$.   6. Reject if $\Sigma X_i^2 > c$.   7. $\bar{X} < c$.

8. $\bar{X} > c$.   9. $\bar{X} > c$.   10. $\Pi X_i < c$.

11. $\bar{X} < c$.   12. Reject H_0 if $\bar{X} > c$.

EXERCISE 8.3

1. (a) Reject H_0 if $|\bar{X} - \mu_0| > c$ (b) Reject H_0 if $\bar{X} - \mu_0 > c$
(c) Reject H_0 if $\bar{X} - \mu_0 < c$.

2. Reject H_0.   5. Reject H_0.   6. Accept H_0.

7. Reject if $\bar{X} < c$.   8. Yes.   9. Yes.

10. Yes.   11. t, $2(n - 1)$ degrees of freedom.

12. Accept H_0.

14. If $p_1 = p_2$, $X/n_1 - Y/n_2$ is approximately normal with mean 0, variance $pq(1/n_1 + 1/n_2)$. Reject $H_0: p_1 = p_2$ if $|X/n_1 - Y/n_2|/\sqrt{\hat{p}\hat{q}(1/n_1 + 1/n_2)} > z_{1-\alpha/2}$ where $\hat{p} = (X + Y)/(n_1 + n_2)$.

15. Reject H_0.   16. Accept H_0.

EXERCISE 8.4

1. Reject.   2. Accept.   4. Accept.

5. Reject.   6. Accept.   7. Accept.

8. Reject.   9. Reject.   10. Accept.

EXERCISE 8.5

1. Reject.   2. Reject.   3. Reject.

4. Reject.   6. $2^k - k$.   7. Reject with $\alpha = .05$

9. Reject.   10. Reject.   11. Reject.

Chapter ⊙9

EXERCISE 9.1

1. $2p, 2(1 - p)$.
2. Given $X = 1$, $2p/(b^2 - a^2)$; Given $X = 0$, $2(1 - p)/[(1 - a)^2 - (1 - b)^2]$.
3. Normal, mean $= (2\bar{x} + 20)/2.2$, variance $= 1/2.2$.
5. $(1100.001)^2 \lambda e^{-1100.001\lambda}/\Gamma(2)$.   6. 4/9 at $\alpha = 9.5$, 5/9 at $\alpha = 9.7$.

7. Given tail:

p	$\frac{1}{4}$	$\frac{1}{2}$	$\frac{3}{4}$
prob.	$\frac{1}{2}$	$\frac{1}{3}$	$\frac{1}{6}$

Given head:

p	$\frac{1}{4}$	$\frac{1}{2}$	$\frac{3}{4}$
Prob.	$\frac{1}{6}$	$\frac{1}{3}$	$\frac{1}{2}$

8. Binomial, $n = 50, p = .05$.
9. Possibly a continuous distribution on $(0,1)$ with a peak at .42, and relatively flat around .42.
10. $\binom{47}{d}(.05)^d(.95)^{47-d}$, $d = 0, 1, 2, \ldots, 47$.

EXERCISE 9.2

4. σ_0^2, $\sigma_0^4/(m - 1)$.   5. λ_0, $\lambda_0^2/(m + 1)$.   7. $\bar{x}$, a or b.
8. $(m\sigma_0^2 + \frac{1}{2} \sum (x_i - \mu_x)^2)/(m + n/2)$.
9. $\lambda_0(n + m + 1)/(n\bar{x}\lambda_0 + m + 1)$.   10. $(m + \sum x_i + 1)\lambda_0/(n\lambda_0 + m + 1)$.
11. $\bar{x}$, if $\bar{x} = (b + a)/2$.   13. $(2.995 + n)/(3992.995 + \sum x)$.
14. Normal.   15. $b/(m + n/2)$; No, since $m > 0$.
16. $\lambda_0(n + m + 1)/(n\bar{x}\lambda_0 + m + 1)$; No, since $\lambda_0 > 0$, $m > -1$.
17. $(m + \sum x_i + 1)\lambda_0/(n\lambda_0 + m + 1)$; No, since $m > -1$.
18. $(a + \sum x_i)/(a + b + n)$; No, since $a > 0$, $b > 0$.
19. $(a + n)/(a + b + \sum x_i)$; No, since $a > 0$, $b > 0$.

EXERCISE 9.3

1. (18.70 minutes, 19.92 minutes).
2. $\int_0^a \dfrac{[10.05]^{231}}{230!} \lambda^{230}e^{-10.05\lambda} \, d\lambda = \sum_{j=0}^{230} \dfrac{(10.05a)^j}{j!} e^{-10.05a} \doteq N_Z\left(\dfrac{230.5 - 10.05a}{\sqrt{10.05a}}\right)$
   $= .005$.
   Then $\dfrac{230.5 - 10.05a}{\sqrt{10.05a}} = -2.57$ and $a = 19.36$; similarly, $b = 27.17$ and the interval goes from 19.36 to 27.17. (See Exercise 9.3.7).

3. $\sigma_0^2 = 10^{-4}$, $m = 1.01$, assume $\mu_X = 2$

$$\int_a^b \frac{(.0005)^{52.01}}{\Gamma(52.01)} (\sigma^2)^{-53.01} e^{-.0005/\sigma^2} \, d\sigma^2 = .95.$$

4. Beta, $a = x_k + 1$, $b = n - x_k + 1$

$$\int_0^a \frac{\Gamma(n+2)}{\Gamma(x_k + 1)(n + 1 - x_k)} p_k^{x_k}(1 - p_k)^{n-x_k} \, dp_k = \sum_{i=x_k+1}^{n+1} \binom{n+1}{i} a^i (1 - a)^{n+1-i}$$

$$\doteq 1 - N_Z\left(\frac{x_k + \frac{1}{2} - (n+1)a}{\sqrt{(n+1)a(1-a)}}\right) = .025,$$

giving $\dfrac{x_k + \frac{1}{2} - (n+1)a}{\sqrt{(n+1)a(1-a)}} = 1.96$, which can be solved for a, given values for

n and x_k. A similar equation holds for b; the interval then is given by (a, b).

5. Gamma with parameters $m = n$ and $\theta = n\bar{x} + \frac{1}{45}$, setting the prior mean equal to 45. The confidence interval is defined by $\int_0^a f(\lambda) \, d\lambda = \int_b^\infty f(\lambda) \, d\lambda = .025$, where $f(\lambda)$ is the posterior density. The result given in Exercise 9.3.7 can be used to evaluate a and b.

9. Limits are $.059 + .118 \sum x_i \pm \sqrt{.00298 + .01191[\Sigma x_i + (\Sigma x_i)^2]}$.

10. Reject H_0 if sample contains 6 or more defectives.

Chapter 10

EXERCISE 10.1

1. $b' = 1$, $a' = 8\frac{1}{4}$, $s'^2 = 1.583$, $v(B') = .00236$, $v(A') = 1.555$.

3. $b' = .742$, $a' = -1.476$, $s'^2 = 8.031$, $v(B') = .104$, $v(A') = 24.615$.

4. $B' = \sum x_i Y_i / \sum x_i^2$, $S^2 = \sum (Y_i - B'x_i)^2/(n - 1)$.

7. $\bar{Y}/\bar{x}$.   8. $\sum Y_i/nx_i$.

10. $A' = \bar{Y} - B'\bar{c}$, $B' = \sum (c_i - \bar{c})(Y_i - \bar{Y})/\sum (c_i - \bar{c})^2$, $c_i = 1/x_i$.

11. (a) $\sum E_i^2 = \sum |E_i| = 0$ for this case (b) No. (c) $y = \frac{1}{3}$.

12. $V(Z_1) = \sigma^2 \sum c_i^2$, $V(Z_2) = \sigma^2 \sum d_i^2$, cov $(Z_1, Z_2) = \sigma^2 \Sigma c_i d_i$.

14. $\sigma^2\left[\dfrac{1}{n} + \dfrac{(x^* - \bar{x})^2}{\sum (x_i - \bar{x})^2}\right]$.   16. No.

17. (b) $V(\bar{Y}_2) = \sigma^2/n_2$; $V(Y_1') = \sigma^2\left[\dfrac{1}{n_1 + n_2 + n_3} + \dfrac{(x_2 - \bar{x})^2}{\sum n_i(x_2 - \bar{x})^2}\right]$;

$V(Y_2') = \sigma^2\left\{\dfrac{1}{9n_1} + \dfrac{1}{9n_2} + \dfrac{1}{9n_3} + \dfrac{(x_2 - \bar{x})^2 \sum (x_i - \bar{x})^2/n_i}{[\sum (x_i - \bar{x})^2]^2}\right\}$; $V(Y_1')$ is smaller.

18. (b) $b' = \sum (w_i - \bar{w})(Z_i - \bar{Z})/\sum (w_i - \bar{w})^2$
    $= \sum (x_i - \bar{x})(Y_i - \bar{Y})/\sum (x_i - \bar{x})^2.$

20. $\bar{Y}/\bar{x}.$

21. $V(\bar{Y}/\bar{x}) = \lambda/n\bar{x}$

$$V\left(\sum \frac{Y_i}{\sqrt{x_i}}\bigg/\sum \sqrt{x_i}\right) = n\lambda/(\sum \sqrt{x_i})^2.$$

EXERCISE 10.2

1. $\hat{B} = (\bar{Z} - \bar{Y})/(x_2 - x_1)$, $\hat{A} = \bar{Z} - \hat{B}x_2$; same as least squares, if the observations are weighted by the reciprocal of the standard deviation.

2. Reject H_0.   3. $5.83 \leq a \leq 10.67, .906 \leq b \leq 1.094.$

4. $\hat{B} = (\bar{Z} - \bar{Y})/(x_2 - x_1)$, $\hat{A} = \bar{Z} - \hat{B}x_2$; these are the same as the weighted least squares estimators.

5. $\hat{b} = .998$, $\hat{a} = .0014$, $\hat{s}^2 = .000125$; accept $a = 0.$

6. Using $E[Y] = bx$, reject $\sigma^2 \leq .00001.$

7. $\hat{\sigma}^2 = \sum (y_i - \hat{a} - \hat{b}x_i)^2/nx_i$; $\hat{a} = \bar{y} - \hat{b}\bar{x}$;
   $\hat{b} = [\sum y_i/x_i - \bar{y} \sum 1/x_i]/[n - \bar{x} \sum 1/x_i].$

8. (a) Use a t-test; Reject H_0 if $\dfrac{|\hat{B}_1 - \hat{B}_2|}{D} > t_{n+m-4}(1 - \alpha/2)$ where $\hat{B}_i$ is the least squares estimator for group i, $i = 1, 2$, and

$$D^2 = S^2\left[\frac{1}{\sum (x_{1i} - \bar{x}_1)^2} + \frac{1}{\sum (x_{2i} - \bar{x}_2)^2}\right],$$

$$S^2 = \frac{\sum (Y_{1i} - \hat{A}_1 - \hat{B}_1x_{1i})^2 + \sum (Y_{2i} - \hat{A}_2 - \hat{B}_2x_{2i})^2}{n_1 + n_2 - 4}$$

(b) Use a t-test; Reject H_0 if $\dfrac{|\hat{A}_1 - \hat{A}_2|}{C} > t_{n+m-4}(1 - \alpha/2)$ where $\hat{A}_i$ is the least squares estimator for group i, $i = 1, 2$, and

$$C^2 = S^2\left[\frac{\sum x_{1i}^2}{n \sum (x_{1i} - \bar{x}_1)^2} + \frac{\sum x_{2i}^2}{m \sum (x_{2i} - \bar{x}_2)^2}\right].$$

S^2 is defined in (a).

(c) $\hat{B} = \dfrac{\sum\sum(Y_{ji} - \bar{Y})(x_{ji} - \bar{x})}{\sum\sum (x_{ji} - \bar{x})^2}$

$\hat{A} = \bar{Y} - \hat{B}\bar{x}$, $\bar{Y} = \dfrac{n\bar{Y}_1 + m\bar{Y}_2}{n + m}$, $\bar{x} = \dfrac{n\bar{x}_1 + m\bar{x}_2}{n + m}.$

9. $\hat{B} = \dfrac{\sum (Y_{1i} - \bar{Y}_1)(x_{1i} - \bar{x}_1) + \sum (Y_{2i} - \bar{Y}_2)(x_{2i} - \bar{x}_2)}{\sum (x_{1i} - \bar{x}_1)^2 + \sum (x_{2i} - \bar{x}_2)^2}$

$\hat{A}_1 = \bar{Y}_1 - \hat{B}\bar{x}_1$, $\hat{A}_2 = \bar{Y}_2 - \hat{B}\bar{x}_2.$

10. $\tilde{B} = (\bar{Y}_2 - \bar{Y}_1)/(x_2 - x_1)$, $\tilde{A} = \bar{Y}_2 - \tilde{B}x_2$

$\Sigma^2 = \sum \sum (Y_{ij} - \tilde{A} - \tilde{B}x_i)^2/(n + m)$.

EXERCISE 10.3

3. $\hat{b}_0 = 8.6182$, $\hat{b}_1 = .9727$, $\hat{b}_1 = .00122$.

5. $\log \hat{a} = -.6962$, $\hat{b} = .5009$, seventh-year profit $= 16.61$.

6. $s^2 = .0001922$, Accept H_0: $b = \frac{1}{2}$.

Index

Absolute value function, 12
Acceptance region, Definition 8.1.1, 286
Algebra of sets, Section 1.2, 4
Alternative hypothesis, 287
Answers to exercises, 403–424
Asymptotic distribution, maximum
    likelihood, 269
Asymptotic property of estimator, 265
Average value of random variable,
    Definition 3.3.2, 96

Bayes estimate, Definition 9.2.1, 344
  estimator, 344
  interval, Definition 9.3.1, 352
  procedures and tests, Example 9.3.4, 356
  theorem, Theorem 2.6.1, 56
Bayesian methods, Chapter 9, 333–358
Bernoulli random variable, Definition
    4.1.3, 130
  factorial moment generating function, 131
  mean, 131
  probability function, 130
  variance, 131
Bernoulli trial, Definition 4.1.1, 125
Best linear unbiased estimator, Definition
    7.2.3, 263
Best test, simple hypotheses, Theorem
    8.2.1, 293
Beta random variable, 343
  density function, 343
  mean, 343
  variance, 343
Binomial random variable, Definition
    4.1.2, 126
  distribution function, Table 1,
    389–395
  mean, 130
  probability function, 126
  variance, 130

Binomial theorem, Theorem 2.4.1, 42
Birthday problem, Example 2.5.3, 49

Cartesian product of two sets, Definition
    1.2.6, 9
Central limit theorem, Theorem 5.6.1, 215
Chebychev inequality, Theorem 5.5.1, 209
  attaining bound, Exercise 5.5.2, 214
Chi-square random variable, density
    function, 240
  distribution function, Table 4, 370, 371
  mean, 241
  one degree of freedom, 164, 166
  variance, 241
Combinations, n things r at a time,
    Definition 2.4.4, 39
Complement of set, Definition 1.2.3, 7
Complete graph, Exercise 2.4.14, 44
Composite hypothesis, Example 8.1.1, 286
Compound experiment, Exercise 2.2.11, 30
Conceptual population, 226
Conditional density function, Definition
    5.3.1, 187
Conditional mean, 190
Conditional probability, events,
    Definition 2.6.1, 53
Conditional probability function,
    Definition 5.3.1, 187
Confidence band for regression line, 378
Confidence interval, Definition 7.3.1, 272
  in least squares, 377
  for exponential λ, Theorem 7.3.5, 279
  for normal mean, Theorem 7.3.4, 277
  for normal variance, Theorem 7.3.1, 274
  relation to tests, 309, 310
Conjugate prior distribution, 343–347
Consistent estimator, Definition 7.2.4,
    264, 265
Constant function, Example 1.3.1, 12

425

Contingency tables, 323–332
Continuity correction, 218
Continuous population, 227
Continuous random variable, Definition
    3.1.2, 76
  density function, Definition 3.2.2, 88
  range, 88
Continuous sample space, Definition
    2.8.2, 67
Correlation coefficient, Definition
    5.2.5, 182
Counting techniques, Section 2.4, 36–46
Covariance, Definition 5.2.4, 181
Critical region, Definition 8.1.1, 286
Cumulant generating function,
    Exercise 3.4.14, 114

Density function, continuous random
    variable, 88
  conditional, 188
*Density functions for standard distributions
    are listed under the name of the
    random variable*
Discrete population, 227
  random variable, Definition 3.1.2, 76
  probability function, Definition
    3.1.3, 78
  sample space, Definition 2.8.1, 65
Distribution function, Definition
    3.2.1, 81
  monotonic function of a random
    variable, Theorem 3.5.1, 115
  nonmonotonic function of a random
    variable, 120
Distributive laws for set operations, 5
Domain of definition, Definition 1.3.1, 11
Douglas, 20, 34, 70, 406

Efficient estimator, Definition 7.2.2, 262
Elements of sample, 230
Equally likely case, 31–35
Estimate, 250
  Bayes, 344
  least squares, 361
  maximum likelihood, 253
  method of moments, 250
Estimator, 250
  asymptotic property of, 265
  Bayes, 344
  best linear unbiased, 263
  consistent, 264

Estimator (*continued*)
  efficient, 262
  least squares, 361
  maximum likelihood, 253
  method of moments, 250
  unbiased, 261
Event, Definitions 2.1.3, 2.1.4, 21
Expected value, Definition 3.3.1, 96
  function of random variables,
    Definition 5.2.1, 176
  properties of, Theorem 3.3.1, 99
Experiment, Definition 2.2.1, 19
  compound, Exercise 2.2.11, 30
Exponential random variable, Definition
    4.4.2, 154
  density function, Theorem 4.4.2, 154
  distribution function, Theorem 4.4.2,
    154
  mean, 155
  variance, 155
  with displaced origin, Exercise 7.1.29, 260

Factorial moment generating function,
    Definition 3.4.4, 110
Fair game, Example 3.3.1, 96
Finitely additive function, Exercise
    1.3.10, 16
Fisher-Neyman factorization criterion,
    Theorem 7.2.3, 267
Folded standard normal random variable,
    Exercise 4.5.3, 165
  density function, Exercise 4.5.11, 166
Function, Definition 1.3.1, 11
  finitely additive, Exercise 1.3.10, 16
  monotonic, Exercise 1.3.10, 16
  monotonic decreasing, 115
  monotonic increasing, 114
  of random variable, Section 3.5, 114–124

Gamma function, 239, 341
  Exercise 4.4.15, 159
Gamma random variable, 341
  density function, 341
  mean, 342
  relation to Poisson process, Exercise
    4.4.13, 159
  variance, 342
Gauss-Markov theorem, Theorem 10.1.5, 370
Generating function, cumulant, Exercise
    3.4.14, 114
  factorial, Definition 3.4.4, 110

Generating function (*continued*)
  moment, Definition 3.4.2, 106
  moments about mean, 109
  probability, 110
Geometric progression, 132
Geometric random variable,
    Definition 4.2.1, 133
  mean, 135
  probability function, Theorem 4.2.1, 133
  variance, 135
Goodness of fit tests, 313–323

Hugh, 20, 34, 71, 289, 406
Hypergeometric random variable,
    Definition 4.2.2, 136
  mean, 139
  probability function, 137
  variance, 139
Hypothesis, 285
  alternative, 287
  composite, 286
  null, 287
  simple, 286

Identically distributed random variables,
    Definition 5.4.5, 203
Identity function, Example 1.3.1, 12
Independent events, Definition 2.7.1, 59
  random variables, 192, 201
  trials, 61
Interquartile range, 100
Intersection of two sets, Definition
    1.2.2, 5
Invariant property, maximum likelihood
    estimation, 268
Inverse function, 115

Joint moments, two random variables,
    Definition 5.2.2, 179
Jointly distributed random variables,
    Chapter 5, 169–224
Josef, 20, 34, 40, 188, 289, 406

Law of large numbers, Theorem 5.5.2, 212
Least squares criterion, Figure 10.1.1, 361
Least squares estimates, Theorem 10.1.1, 361
Least squares estimators, 364
  and maximum likelihood, Theorem
    10.2.1, 375
  covariance, Theorem 10.1.3, 365
  means, Theorem 10.1.2, 364

Least squares estimators (*continued*)
  variances, Theorem 10.1.3, 365
Likelihood function of sample, Definition
    7.1.1, 253
Likelihood ratio test criterion, Definition
    8.3.1, 303
  examples, 304, 306, 308
  large sample distribution, Theorem
    8.3.3, 308
Linear statistical model, 384
Lognormal random variable, Exercise
    10.3.4, 386

Marginal density function, 173
Marginal of sample, 337
Marginal probability function, 172
Marie, v, 25, 141, 405
Maximum likelihood estimator, Definition
    7.1.2, 253
  asymptotic distribution, Theorem
    7.2.4, 269
  invariant property, 268
Maximum sample value, Definition
    6.2.3, 235
  distribution, 244
Mean of a random variable, Definition
    3.3.2, 96
  of a linear function, Theorem 5.4.6, 205
Mean square error, Exercise 7.2.15, 271
Median, 100
Method of moments estimator, 250
Minimum sample value, Definition
    6.2.3, 235
  distribution, 244
Mixture random variable, Exercise
    3.2.11, 92
  Example 9.1.3, 337
Moment of a random variable, Definition
    3.4.1, 104
  about mean, Definition 3.4.3, 108
  factorial, 110
Moment generating function, Definition
    3.4.2, 106
  cumulant, 114
  factorial, 108
  of sum of random variables, Theorem
    5.4.3, 204
  two dimensional random variables,
    Definition 5.2.3, 179
Monotonic decreasing function, 115
Monotonic increasing function, 114

Most powerful test, 296, 298
Multinomial random variable, Definition
    5.4.3, 199
  probability function, Theorem
    5.4.1, 199
Multinomial trial, Definition 5.4.2, 199
Multiplication principle, Definition 2.4.1, 37
Mutually exclusive events, Definition
    2.7.2, 60

n-dimensional random variables,
    Definition 5.4.1, 198
n-factorial, 38
n-tuple, Definition 1.2.5, 8
Negative binomial random variable,
    Exercise 4.2.11, 142
  mean, 142
  probability function, 142
  variance, 142
Neyman-Pearson lemma, Theorem 8.2.1, 293
Nonlinear statistical model, 385
Normal approximation to binomial,
    Example 5.6.1, 217–219
  with continuity correction, Example
    5.6.2, 219
Normal approximation to negative
    binomial, Exercise 5.6.14, 224
Normal approximation to Poisson,
    Example 5.6.3, 220
Normal equations, 362
Normal random variable, Definition
    4.5.1, 160
  distribution function, 162
  mean, 162
  moment generating function, Theorem
    4.5.1, 161
  variance, 162
Null hypothesis, 287
Null set, 8

One-sided alternative hypotheses,
    Example 8.2.5, 300
One-sided confidence interval, 280
Operating characteristic, Definition
    8.1.3, 288
Operation, associative, 4
  commutative, 4
Order statistics, Definition 6.2.3, 235
  distribution function, Theorem 6.2.6, 244

Pairwise independent events, Example
    2.7.3, 61
Partition, 56
Pascal random variable, Exercise
    4.2.11, 142
Percentiles, Definition 3.3.4, 100
Permutation, Definition 2.4.2, 37
  n things r at a time, Definition
    2.4.3, 38
Poisson approximation to binomial,
    Example 4.3.4, 149
Poisson process, Definition 4.3.1, 143
Poisson random variable, Definition
    4.3.2, 144
  distribution function, Table 2, 396–397
  mean, 148
  probability function, Theorem 4.3.1, 144
  variance, 148
Population, 226
  conceptual, 226
  continuous, 227
  discrete, 227
Population random variable, 228
Positive random variable, Exercise
    3.5.19, 123
Posterior density function, Definition
    9.1.2, 337
  Table 9.1, 348
Power function, Definition 8.1.3, 288
Prior distribution, Definition 9.1.1, 334
  conjugate, 343–347
Prior mean, Example 9.1.2, 335
  variance, Example 9.1.2, 336
Probability, elementary theorems, 28–29
  relative frequency, 17
  subjective, 333
Probability axioms, 26, 27
Probability density function, Definition
    3.2.2, 88
  conditional, 187
  domain, 88
Probability function, 14
  conditional, Definition 2.6.1, 53
Probability function for discrete random
    variables, Definition 3.1.3, 78
  conditional, Definition 5.3.1, 187
*Probability functions for standard
    distributions are listed under the
    name of the random variable*
Probability generating function, 110

Probability integral transform, Exercise 3.5.16, 123
Probability measure, 27

Quartiles, 100

Rachel, 34, 113, 406
Random interval, 272
Random sample from a population, 229
Random sample of a random variable, Definition 6.1.1, 229
Random variable, 14
  Definition 3.1.1, 76
  continuous, Definition 3.1.2, 76
  discrete, Definition 3.1.2, 76
  domain, 76
  range, 76
Range of function, Definition 1.3.1, 11
References, 387
Regression of Y on X, 190, 361
Rejection region, Definition 8.1.1, 286
Relation, reflexive, 3
  symmetric, Exercise 1.1.2, 3
  transitive, Exercise 1.1.1, 3
Reliability, Exercise 4.4.17, 159
  Example 7.3.6, 280
Reproductive property of binomial, Exercise 5.4.4, 207
  of chi-square, Exercise 6.2.2, 247
  of normal, Theorem 5.4.5, 204
  of Poisson, Exercise 5.4.5, 207
Riley, Mrs., 49, 52
Rounding error, Example 5.6.4, 221

Sample maximum value, Definition 6.2.3, 235
  mean, Definition 6.2.2, 233
  median, Definition 6.2.4, 236
  minimum, Definition 6.2.3, 235
  moment, Definition 6.2.2, 233
  range, Definition 6.2.4, 236
Sample space, Definition 2.1.2, 19
  discrete, Definition 2.8.1, 65
  continuous, Definition 2.8.2, 67
Sample standard deviation, Definition 6.2.2, 233
Sample variance, Definition 6.2.2, 233
  computational formula, Theorem 6.2.1, 234
Sampling distribution, 236
  for sample moments, Theorem 6.2.2, 236

Saint Petersburg paradox, Exercise 3.3.15, 104
Sandy, 25, 405
Schwarz inequality, 183
Set, 1
Set equality, Definition 1.1.1, 2
Simple hypothesis, Example 8.1.1, 286
  best test, Theorem 8.2.1, 293
Single-element event, Definition 2.3.1, 31
Square function, Example 1.3.1, 12
Standard deviation of random variable, Definition 3.3.3, 98
Standard form for a random variable, Example 3.5.5, 119
Standard normal random variable, Definition 4.5.2, 162
  distribution function, Table 3, 398–399
Statistic, Definition 6.2.1, 232
Student's t-distribution, Theorem 7.3.2, 276
  distribution function, Table 5, 402
Subjective probability, 333
Subset, Definition 1.1.2, 2
Sue, 141
Sufficient statistic, Definition 7.2.5, 266
  and tests, Theorem 8.2.4, 301

t-distribution, density, 276
  percentiles, Table 5, 402
Taylor series approximations, Exercise 3.5.23, 124
Test of a hypothesis, Definition 8.1.1, 286
  in Bayes procedure, 356
  in regression, 379
Test statistic, 297
Tina, 25, 405
Trial, Bernoulli, 125
  multinomial, 199
Two dimensional random variable, Definition 5.1.1, 169
  density function, Definition 5.1.3, 172
  distribution function, Definition 5.1.4, 174
  probability function, Definition 5.1.2, 171
Type I error, Definition 8.1.2, 286
Type II error, Definition 8.1.2, 286

Unbiased estimator, Definition 7.2.1, 261
Uniform random variable, Definition 4.4.1, 152
  density function, Theorem 4.4.1, 152
  distribution function, 153
  mean, 153

Uniform random variable (*continued*)
  variance, 153
Uniformly most powerful test, Example
  8.2.5, 300
Union of two sets, Definition 1.2.1, 4
Universal set, 6

Variance of a random variable,
  Definition 3.3.3, 98
  computational formula, 99
Variance of a linear function, Theorem
  5.4.6, 205
Venn diagram, 5

Applied Probability and Statistics (*Continued*)

OTNES and ENOCHSON · Digital Time Series Analysis

RAO and MITRA · Generalized Inverse of Matrices and Its Applications

SARD and WEINTRAUB · A Book of Splines

SARHAN and GREENBERG · Contributions to Order Statistics

SEAL · Stochastic Theory of a Risk Business

SEARLE · Linear Models

THOMAS · An Introduction to Applied Probability and Random Processes

WHITTLE · Optimization under Constraints

WILLIAMS · Regression Analysis

WONNACOTT and WONNACOTT · Econometrics

YOUDEN · Statistical Methods for Chemists

ZELLNER · An Introduction to Bayesian Inference in Econometrics

Tracts on Probability and Statistics

BILLINGSLEY · Ergodic Theory and Information

BILLINGSLEY · Convergence of Probability Measures

CRAMÉR and LEADBETTER · Stationary and Related Stochastic Processes

JARDINE and SIBSON · Mathematical Taxonomy

KINGMAN · Regenerative Phenomena

RIORDAN · Combinatorial Identities

TAKACS · Combinatorial Methods in the Theory of Stochastic Processes